STATE SPACE AND UNOBSERVED COMPONENT MODELS

Theory and Applications

This volume offers a broad overview of the state-of-the-art developments in the theory and applications of state space modelling. With fourteen chapters from twenty three contributors, it offers a unique synthesis of state space methods and unobserved component models that are important in a wide range of subjects, including economics, finance, environmental science, medicine and engineering. The book is divided into four sections: introductory papers, testing, Bayesian inference and the bootstrap, and applications. It will give those unfamiliar with state space models a flavour of the work being carried out as well as providing experts with valuable state-of-the-art summaries of different topics. Offering a useful reference for all, this accessible volume makes a significant contribution to the advancement of time series analysis.

ANDREW HARVEY is Professor of Econometrics and Fellow of Corpus Christi College, University of Cambridge. He is the author of *The Econometric Analysis of Time Series* (1981), *Time Series Models* (1981) and *Forecasting, Structural Time Series Models and the Kalman Filter* (1989).

SIEM JAN KOOPMAN is Professor of Econometrics at the Free University Amsterdam and Research Fellow of Tinbergen Institute, Amsterdam. He has published in international journals and is coauthor of *Time Series Anaysis by State Space Models* (with J. Durbin, 2001).

NEIL SHEPHARD is Professor of Economics and Official Fellow, Nuffield College, Oxford University.

STATE SPACE AND UNOBSERVED COMPONENT MODELS

Theory and Applications

Edited by

Andrew Harvey
Faculty of Economics and Politics, University of Cambridge

Siem Jan Koopman
*Department of Econometrics and Operations Research,
Free University Amsterdam*

Neil Shephard
Nuffield College, University of Oxford

PUBLISHED BY THE PRESS SYNDICATE OF THE UNIVERSITY OF CAMBRIDGE
The Pitt Building, Trumpington Street, Cambridge, United Kingdom

CAMBRIDGE UNIVERSITY PRESS
The Edinburgh Building, Cambridge CB2 2RU, UK
40 West 20th Street, New York, NY 10011–4211, USA
477 Williamstown Road, Port Melbourne, VIC 3207, Australia
Ruiz de Alarcón 13, 28014 Madrid, Spain
Dock House, The Waterfront, Cape Town 8001, South Africa

http://www.cambridge.org

© Cambridge University Press 2004

This book is in copyright. Subject to statutory exception
and to the provisions of relevant collective licensing agreements,
no reproduction of any part may take place without
the written permission of Cambridge University Press.

First published 2004

Printed in the United Kingdom at the University Press, Cambridge

Typeface Computer Modern 11/14 pt.   *System* LaTeX 2_ε [TB]

A catalogue record for this book is available from the British Library

Library of Congress Cataloguing in Publication data

State space and unobserved component models : theory and applications :
proceedings of a conference in honour of James Durbin / edited by
Andrew C. Harvey, Siem Jan Koopman, Neil Shephard.
p. cm.
Includes bibliographical references and indexes.
ISBN 0 521 83595 X
1. State-space methods–Congresses.   2. System analysis–Congresses.
I. Durbin, J. (James), 1923–   II. Harvey, A. C. (Andrew C.)
III. Koopman, S. J. (Siem Jan)   IV. Shephard, Neil.
QA402.S835 2004   2003063508

ISBN 0 521 83595 X hardback

The publisher has used its best endeavours to ensure that the URLs for external websites
referred to in this book are correct and active at the time of going to press. However, the
publisher has no responsibility for the websites and can make no guarantee that a site will
remain live or that the content is or will remain appropriate.

Contents

Preface		*page* vii
Acknowledgements		xiv
Part I State space models		1
1	Introduction to state space time series analysis *James Durbin*	3
2	State structure, decision making and related issues *Peter Whittle*	26
3	An introduction to particle filters *Simon Maskell*	40
Part II Testing		73
4	Frequency domain and wavelet-based estimation for long-memory signal plus noise models *Katsuto Tanaka*	75
5	A goodness-of-fit test for AR(1) models and power against state space alternatives *T.W. Anderson and Michael A. Stephens*	92
6	Tests for cycles *Andrew C. Harvey*	102
Part III Bayesian inference and bootstrap		121
7	Efficient Bayesian parameter estimation *Sylvia Frühwirth-Schnatter*	123
8	Empirical Bayesian inference in a nonparametric regression model *Gary Koop and Dale Poirier*	152
9	Resampling in state space models *David S. Stoffer and Kent D. Wall*	171
Part IV Applications		203
10	Measuring and forecasting financial variability using realised variance *Ole E. Barndorff-Nielsen, Bent Nielsen, Neil Shephard and Carla Ysusi*	205
11	Practical filtering for stochastic volatility models *Jonathon R. Stroud, Nicholas G. Polson and Peter Müller*	236

12	On RegComponent time series models and their applications *William R. Bell*	248
13	State space modelling in macroeconomics and finance using SsfPack in S+Finmetrics *Eric Zivot, Jeffrey Wang and Siem Jan Koopman*	284
14	Finding genes in the human genome with hidden Markov models *Richard Durbin*	336

References	351
Author index	373
Subject index	377

Preface

State space methods and unobserved component models are important in a wide range of subjects, including economics, finance, environmental science, medicine and engineering. The conference 'State space and unobserved component models', part of the Academy Colloquium programme of the Royal Netherlands Academy of Arts and Sciences, held in Amsterdam from 29 August to 3 September, 2002 brought together researchers from many different areas, but all pursuing a common statistical theme. The papers selected for this volume will give people unfamiliar with state space models a flavour of the work being carried out as well as providing experts with valuable state-of-the-art summaries of different topics.

The conference on state space methods afforded an ideal opportunity to honour Jim Durbin. Jim has been an active researcher in statistics for over fifty years. His first paper, published in 1950, set out the theory of what was to become known as the Durbin–Watson test. He subsequently published in many other areas of statistics, including sampling theory and regression, but over the last fifteen years or so his attention has again been focussed on time series, and in particular on state space models. A steady stream of work has appeared, beginning with the study of the British seat belt law with Andrew Harvey and culminating in the book with Siem Jan Koopman. It is entirely fitting that the first article in the volume should be by Jim. His clear and lucid style has been an inspiration to generations of students at the London School of Economics – including all three editors of this book – and his paper provides an ideal introduction to unobserved components models. We write some words about Jim's career at the end of this Preface.

The introductory section of the book has two other papers. The first, by Peter Whittle, explores various aspects of the structure of state space models, particularly as they pertain to decision making, from a control engineering perspective. The second, by Simon Maskell, is an introduction to the use of particle filtering for readers who are familiar with Kalman filtering. Particle filters have proved of enormous value in dealing with nonlinear state space models. The potential for using particle filters in areas such as economics and finance is well illustrated by the contribution made by Jonathan Stroud, Nicholas Polson and Peter Müller in the applications section of the book.

The second section deals with testing. Katsuto Tanaka's article is on the use of wavelet-based methods in the estimation of long memory models with additive noise. Like particle filters, wavelets have had significant impact on the engineering literature and it is interesting to see whether they are able to help in the statistical treatment of an unobserved component model. Andrew Harvey's article brings together various tests concerning cycles. Many of the test statistics are linked by the Cramér–von Mises distribution, a distribution that also plays an important role in the paper by Ted Anderson and Michael Stephens. They examine some goodness-of-fit statistics, a field to which Jim Durbin has made major contributions.

The third section contains two papers on Bayesian inference and one on the bootstrap. Sylvia Frühwirth-Schnatter provides the first systematic treatment of the effect of parameterising state space models on the effectiveness of Markov chain Monte Carlo algorithms. The econometricians Gary Koop and Dale Poirier carefully study a Bayesian treatment of a spline model, focusing on the effect of priors on hyperparameters and initial values of states. David Stoffer and Kent Wall provide a survey of the recent work on the use of bootstrap to provide accurate classical inference on state space models.

The last section contains applications. The papers by Ole Barndorff-Nielsen, Bent Nielsen, Neil Shephard and Carla Ysusi and by Stroud *et al.* both deal with stochastic volatility, a key area in financial econometrics. Eric Zivot, Jeffrey Wang and Siem Jan Koopman provide details of a port of `SsfPack` to `S-PLUS` as part of the S+FinMetrics module. `SsfPack` is a library of flexible functions for routine Kalman filtering, smoothing and simulation smoothing for general Gaussian, linear state space models. They are sufficiently flexible that they can also be used as components in the analysis of various non-Gaussian and nonlinear models when combined with importance sampling or Markov chain Monte Carlo techniques. The authors illustrate some of the possibilities with applications of state space methods in finance and macroeconomics. William Bell's paper is on regression component models, with particular emphasis on applications within the US Bureau of the Census. Finally, Richard Durbin, Jim's son, shows how hidden Markov chains can be used to find genes in the human genome.

About the conference

The editors of this volume, with the support of the Royal Netherlands Academy of Arts and Sciences, organised the conference 'State space and unobserved component models' at the end of summer 2002 (29 August–3 September). The venue was 'Het Trippenhuis' in the centre of Amsterdam.

The conference focused on the following topics:

- modern computing tools for nonlinear and non-Gaussian models;
- unobserved components: estimation and testing;
- signal extraction and dynamic factor models;
- methods for official statistics;
- applications in areas like biostatistics, economics, engineering and finance.

The first part of the conference (the Academy Colloquium, 29–31 August) was designed as a workshop for about fifty invited researchers in the field. The participants and the programme of the Colloquium are listed below. The second part was the Masterclass that was attended by an international audience of more than 100 researchers and Ph.D. students. The programme of the masterclass is also presented.

We were delighted that many participated in this conference by presenting a paper on a topic related to the overall theme of the conference or by discussing papers or by attending the conference. A selection of the papers

Conference participants

T.W. Anderson,	Stanford University
Richard T. Baillie,	Michigan State University
William R. Bell,	Statistical Research Division, US Bureau of Census
Charles S. Bos,	Free University Amsterdam
Fabio Busetti,	Bank of Italy, Rome
Richard A. Davis,	Colorado State University
Herman van Dijk,	Erasmus University Rotterdam
Arnaud Doucet,	University of Cambridge
Catherine Doz,	University of Cergy-Pontoise, France
J. Durbin,	London School of Economics
Richard M. Durbin,	Sanger Centre, Cambridge
Cristiano A.C. Fernandes,	Pontificada Universidad Catolica, Rio de Janeiro
David F. Findley,	Statistical Research Division, US Bureau of Census
Sylvia Frühwirth-Schnatter,	Johannes Kepler University, Linz
Andrew C. Harvey,	University of Cambridge
Richard H. Jones,	University of Colorado
Genshiro Kitagawa,	Institute of Statistics and Mathematics, Tokyo
Gary Koop,	University of Leicester

Siem Jan Koopman, Free University, Amsterdam
Hans Rudolf Künsch, ETH Zurich
Kai Ming Lee, Tinbergen Institute Amsterdam
Rob E. Luginbuhl, Free University Amsterdam
Jan R. Magnus, CentER for Economic Research, Tilburg University
Simon Maskell, QinetiQ, Malvern
Brendan McCabe, University of Liverpool
Filippo Moauro, National Institute of Statistics, Italy
Roderick Molenaar, ABP Investments and Research
Charles Nelson, University of Washington, Seattle
Jukka Nyblom, Stockholm School of Economics
Marius Ooms, Free University Amsterdam
Keith Ord, University of Georgetown, Washington DC
Daniel Peña, Carlos III University, Madrid
Danny Pfeffermann, Hebrew University, Tel Aviv
Mike K. Pitt, Warwick University, Coventry
Tommaso Proietti, Udine University
Thomas Rothenberg, University of California, Berkeley
Neil Shephard, Nuffield College, University of Oxford
Robert H. Shumway, University of California
Pieter Jelle van der Sluis, ABP Investments and Research
David Stoffer, University of Pittsburgh
Jonathan R. Stroud, University of Chicago
Katsuto Tanaka, Hitotsubashi University
A.M. Robert Taylor, University of Birmingham
Richard Tiller, US Bureau of Labour Statistics
Pedro Valls, IBMEC Business School, Sao Paulo
Aart de Vos, Free University Amsterdam
Peter Whittle, University of Cambridge
Peter C. Young, University of Lancaster
Kenneth F. Wallis, Warwick University, Coventry
Eric Zivot, University of Washington, Seattle

Papers presented at the conference

T.W. Anderson, A goodness-of-fit test for AR(1) models and power against state space alternatives
Richard T. Baillie, A high frequency perspective on the forward premium anomaly
William R. Bell, On some applications of RegComponent time series models
Richard A. Davis, Observation driven models for Poisson counts
Arnaud Doucet, Optimisation of particle methods using stochastic approximation

Richard M. Durbin,	Finding genes in the human genome with hidden Markov models
Sylvia Frühwirth-Schnatter,	Efficient Bayesian parameter estimation for state space models based on reparametrisations
Andrew C. Harvey,	Testing for cycles
Richard H. Jones,	Binary longitudinal data
Genshiro Kitagawa,	State space modelling for signal extraction problems in seismology
Gary Koop,	Empirical Bayesian inference in nonparametric model
Siem Jan Koopman,	Modelling economic convergence using unobserved converging components
Hans Rudolf Künsch,	Variations on the particle filter
Charles Nelson,	The structural break in the equity premium
Keith Ord,	The single source of error specification for state space models: an appraisal
Daniel Peña,	Dimension reduction in multivariate time series
Danny Pfeffermann,	State space modelling with correlated measurement errors with application to small area estimation under benchmark constraints
Neil Shephard,	High frequency financial econometrics: extracting information from realised variances
Robert H. Shumway,	Dynamic mixed models for merging multiple time series fragments
David Stoffer,	Resampling in state space models
Jonathan Stroud,	Practical filtering with parameter learning
Katsuto Tanaka,	Wavelet methods for inference problems associated with long-memory signal plus noise models
A.M.Robert Taylor,	Variance shifts, structural breaks and stationarity tests
Peter Whittle,	State structure, decision making and related issues
Peter C. Young,	Data-based mechanistic modelling and state dependent parameter models

Masterclass

Richard A. Davis,	The innovations algorithm and parameter driven models
Jim Durbin,	Introduction to state space models
Simon Maskell,	Monte Carlo filtering
Jukka Nyblom,	Testing for unobserved components models
Keith Ord,	A new look at models for exponential smoothing
Richard Tiller,	Application of state space modeling for official labor force statistics in the USA
Eric Zivot,	State space modelling in macroeconomics and finance using S+FinMetrics

is included in this volume. We would particularly like to thank those scholars who helped us to referee these papers.

About Professor James Durbin

James Durbin was born 1923 in England. He was educated at St John's College, Cambridge. From 1950 he worked at the London School of Economics and Political Science (LSE) until his retirement in 1988; he remains an Emeritus Professor of Statistics there. He visited research institutions in many corners of the world during his period at the LSE and after his retirement. His research areas include serial correlation (13 publications), time series (>30), sample survey methodology (9), goodness-of-fit tests and sample distribution functions (13) and probability and statistical theory (20). He was deputy editor of *Biometrika* (1962–4) and associate editor of *Biometrika* (1960–2, 1964–7), *Annals of Statistics* (1973–5) and *Journal of the Royal Statistical Society, Series B* (1978–81).

He was the President of the Royal Statistical Society for the period 1986–7 and the President of the International Statistical Institute for the period of 1983–5. Further, he is a Fellow of the Econometric Society (since 1967), the Institute of Mathematical Statistics (since 1958) and the American Statistical Society (since 1960). In 2001 he became a Fellow of The British Academy.

More details about the career of Professor Durbin can be found *Econometric Theory* (1988, volume 4, issued) in which the interview with Professor J. Durbin by Professor Peter C. B. Phillips was published. His publications between 1950 and 1988 are listed at the end of the interview. Below we report his publications from 1988 onwards.

Publications of Professor James Durbin since 1988

(i) Statistics and statistical science (presidential address). *Journal of the Royal Statistical Society, Series A*, 150, 177–191, 1988.

(ii) Is a philosophical consensus for statistics attainable? *Journal of Econometrics*, 37, 51–61, 1988.

(iii) Maximum likelihood estimation of the parameters of a system of simultaneous regression equations. *Econometric Theory*, 4, 159–170, 1988.

(iv) A reconciliation of two different expressions for the first passage density of Brownian motion to a curved boundary. *Journal of Applied Probability*, 25, 829–832, 1988.

(v) First passage densities of Gaussian and point processes to general boundaries with special reference to Kolmogorov–Smirnov tests when parameters are estimated. *Proceedings of First World Conference of the Bernoulli Society, Tashkent, USSR*, 1988.

(vi) Extensions of Kalman modelling to non-Gaussian observations. *Quadermi di Statistica e Mathematica applicata*, 12, 3–12, 1990.

(vii) The first passage density of the Brownian motion process to a curved boundary. *Journal of Applied Probability*, 29, 291–304, 1992.

(viii) On a test of serial correlation for regression models with lagged dependent variables. *The Art of Statistical Science*, edited K. V. Mardia, pp. 27–32, New York: Wiley, 1992.

(ix) Optimal estimating equations for state vectors in non-Gaussian and nonlinear state space time series models. *Selected Proceedings of Athens, Georgia, Symposium on Estimating Functions*. Edited I. V. Basawa, V. P. Godambe and R. L. Taylor, 1997.

(x) (with S. J. Koopman) Monte Carlo maximum likelihood estimation for non-Gaussian state space models. *Biometrika*, 84, 669–684, 1997.

(xi) (with B. Quenneville) Benchmarking by state space models. *International Statistical Review*, 65, 23–48, 1997.

(xii) (with J. R. Magnus) Estimation of regression coefficients of interest when other regression coefficients are of no interest. *Econometrica*, 67, 639–643, 1999.

(xiii) The state space approach to time series analysis and its potential for official statistics (The Foreman Lecture). *Australian and New Zealand Journal of Statistics*, 42, 1–23, 2000.

(xiv) (with S. J. Koopman) Time series analysis of non-Gaussian observations based on state space models from both classical and Baysian perspectives (with discussion). *Journal of the Royal Statistical Society, Series B*, 62, 3–56, 2000.

(xv) (with S. J. Koopman) Fast filtering and smoothing for multivariate state space models. *Journal of Time Series Analysis*, 21, 281–296, 2000.

(xvi) (with S. J. Koopman) *Time Series Analysis by State Space Methods*. Oxford: Oxford University Press, 2001.

(xvii) (with S. J. Koopman) A simple and efficient simulation smoother for state space time series analysis. *Biometrika*, 89, 603–616, 2002.

(xviii) (with S. J. Koopman) Filtering and smoothing of state vector for diffuse state space models. *Journal of Time Series Analysis*, 24, 85–98, 2003.

Acknowledgements

We would like to thank the Royal Netherlands Academy of Arts and Sciences (KNAW, http://www.knaw.nl) for their financial support and for dealing with the organisation of the conference. Further financial assistance was provided by the Netherlands Organisation for Scientific Research (NWO, http://www.nwo.nl), the Tinbergen Institute for Economic Research (TI Amsterdam and Rotterdam, http://www.tinbergen.nl), and ABP Investments and Research (Schiphol, Amsterdam, http://www.abp.nl). Particular thanks are due to Martine Wagenaar of KNAW for being the driving force of the organisation of the conference and for her cooperation before, during and after the conference. Further we would like to thank Charles Bos, Irma Hindrayanto and Marius Ooms of Free University, Amsterdam for their assistance during various stages of the organisation of the conference and the preparation of this volume. We would also like to thank all the participants, particularly those who gave papers or acted as discussants. Finally, Neil Shephard thanks the ESRC for supporting his research.

Part I

State space models

1
Introduction to state space time series analysis

James Durbin

Department of Statistics, London School of Economics and Political Science

Abstract

The paper presents a broad general review of the state space approach to time series analysis. It begins with an introduction to the linear Gaussian state space model. Applications to problems in practical time series analysis are considered. The state space approach is briefly compared with the Box–Jenkins approach. The Kalman filter and smoother and the simulation smoother are described. Missing observations, forecasting and initialisation are considered. A representation of a multivariate series as a univariate series is displayed. The construction and maximisation of the likelihood function are discussed. An application to real data is presented. The treatment is extended to non-Gaussian and nonlinear state space models. A simulation technique based on importance sampling is described for analysing these models. The use of antithetic variables in the simulation is considered. Bayesian analysis of the models is developed based on an extension of the importance sampling technique. Classical and Bayesian methods are applied to a real time series.

State Space and Unobserved Component Models: Theory and Applications, eds. Andrew C. Harvey, Siem Jan Koopman and Neil Shephard. Published by Cambridge University Press. © Cambridge University Press 2004

1.1 Introduction to state space models

1.1.1 Basic ideas

The organisers have asked me to provide a broad, general introduction to state space time series analysis. In the pursuit of this objective I will try to make the exposition understandable for those who have relatively little prior knowledge of the subject, while at the same time including some results of recent research. My starting point is the claim that state space models provide an effective basis for practical time series analysis in a wide range of fields including statistics, econometrics and engineering.

I will base my exposition on the recent book by Durbin and Koopman (2001), referred to from now on as the DK book, which provides a comprehensive treatment of the subject. Readers may wish to refer to the website http://www.ssfpack.com/dkbook/ for further information about the book. Other books that provide treatments of state space models and techniques include Harvey (1989), West and Harrison (1997), Kitagawa and Gersch (1996) and Kim and Nelson (1999). More general books on time series analysis with substantial treatments of state space methods are, for example, Brockwell and Davis (1987), Hamilton (1994) and Shumway and Stoffer (2000).

I will begin with a particular example that I will use to introduce the basic ideas that underlie state space time series analysis. This refers to logged monthly numbers of car drivers who were killed or seriously injured in road accidents in Great Britain, 1969–84. These data come from a study by Andrew Harvey and me, undertaken on behalf of the British Department of Transport, regarding the effect on road casualties of the seat belt law that was introduced in February 1983; for details see Durbin and Harvey (1985) and Harvey and Durbin (1986).

Inspection of Figure 1.1 reveals that the series is made up of a trend which initially is increasing, then decreases and subsequently flattens out, plus a seasonal effect which is high in the winter and low in the summer, together with a sudden drop in early 1983 seemingly due to the introduction of the seat belt law. Other features that could be present, though they are not apparent from visual inspection, include cycles and regression effects due to the influence of such factors as the price of petrol, weather variations and traffic density. Thus we arrive at the following model:

$$y_t = \mu_t + \gamma_t + c_t + r_t + i_t + \varepsilon_t, \tag{1.1}$$

where

y_t = observation (often logged, possibly a vector)
μ_t = trend (slow change in level)
γ_t = seasonal (pattern can change over time)
c_t = cycle (of longer period than seasonal)
r_t = regression component (coefficients can vary over time)
i_t = intervention effect (e.g. seat belt law)
ε_t = random error or disturbance or noise.

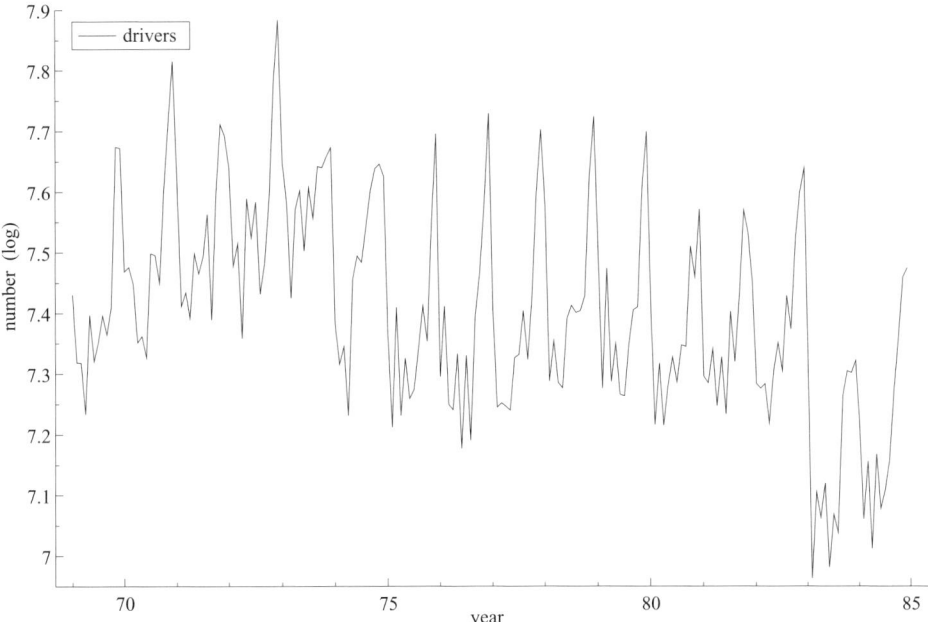

Fig. 1.1. Monthly numbers (logged) of car drivers who were killed or seriously injured in road accidents in Great Britain.

In the state space approach we construct submodels designed to model the behaviour of each component such as trend, seasonal, etc. separately and we put these submodels together to form a single matrix model called a *state space model*. The model used by Harvey and Durbin for the analysis of the data of Figure 1.1 included all the components of (1.1) except c_t; some of the results of their analysis will be presented later.

1.1.2 Special cases of the basic model

We consider the following two special cases.

1.1.2.1 The local level model

This is specified by

$$\begin{aligned} y_t &= \mu_t + \varepsilon_t, & \varepsilon_t &\sim N(0, \sigma_\varepsilon^2), \\ \mu_{t+1} &= \mu_t + \eta_t, & \eta_t &\sim N(0, \sigma_\eta^2), \end{aligned} \quad (1.2)$$

for $t = 1, \ldots, n$, where the ε_ts and η_ts are all mutually independent and are also independent of μ_1.

The objective of this model is to represent a series with no trend or seasonal whose level μ_t is allowed to vary over time. The second equation of the model is a *random walk*; random walks are basic elements in many state space time series models. Although it is simple, the local level model is not an artificial model and it provides the basis for the treatment of important series in practice. It is employed to explain the ideas underlying state space time series analysis in an elementary way in Chapter 2 of the DK book. The properties of time series that are generated by a local level model are studied in detail in Harvey (1989).

1.1.2.2 The local linear trend model

This is specified by

$$\begin{aligned} y_t &= \mu_t + \varepsilon_t, & \varepsilon_t &\sim N(0, \sigma_\varepsilon^2), \\ \mu_{t+1} &= \mu_t + \nu_t + \xi_t, & \xi_t &\sim N(0, \sigma_\xi^2), \\ \nu_{t+1} &= \nu_t + \zeta_t, & \zeta_t &\sim N(0, \sigma_\zeta^2). \end{aligned} \quad (1.3)$$

This extends the local level model to the case where there is a trend with a slope ν_t where both level and slope are allowed to vary over time. It is worth noting that when both ξ_t and ζ_t are zero, the model reduces to the classical linear trend plus noise model, $y_t = \alpha + \beta t +$ error. It is sometimes useful to smooth the trend by putting $\xi_t = 0$ in (1.3). Details of the model and its extensions to the general class of structural time series models are given in the DK book Section 3.2 and in Harvey (1989).

The matrix form of the local linear trend model is

$$y_t = \begin{pmatrix} 1 & 0 \end{pmatrix} \begin{pmatrix} \mu_t \\ \nu_t \end{pmatrix} + \varepsilon_t,$$

$$\begin{pmatrix} \mu_{t+1} \\ \nu_{t+1} \end{pmatrix} = \begin{pmatrix} 1 & 1 \\ 0 & 1 \end{pmatrix} \begin{pmatrix} \mu_t \\ \nu_t \end{pmatrix} + \begin{pmatrix} \xi_t \\ \zeta_t \end{pmatrix}.$$

By considering this and other special cases in matrix form we are led to the following general model which provides the basis for much of our further treatment of state space models.

1.1.3 The linear Gaussian state space model

This has the form

$$\begin{aligned}
y_t &= Z_t \alpha_t + \varepsilon_t, & \varepsilon_t &\sim N(0, H_t), & & \\
\alpha_{t+1} &= T_t \alpha_t + R_t \eta_t, & \eta_t &\sim N(0, Q_t), & t &= 1, \ldots, n, \quad (1.4) \\
\alpha_1 &\sim N(a_1, P_1). & & & &
\end{aligned}$$

Matrices Z_t, H_t, T_t, R_t and Q_t are assumed known. Initially, a_1 and P_1 are assumed known; we will consider later what to do when some elements of them are unknown. The $p \times 1$ vector y_t is the observation. The unobserved $m \times 1$ vector α_t is called the *state*. The disturbances ε_t and η_t are independent sequences of independent normal vectors. The matrix R_t, when it is not the identity, is usually a selection matrix, that is, a matrix whose columns are a subset of the columns of the identity matrix; it is needed when the dimensionality of α_t is greater than that of the disturbance vector η_t. The first equation is called the *observation equation* and the second equation is called the *state equation*.

The structure of model (1.4) is a natural one for representing the behaviour of many time series as a first approximation. The first equation is a standard multivariate linear regression model whose coefficient vector α_t varies over time; the development over time of α_t is determined by the first-order vector autoregression given in the second equation. The Markovian nature of the model accounts for many of its elegant properties.

In spite of the conceptual simplicity of this model it is highly flexible and has a remarkably wide range of applications to problems in practical time series analysis. I will mention just a few.

(i) Structural time series models. These are models of the basic form (1.1) where the submodels for the components are chosen to be compatible with the state space form (1.4). The local level model and the local linear trend model are simple special cases. The models are sometimes called *dynamic linear models*.

(ii) ARMA and Box–Jenkins (BJ) ARIMA models. These can be put in state space form as described in the DK book, Section 3.3. This means that ARIMA models can be treated as special cases of state space models. I will make a few remarks at this point on the relative merits of the BJ approach and the state space approach for practical time series analysis.

   (a) BJ is a 'black box' approach in which the model is determined purely by the data without regard to the structure underlying the

data, whereas state space fits the data to the structure of the system which generated the data.
   (b) BJ eliminates trend and seasonal by differencing. However, in many cases these components have intrinsic interest and in state space they can be estimated directly. While in BJ estimates can be 'recovered' from the differenced series by maximising the residual mean square, this seems an artificial procedure.
   (c) The BJ identification procedure need not lead to a unique model; in some cases several apparently quite different models can appear to fit the data equally well.
   (d) In BJ it is difficult to handle regression effects, missing observations, calendar effects, multivariate observations and changes in coefficients over time; these are all straightforward in state space.

   A fuller discussion of the relative merits of state space and BJ is given in the DK book, Section 3.5. The comparison is strongly in favour of state space.
(iii) Model (1.4) handles time-varying regression and regression with autocorrelated errors straightforwardly.
(iv) State space models can deal with problems in spline smoothing in discrete and continuos time on a proper modelling basis in which parameters can be estimated by standard methods, as compared with customary *ad hoc* methods.

1.2 Basic theory for state space analysis

1.2.1 Introduction

In this section we consider the main elements of the methodology required for time series analysis based on the linear Gaussian model (1.4). Let $Y_t = \{y_1, \ldots, y_t\}$, $t = 1, \ldots, n$. We will focus on the following items:

- *Kalman filter.* This recursively computes $a_{t+1} = E(\alpha_{t+1}|Y_t)$ and $P_{t+1} = Var(\alpha_{t+1}|Y_t)$ for $t = 1, \ldots, n$. Since distributions are normal, these quantities specify the distribution of α_{t+1} given data up to time t.
- *State smoother.* This estimates $\hat{\alpha}_t = E(\alpha_t \,|\, Y_n)$ and $V_t = \text{Var}(\alpha_t \,|\, Y_n)$ and hence the conditional distribution of α_t given all the observations for $t = 1, \ldots, n$.
- *Simulation smoother.* An algorithm for generating draws from

$$p(\alpha_1, \ldots, \alpha_n | Y_n).$$

This is an essential element in the analysis of non-Gaussian and nonlinear models as described in Section 1.4.
- *Missing observations.* We show that the treatment of missing observations is particularly simple in the state space approach.
- *Forecasting* is simply treated as a special case of missing observations.
- *Initialisation.* This deals with the case where some elements of $a_1 = E(\alpha_1)$ and $V_1 = \text{Var}(\alpha_1)$ are unknown.
- *Univariate treatment of multivariate series.* This puts a multivariate model into univariate form, which can simplify substantially the treatment of large complex models.
- *Parameter estimation.* We show that the likelihood function is easily constructed using the Kalman filter.

1.2.2 Kalman filter

We calculate
$$a_{t+1} = E(\alpha_{t+1}|Y_t), \qquad P_{t+1} = Var(\alpha_{t+1}|Y_t),$$
by the recursion
$$\begin{aligned}
v_t &= y_t - Z_t a_t, \\
F_t &= Z_t P_t Z_t' + H_t, \\
K_t &= T_t P_t Z_t' F_t^{-1}, \\
L_t &= T_t - K_t Z_t, \\
a_{t+1} &= T_t a_t + K_t v_t, \\
P_{t+1} &= T_t P_t L_t' + R_t Q_t R_t' \qquad t = 1, \ldots, n,
\end{aligned} \tag{1.5}$$
with a_1 and P_1 as the mean vector and the variance matrix of α_1.

1.2.3 State smoother

We calculate
$$\hat{\alpha}_t = E(\alpha_t \mid Y_n), \quad V_t = Var(\alpha_t \mid Y_n),$$
by the backwards recursion
$$\begin{aligned}
r_{t-1} &= Z_t' F_t^{-1} v_t + L_t' r_t, \\
N_{t-1} &= Z_t' F_t^{-1} Z_t + L_t' N_t L_t, \\
\hat{\alpha}_t &= a_t + P_t r_{t-1}, \\
V_t &= P_t - P_t N_{t-1} P_t \qquad t = n, \ldots, 1,
\end{aligned} \tag{1.6}$$

with $r_n = 0$ and $N_n = 0$. The recursive nature of formulae (1.5) and (1.6), which arises from the Markovian nature of model (1.4), implies that calculations based on them are very fast on modern computers.

The proofs of these and many related results in state space theory can be derived very simply by the use of the following elementary lemma in multivariate normal regression theory. Suppose that x, y and z are random vectors of arbitrary orders that are jointly normally distributed with means μ_p and covariance matrices

$$\Sigma_{pq} = E[(p - \mu_p)(q - \mu_q)']$$

for $p, q = x, y$ and z with $\mu_z = 0$ and $\Sigma_{yz} = 0$. The symbols x, y, z, p and q are employed for convenience and their use here is unrelated to their use in other parts of the paper.

Lemma

$$\begin{aligned} E(x|y, z) &= E(x|y) + \Sigma_{xz}\Sigma_{zz}^{-1} z, \\ \operatorname{Var}(x|y, z) &= \operatorname{Var}(x|y) - \Sigma_{xz}\Sigma_{zz}^{-1}\Sigma_{xz}'. \end{aligned}$$

The proof of this familiar lemma can be obtained straightforwardly from elementary multivariate normal regression theory; see, for example, the DK book Section 2.13 for details. Proofs of the Kalman filter and smoother are given in the DK book, Sections 4.2 and 4.3. The elementary nature of this lemma drives home the point that the theoretical basis of state space analysis is very simple.

1.2.4 Simulation smoothing

A simulation smoother in Gaussian state space time series analysis draws samples from the Gaussian conditional distribution of state or disturbance vectors given the observations. This has proved important in practice for the analysis of non-Gaussian models and for carrying out Bayesian inference. Recently a new technique for implementing this has been proposed by Durbin and Koopman 2002 which is both simple and computationally efficient and which we now describe.

The construction of a simulation smoother for the state vector α_t is relatively simple given the lemma in Section 1.2.3. Since the state space model (1.4) is linear and Gaussian, the density $p(\alpha_1, \ldots, \alpha_n|Y_n)$ is multivariate normal. Its variance matrix has the important property that it does not depend upon Y_n; this follows immediately from the general result that in a multivariate normal distribution the conditional variance matrix of a vector given that a second vector is fixed does not depend on the second vector. These observations lead to a straightforward derivation of the following algorithm for drawing random vectors $\tilde{\alpha}_t$ from $p(\alpha|Y_n)$:

Step 1. Obtain random draws ε_t^+ and η_t^+ from densities $N(0, H_t)$ and $N(0, Q_t)$, respectively, for $t = 1, \ldots, n$. Generate α_t^+ and y_t^+ by means of recursion (1.4) with ε_t, η_t replaced by ε_t^+, η_t^+ where the recursion is initialised by the draw $\alpha_1^+ \sim N(a_1, P_1)$.

Step 2. Compute $\hat{\alpha}_t = E(\alpha_t|Y_n)$ and $\hat{\alpha}_t^+ = E(\alpha_t|Y_n^+)$ where $Y_n^+ = \{y_1^+, \ldots, y_n^+\}$ by means of standard filtering and smoothing using (1.5) forwards and (1.6) backwards.

Step 3. Take

$$\tilde{\alpha}_t = \hat{\alpha}_t - \hat{\alpha}_t^+ + \alpha_t^+,$$

for $t = 1, \ldots, n$.

This algorithm for generating $\tilde{\alpha}_t$ only requires standard Kalman filtering and state smoothing applied to the constructed series $y_1^+, \ldots, y_n^+$ and is therefore easy to incorporate in new software; special algorithms for simulation smoothing such as the ones developed by Frühwirth-Schnatter (1994c), Carter and Kohn (1994) and de Jong and Shephard (1995) are not required. The algorithm and similar ones for the disturbances are intended to replace those given in Section 4.7 of the DK book.

1.2.5 Missing observations

These are easy to handle in state space analysis. If observation y_j is missing for any j from 2 to $n-1$, all we have to do is put $v_j = 0$ and $K_j = 0$ in equations (1.5) and (1.6). The proof is given in Section 4.8 of the DK book.

1.2.6 Forecasting

This also is very easy in state space analysis. Suppose we want to forecast $y_{n+1}, \ldots, y_{n+k}$ given $y_1, \ldots, y_n$ and calculate mean square forecast errors. We merely treat $y_{n+1}, \ldots, y_{n+k}$ as missing observations and proceed using (1.5) as in Section 1.2.5. We use

$$Z_{n+1}a_{n+1}, \ldots, Z_{n+k}a_{n+k}$$

as the forecasts and use

$$V_{n+1}, \ldots, V_{n+k}$$

to provide mean square errors; for details see the DK book, Section 4.9.

1.2.7 Initialisation

So far we have assumed for convenience that the initial state $\alpha_1 \sim N(a_1, P_1)$, where a_1 and P_1 are known. In practice, however, it will usually be the case that some or all elements of a_1 and P_1 are unknown. In this situation we have two choices: we can either assume that the corresponding elements of α_1 are fixed and estimate them by maximum likelihood or we can assume that they have zero means and variance matrix κI, where I is the identity matrix and $\kappa \to \infty$. In the latter case we call the distribution a *diffuse prior*. It turns out that the two approaches give very similar results. The resulting filtering and smoothing recursions have similar structure to those with a_1 and P_1 known. A full treatment of the initialisation problem is given in Chapter 5 of the DK book.

1.2.8 Univariate treatment of multivariate series

If a multivariate state space model is not small and simple it is often advantageous computationally to rewrite the model in univariate form. This is straightforward in most cases and can result in large computational savings. It can also assist in sorting out problems or complexities that refer only to particular elements of observational vectors. Suppose that in contrast to the specification in model (1.4), y_t has dimensionality p_t and H_t is diagonal. Write the observation and disturbance vectors as

$$y_t = \begin{pmatrix} y_{t,1} \\ \vdots \\ y_{t,p_t} \end{pmatrix}, \quad \varepsilon_t = \begin{pmatrix} \varepsilon_{t,1} \\ \vdots \\ \varepsilon_{t,p_t} \end{pmatrix},$$

and the observation equation matrices as

$$Z_t = \begin{pmatrix} Z_{t,1} \\ \vdots \\ Z_{t,p_t} \end{pmatrix}, \quad H_t = \begin{pmatrix} \sigma_{t,1}^2 & 0 & 0 \\ 0 & \ddots & 0 \\ 0 & 0 & \sigma_{t,p_t}^2 \end{pmatrix},$$

where $y_{t,i}$, $\varepsilon_{t,i}$, $\sigma_{t,i}^2$ are scalars and $Z_{t,i}$ is a row vector. The univariate observation equation is

$$y_{t,i} = Z_{t,i}\alpha_{t,i} + \varepsilon_{t,i}, \quad i = 1, \ldots, p_t, \quad t = 1, \ldots, n,$$

where $\alpha_{t,i} = \alpha_t$. The univariate state equation is

$$\alpha_{t,i+1} = \alpha_{t,i}, \quad i = 1, \ldots, p_t - 1,$$
$$\alpha_{t+1,1} = T_t\alpha_{t,p_t} + R_t\eta_t, \quad t = 1, \ldots, n,$$

with $\alpha_{1,1} = \alpha_1 \sim N(a_1, P_1)$.

The Kalman filter and smoother for this univariate representation follow directly from (1.5) and (1.6) and are given explicitly in Section 6.4 of the DK book; the case where the matrix H_t is nondiagonal is also considered.

1.2.9 Estimation of unknown parameters by maximum likelihood

Given the unknown parameter vector ψ and $\alpha_1 \sim N(a_1, P_1)$, where a_1 and P_1 are known, the likelihood is

$$L(\psi) = p(y_1, \ldots, y_n) = p(y_1) \prod_{t=2}^{n} p(y_t \mid Y_{t-1}).$$

Now,

$$\begin{aligned} E(y_t \mid Y_{t-1}) &= E(Z_t \alpha_t + \varepsilon_t \mid Y_{t-1}) = Z_t a_t, \\ v_t &= y_t - Z_t a_t, \\ Var(y_t \mid Y_{t-1}) &= Var(v_t) = F_t. \end{aligned}$$

Thus,

$$L(\psi) = \prod_{t=1}^{n} p(v_t), \qquad v_t \sim N(0, F_t).$$

This can be easily calculated from the output of the Kalman filter. Similar expressions are obtained when some elements of α_1 are diffuse or estimated by maximum likelihood.

Log $L(\psi)$ is maximised numerically using the score vector or the EM algorithm, both of which emerge in computationally convenient forms for the state space model. Details are given in Chapter 7 of the DK book.

1.2.10 Bayesian analysis

The results in the previous sections provide the basis for the application of classical inference to the linear Gaussian state space model. We now consider a Bayesian analysis based on the Gibbs sampler as proposed by Frühwirth-Schnatter (1994c). For example, consider the local level model (1.2) where the variances σ_ε^2 and σ_η^2 are treated as random variables. A model for a variance σ^2 can be based on the inverse gamma distribution with logdensity

$$\log p(\sigma^2 \mid c, s) = -\log \Gamma\left(\frac{c}{2}\right) - \frac{c}{2} \log \frac{s}{2} - \frac{c+2}{2} \log \sigma^2 - \frac{s}{2\sigma^2}, \qquad \text{for } \sigma^2 > 0,$$

and $p(\sigma^2|c,s) = 0$ for $\sigma^2 \leq 0$; see, for example, Poirier (1995, Table 3.3.1). We denote this density by $\sigma^2 \sim IG(c/2, s/2)$, where c determines the shape and s determines the scale of the distribution. It has the convenient property that if we take this as the prior density of σ^2 and we take a sample $u_1, \ldots, u_n$ of independent $N(0, \sigma^2)$ variables, the posterior density of σ^2 is

$$p(\sigma^2|u_1, \ldots, u_n) = IG\left[(c+n)/2, \left(s + \sum_{i=1}^{n} u_i^2\right)/2\sigma^2\right]. \quad (1.7)$$

The posterior means of $\mu = (\mu_1, \ldots, \mu_n)'$ and of the variances $\psi = (\sigma_\varepsilon^2, \sigma_\eta^2)'$ can be obtained by simulating from the joint density $p(\psi, \mu|y)$. In a Markov chain Monte Carlo (MCMC) procedure, the sampling from this joint density is implemented as a Markov chain. After the initialisation $\psi = \psi^{(0)}$, we repeat the following simulation steps M^* times:

(i) sample $\mu^{(i)}$ from $p(\mu|y, \psi^{(i-1)})$ using the simulation smoother of Section 1.2.4;

(ii) sample $\psi^{(i)}$ from $p(\psi|y, \mu^{(i)})$ using the inverse gamma density;

for $i = 1, \ldots, M^*$. After the process has stabilised, we treat the last M samples from Step (ii) as being generated from the density $p(\mu, \psi|y)$, so if we discard the sampled μs, the resulting draws of ψ can be thought of as samples from $p(\psi|y)$. Usually, sampling directly from conditional densities is easier than sampling from the marginal density $p(\psi|y)$. For the implementation of Step (ii) a sample value of an element of ψ is chosen from the posterior density (1.7). We can take u_t in (1.7) as an element of ε_t or η_t obtained from the simulation smoother in Step (i). Similar methods can be applied to the general linear Gaussian model (1.4). An alternative to MCMC simulation can be developed based on importance sampling. Details of both approaches are given in Chapter 8 of the DK book.

1.3 Application to seat belt data

1.3.1 Introduction

We now consider the application of methods based on the linear Gaussian model to the data introduced in Section 1.1.1 on the effect of the seat belt law on death and serious injuries of car drivers in Great Britain. The data are displayed in Figure 1.1; we have monthly observations from January 1969 to December 1984. The seat belt law was introduced on 31 January 1983.

We use the model

$$y_t = \mu_t + \gamma_t + \beta x_t + \delta i_t + \varepsilon_t, \quad \varepsilon_t \sim N(0, \sigma_\varepsilon^2),$$
$$\mu_t = \mu_{t-1} + \eta_t, \quad \eta_t \sim N(0, \sigma_\eta^2), \quad (1.8)$$
$$\gamma_t = -\gamma_{t-1} - \cdots - \gamma_{t-11} + \omega_t, \quad \omega_t \sim N(0, \sigma_\omega^2),$$

where

y_t = log of number car drivers killed or seriously injured,
μ_t = trend,
γ_t = seasonal,
x_t = price of petrol,
i_t = 0 before February 1983 and 1 afterwards,
$\varepsilon_t, \eta_t, \omega_t$ = mutually and serially independent disturbances.

We shall give illustrations of both classical and Bayesian inference methods.

1.3.2 Classical inference

The classical estimation results are given in Table 1.1.

Table 1.1. *Classical estimation results*

parameter	estimates	std error	
β	−0.29140	0.098318	
δ	−0.23773	0.046317	
σ_ε^2	0.00378		
σ_η^2	0.00027		$\sigma_\eta^2/\sigma_\varepsilon^2 = 0.0707$
σ_ω^2	1.1620×10^{-6}		$\sigma_\omega^2/\sigma_\varepsilon^2 = 0.0003$

The t-values of estimates of β and δ are −2.96 and −5.13, respectively. There was a drop of 21% $(1 - \exp(-0.24))$ in drivers killed and seriously injured due to the law. The trend plus original data, seasonal and irregular components of the series are shown in Figure 1.2. The sharp drop in the trend in February 1983 is very evident. The graph of the seasonal pattern was constant throughout the period.

1.3.3 Bayesian inference

Following Frühwirth-Schnatter 1994a we carry out a Bayesian analysis of the data based on the same structural time series model but to keep the

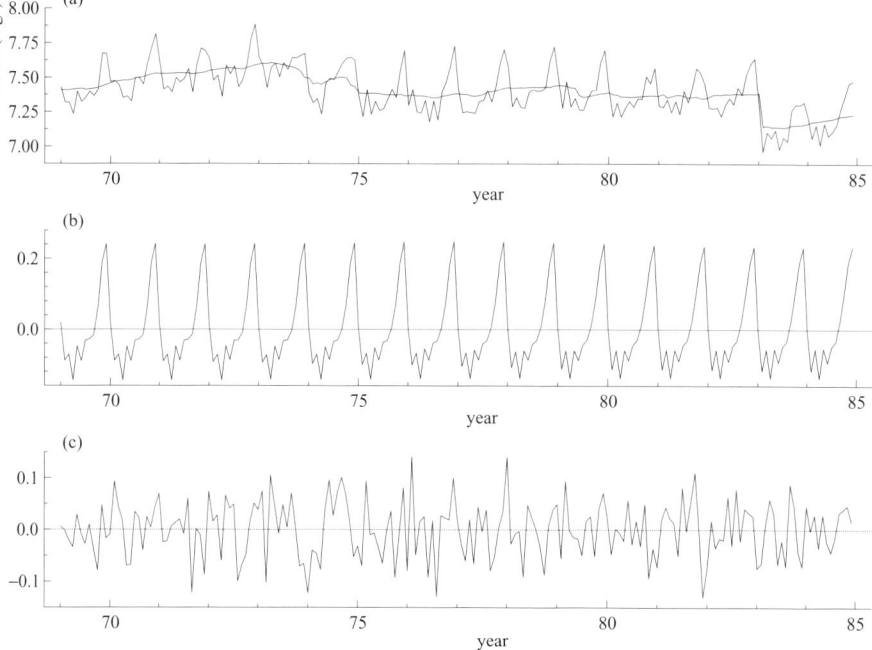

Fig. 1.2. Estimated components for basic time series model with intervention and regression components: (a) trend; (b) seasonal; (c) irregular.

illustration simple we treat the seasonal term γ_t as constant through time using fixed dummies and we ignore the regression term βx_t.

Table 1.2. *Posterior results for parameters based on Gibbs sampler with* $M = 2000$

parameter	posterior mean	posterior std deviation	quantiles	
			0.025	0.975
σ_ε^2	0.00380	0.000543	0.00281	0.00492
σ_η^2	0.00053	0.000253	0.00021	0.00126
δ	-0.259	0.0803	-0.420	-0.103

We have applied the Gibbs sampler as described in Section 1.2.10 with $M = 2000$. For this model we have $\psi = (\sigma_\varepsilon^2, \sigma_\eta^2, \delta)'$. Figure 1.3 shows the realised draws from $p(\psi|y)$, the correlogram of the series of draws and a histogram of the realised draws which can be treated as an estimate of the posterior density of ψ. The estimated posterior means and standard deviations of elements of ψ together with posterior quantiles are reported in Table 1.2. These may be compared with the relevant classical estimates in Section 1.3.2.

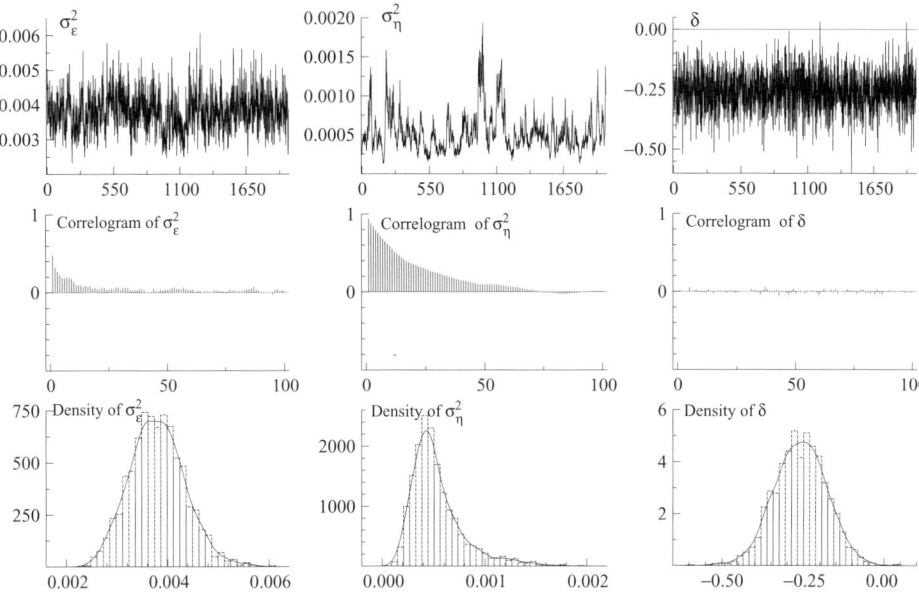

Fig. 1.3. Gibbs sampler results for the seat belt model: columns represent elements of ψ and rows represent realised draws of ψ, correlogram of series of draws and histogram of draws, respectively.

1.4 Non-Gaussian and nonlinear state space models

1.4.1 Introduction

The linear Gaussian state space model is adequate for a wide range of applications in practical time series analysis, possibly after transformations such as the log transformation. However, there are many practical situations where the linear Gaussian model is inadequate and more appropriate models are required. For example, if the observed data are the monthly numbers of people killed in road accidents in a particular region and if the numbers concerned are small, then the Poisson distribution will usually provide a better fit to the data than the normal distribution. We therefore need to develop a form of the state space model which represents the development over time of a Poisson variable rather than a normal variable. Other types of non-Gaussian models are needed to deal with such things as outliers in the observation equation or structural shifts in the state equation. These cases can be handled by employing heavy-tailed distributions for the disturbances in the observation or state equation, such as the t-distribution or a Gaussian mixture distribution.

Another class of cases we need to consider is where the appropriate state space model is nonlinear. For example, suppose that the trend and seasonal of a series combine multiplicatively but the disturbance term is additive. This situation cannot be handled by a linear model. Another case occurs in stochastic volatility models in financial statistics where we assume that the disturbance variance in the observation equation varies over time according to a specified submodel.

1.4.2 The non-Gaussian model

Suppose that the observational vectors y_t are determined by

$$p(y_t|\alpha_1,\ldots,\alpha_t,y_1,\ldots,y_{t-1}) = p(y_t|Z_t\alpha_t)$$

and that state vectors are determined, independently of Y_t, by

$$\alpha_{t+1} = T_t\alpha_t + R_t\eta_t, \qquad \eta_t \sim p(\eta_t),$$

for $t = 1,\ldots,n$, where the η_ts are serially independent. Densities $p(y_t|Z_t\alpha_t)$ or $p(\eta_t)$ or both can be non-Gaussian. Denote the signal $Z_t\alpha_t$ by θ_t. Consider the two special cases:

(i) Observations come from exponential family distributions

$$p(y_t \mid \theta_t) = \exp\left[y_t'\theta_t - b_t(\theta_t) + c_t(y_t)\right],$$

where $b_t(\theta_t)$ is twice differentiable.

(ii) Observations are generated by the relation

$$y_t = \theta_t + \varepsilon_t, \qquad \varepsilon_t \sim p(\varepsilon_t),$$

where ε_ts are non-Gaussian and serially independent.

We shall consider below how to deal with these models by simulation.

1.4.3 The nonlinear model

Consider the model

$$\begin{aligned} y_t &= Z_t(\alpha_t) + \varepsilon_t, & \varepsilon_t &\sim N(0, H_t), \\ \alpha_{t+1} &= T_t(\alpha_t) + R_t\eta_t, & \eta_t &\sim N(0, Q_t), \end{aligned}$$

for $t = 1,\ldots,n$. The functions $Z_t(\cdot)$ and $T_t(\cdot)$ are differentiable vector functions of α_t with dimensions p and m respectively. Again, we shall discuss below how to deal with this model by simulation.

1.4.4 Choice of simulation method

In Chapter 11 of the DK book we develop a comprehensive methodology for the treatment of non-Gaussian and nonlinear models. Since no analytical techniques are available the methodology is based on simulation. Most previous work using simulation for state space models has been based on MCMC techniques. Some of the those who have made contributions in this area are Fruwirth-Schnatter, Carlin and Carter, Kohn, Shephard, Pitt, Gamerman and Cargnoni (references are given in the DK book). Siem Jan Koopman and I looked at this work and we thought that the MCMC approach looked too complicated for the types of time series applications we were interested in. We were looking for a methodology that could be applied as part of a package that applied workers could use in a routine way, possibly for a large number of series, without having to scrutinise the development of the calculations for individual series in the way that is usual for MCMC. We decided to investigate whether methods could be developed based on traditional simulation techniques of importance sampling and antithetic variables, as discribed, for example, by Ripley (1987), that would be more suitable for our purpose than MCMC. We believe that we succeeded in this. I will now describe the ideas behind the methodology beginning with importance sampling.

1.4.5 Importance sampling

Denote the stacked vectors of states and observations over the whole series with $t = 1, \ldots, n$ by α and y. We consider the estimation of the conditional mean

$$\bar{x} = E[x(\alpha)|y] = \int x(\alpha) p(\alpha|y) d\alpha,$$

of an arbitrary function $x(\alpha)$ of α given observation vector y. This includes estimates of the mean vector $E(\alpha_t|y)$ and its conditional variance matrix $\text{Var}(\alpha_t|y)$; it also includes estimates of the conditional density and the distribution function of $x(\alpha)$ given y for scalar $x(\alpha)$. The conditional density $p(\alpha|y)$ usually depends on an unknown parameter vector ψ. In applications from a classical perspective, vector ψ is replaced by its maximum likelihood estimate $\hat{\psi}$, while in Bayesian analysis ψ is treated as a random vector.

For given ψ, let $g(\alpha|y)$ be a Gaussian density which is chosen to resemble $p(\alpha|y)$ as closely as is reasonably possible; we have

$$\bar{x} = \int x(\alpha) \frac{p(\alpha|y)}{g(\alpha|y)} g(\alpha|y) d\alpha = E_g \left[x(\alpha) \frac{p(\alpha|y)}{g(\alpha|y)} \right],$$

where E_g denotes expectation with respect to the so-called *importance density* $g(\alpha|y)$.

After some algebra we have

$$\bar{x} = \frac{E_g\left[x(\alpha)w(\alpha,y)\right]}{E_g\left[w(\alpha,y)\right]}, \quad \text{where} \quad w(\alpha,y) = \frac{p(\alpha,y)}{g(\alpha,y)}.$$

We reduce the expression for $\bar{x}$ to this form since the joint density $p(\alpha,y)$ is easier to deal with than the conditional density $p(\alpha|y)$. Choose a series of independent draws $\alpha^{(1)}, \ldots, \alpha^{(N)}$ from the distribution with density $g(\alpha \mid y)$ and take

$$\hat{x} = \frac{\sum_{i=1}^{N} x_i w_i}{\sum_{i=1}^{N} w_i}, \quad x_i = x(\alpha^{(i)}) \quad w_i = w(\alpha^{(i)}, y).$$

Since draws are independent, $\hat{x}$ converges to $\bar{x}$ probabilistically as $N \to \infty$. This estimate, however, is numerically inefficient and can be improved substantially as I shall describe shortly.

We obtain $g(\alpha|y)$ from a linear Gaussian approximating model which is chosen so that $g(\alpha|y)$ and $p(\alpha|y)$ have the same mode. Details are given in Chapter 11 of the DK book. In practice we do the computing with the disturbance vector η_t since in the approximating model

$$\alpha_{t+1} = T_t \alpha_t + R_t \eta_t,$$

where η_t has usually a smaller dimension than α_t, so we can easily get α_ts from η_ts. We therefore draw simulation samples from $g(\eta|y)$, where η is a stacked vector of the η_ts. This is done by the substitution of independent $N(0,1)$ variables into simple recursive formulae. The operation of drawing η from $g(\eta|y)$ is called simulation smoothing and is efficiently performed by an algorithm similar to that considered in Section 1.2.4; details are given in Durbin and Koopman 2002.

I have described here the basic ideas underlying the simulation techniques that we use for handling non-Gaussian and nonlinear models. However, we can obtain considerable improvements in efficiency by using antithetic variables.

1.4.6 Antithetic variables

An *antithetic variable* in this context is a function of a random draw of η which is equi-probable with η and which, when included together with η in the estimate of $\bar{x}$, increases the efficiency of the estimation. We employ two types of antithetic variables. The first is the standard one given by

$\check{\eta} = 2\hat{\eta} - \eta$ where $\hat{\eta} = E_g(\eta|y)$. Since $\check{\eta} - \hat{\eta} = -(\eta - \hat{\eta})$ and η is normal, the two vectors η and $\check{\eta}$ are equi-probable. Thus we obtain two simulation samples from each draw of the simulation smoother; moreover, values of conditional means calculated from the two samples are negatively correlated, giving further efficiency gains. When this antithetic is used the simulation sample is said to be *balanced for location*.

The second antithetic *balances for scale*. Let c be the sum of squares of the k $N(0,1)$ values drawn for a particular simulated value of η and let $q = Pr(\chi_k^2 < c)$. Let c^* be the value of χ_k^2 corresponding to probability $1 - q$. Then as c varies, c and c^* have the same distribution. Now take, $\eta^* = \hat{\eta} + \sqrt{c^*/c}(\eta - \hat{\eta})$. Then η^* has the same distribution as η. Do the same for $\check{\eta}$ above. Thus we have four draws of η for each set of $N(0,1)$ draws. Because of the balance in the four draws they are substantially more efficient than four independent draws would be.

We use straightforward developments of this technique to construct estimates of the likelihood function which we can maximise numerically to obtain estimates of unknown parameters. Dealing with missing observations is simple: we just treat these observations as missing values in the linear Gaussian approximating model in the way described earlier for the linear Gaussian model. Forecasts are obtained simply by treating future values as missing observations Further details are given in Chapters 11 and 12 of the DK book.

1.4.7 Bayesian analysis

We treat the parameter vector ψ as random with prior density $p(\psi)$ which initially we assume is a proper prior. As in the classical case suppose we wish to calculate the posterior mean

$$\bar{x} = E[x(\alpha)|y]$$

of a function $x(\alpha)$ of the stacked state vector α given the stacked observation vector y. This is a general formulation which enables us not only to estimate the posterior means of quantities of interest such as trend or seasonal, but also posterior variance matrices and posterior distribution functions and densities of scalar functions of the state.

We have

$$\begin{aligned}\bar{x} &= \int x(\alpha)p(\psi,\alpha|y)\mathrm{d}\psi\mathrm{d}\alpha \\ &= \int x(\alpha)p(\psi|y)p(\alpha|\psi,y)\mathrm{d}\psi\mathrm{d}\alpha.\end{aligned}$$

As an importance density for $p(\psi|y)$ we take its large sample normal approximation

$$g(\psi|y) = N(\hat{\psi}, \hat{V}),$$

where $\hat{\psi}$ is the solution of the equation

$$\frac{\partial \log p(\psi|y)}{\partial \psi} = \frac{\partial \log p(\psi)}{\partial \psi} + \frac{\partial \log p(y|\psi)}{\partial \psi} = 0,$$

and

$$\hat{V}^{-1} = -\left.\frac{\partial^2 \log p(\psi)}{\partial \psi \partial \psi'} - \frac{\partial^2 \log p(y|\psi)}{\partial \psi \partial \psi'}\right|_{\psi=\hat{\psi}}.$$

This large sample approximation is well known; see for example Bernardo and Smith 1994, Section 5.3). Let $g(\alpha|\psi,y)$ be a Gaussian importance density for α given ψ and y obtained as in Section 1.4.5 from an approximating linear Gaussian model. It follows that

$$\bar{x} = \int x(\alpha) \frac{p(\psi|y)p(\alpha|\psi,y)}{g(\psi|y)g(\alpha|\psi,y)} g(\psi|y) g(\alpha|\psi,y) \mathrm{d}\psi \mathrm{d}\alpha.$$

After some algebra, we obtain

$$\bar{x} = \frac{E_g[x(\alpha)z(\psi,\alpha,y)]}{E_g[z(\psi,\alpha,y)]},$$

where

$$z(\psi,\alpha,y) = \frac{p(\psi)g(y|\psi)}{g(\psi|y)} \frac{p(\alpha,y|\psi)}{g(\alpha,y|\psi)}.$$

We estimate this from N random draws of ψ from $g(\psi|y)$ and of α from $g(\alpha|\psi,y)$. However, for practical calculations we work with η rather than α directly, obtaining results for α via the state space equation recursion. For cases in which a proper prior is not available let $p(\psi)$ denote a noninformative prior. It is found that the resulting formula for $\bar{x}$ is exactly the same as for the proper prior. Further details are given in Chapter 13 of the DK book.

1.4.8 Illustration of an application to Poisson data

We consider the effect of the seat belt law on deaths of van drivers. The data are from the same study by Harvey and Durbin of road casualties as considered in Section 1.3. The numbers observed are too small for the normal approximation used in Section 1.3 to be valid. Let y_t be the number of van drivers killed. For the Poisson model let $\theta_t = Z_t \alpha_t$. The model is

$$p(y_t|\alpha_t) = \exp[\theta_t y_t - \exp \theta_t - \log y_t!]$$

with $E(y_t|\alpha_t) = \exp \theta_t$, analogously to the standard treatment in generalised linear modelling for non-time-series data. For the van driver data we take

$$\theta_t = \mu_t + \gamma_t + \delta i_t,$$

where μ_t, γ_t and i_t are as in Section 1.3. We perform both classical and Bayesian analyses. The results are given in Table 1.3.

Table 1.3. *Estimated effect of seat belt law on deaths of van drivers (δ)*

classical	$\hat{\delta}$	standard error of $\hat{\delta}$
	−0.278	0.114
Bayesian	posterior mean of δ	posterior std dev. of δ
	−0.280	0.126

These results are closely similar. Both give a reduction in deaths of around 24%. Standard errors due to simulation were 0.0036 and 0.0040. Calculations were based on 500 draws of $\eta|y$ with four antithetics per draw. These simulation standard errors are very small and they demonstrate that accurate analyses can be obtained using these techniques based on relatively small simulation samples.

Estimated standard errors of $\hat{\mu}_t$ in the classical analysis and posterior standard deviations in the Bayesian analysis are compared in Figure 1.4. There are two things to be noted about these graphs. The first is that the Bayesian standard deviation is always greater than the classical standard error. This is to be expected since the classical analysis is based on the assumption that the maximum likelihood parameter estimate $\hat{\psi}$ is the true value of the parameter vector ψ. A technique for estimating the bias in estimates of variance due to this assumption is given for the linear Gaussian model in Section 7.3.7 of the DK book; exactly the same procedure can be used for non-Gaussian and nonlinear models. The second point to note about Figure 1.4 is that, as was to be expected, there is a large increase in both graphs at the intervention point; however, it is perhaps surprising that there is a large drop below the general level immediately after the intervention point. Figure 1.5 presents a histogram estimate of the posterior density of the intervention effect δ.

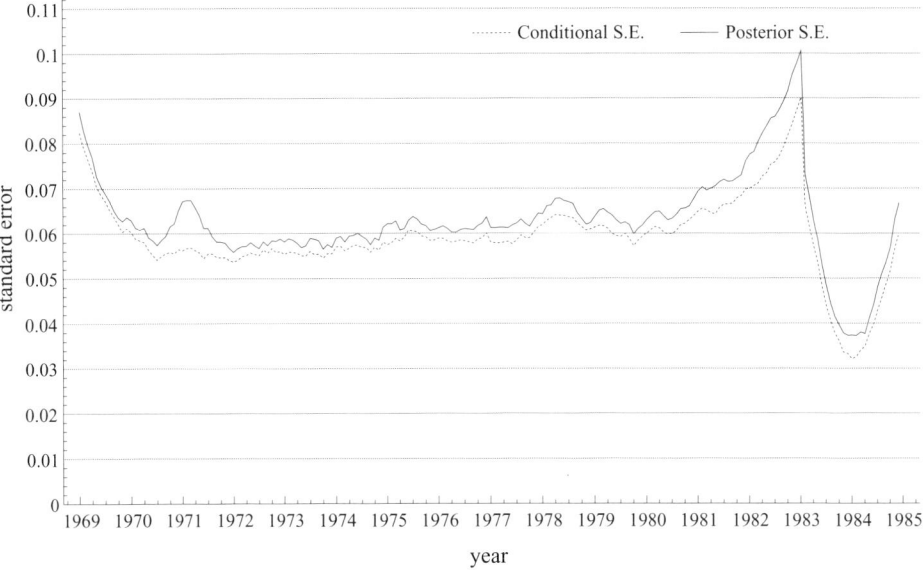

Fig. 1.4. Standard errors for trend level including intervention.

1.5 Software

Software for the application of the methodology considered in this paper is freely available on the internet. It is described in the DK book, Section 6.6 and 14.6. Software implementations of state space methods become more and more available as standard options in statistical packages. The STAMP program by Koopman, Harvey, Doornik and Shephard (2000) has implemented state space techniques for structural time series models as described in Section 1.1; it is a menu-driven user-friendly Windows package. Micro-CAPTAIN of Young (1998) is a similar package but is implemented for MATLAB. Software tools for the standard state space model are available in packages such as Eviews, SAS and TSP. A complete implementation for a general class of state space models with time-varying system matrices is the `SsfPack` package developed by Koopman, Shephard and Doornik (1999) for the Ox programming system and documented for the S-PLUS package by Zivot, Wang, and Koopman (see Chapter 13).

1.6 Conclusion

In this paper I have reviewed the use of linear Gaussian, non-Gaussian and nonlinear state space models for problems in practical time series analysis.

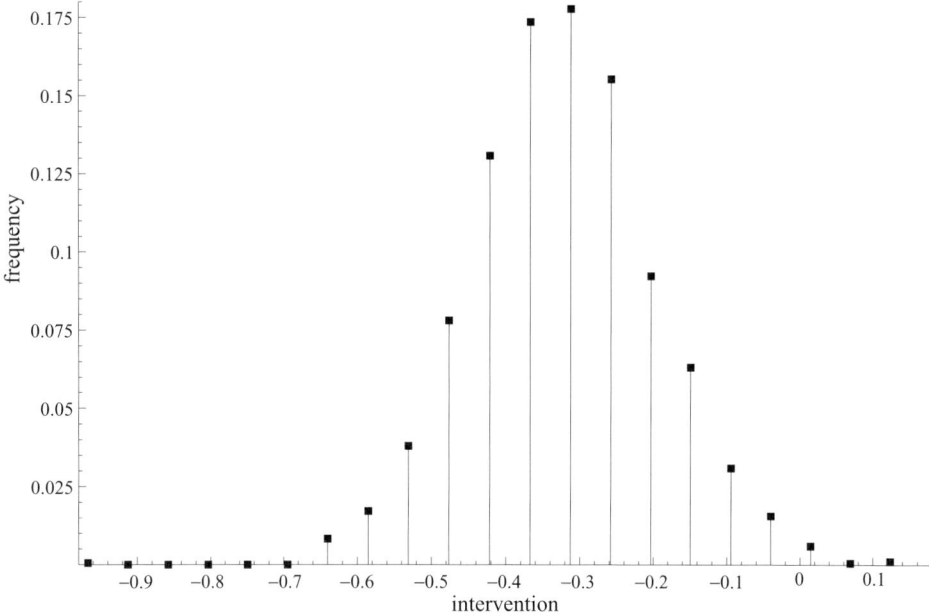

Fig. 1.5. Posterior distribution of intervention effect.

My contention is that the state space approach provides a convenient and efficient methodology for handling problems in this area.

Acknowledgements

I am indebted to Siem Jan Koopman for the calculations and diagrams in this paper and to Irma Hindrayanto for assistance.

2
State structure, decision making and related issues

Peter Whittle

Statistical Laboratory, Centre for Mathematical Sciences, University of Cambridge

Abstract

The article explores the extended and strengthened role of state structure when statistical analysis is coupled with the optimisation of decisions. The LQG theory is of course well developed, with its explicit algorithms and complete formal duality of estimation and control. However, the path-integral formulation which is so natural for the state-structured case leads to an elegant formalism which is less well known. The risk-sensitive models give a mild but significant generalisation of the LQG case, with a complete theory and a special simplification in the state-structured case. The application of large-deviation methods, when these are applicable, leads to a direct but radical generalisation of the LQG theory.

2.1 State structure in time series analysis

Durbin and Koopman (2001) (and references quoted therein) have eloquently demonstrated the importance of the concept of state in time series analysis. If the underlying model has state structure then this greatly eases inference, but a central (and well-recognised) thesis of this paper is that it also eases decision-making. This enhanced role also requires an enhancement of the concept of state.

The simplest state-structured model is the first-order scalar linear autoregression

$$x_t = ax_{t-1} + \epsilon_t, \qquad (2.1)$$

where the residuals ('noise variables') ϵ_t are supposed $NID(0, v)$. From this one obtains an immediate and simple evaluation of the quantity certainly required for inference: the likelihood based on the sample $(x_1, x_2, \ldots, x_n)$. If this likelihood conditional on the value of x_0 is written $\exp(-\mathcal{D})$ then $\mathcal{D}$ can be immediately evaluated as

$$\mathcal{D} = \text{const.} + \tfrac{1}{2}\sum_{t=1}^{n}(x_t - ax_{t-1})^2 + \tfrac{1}{2}n\log v, \qquad (2.2)$$

and the values of a and v minimising this expression are the maximum likelihood estimators of these parameters. Expression (2.2) implies an evaluation of both the inverse and the determinant of the covariance matrix of the conditioned sample. The evaluation is strikingly simple: the recursive nature of (2.1) implies that the quadratic form (2.2) involves only the sample autocovariances of lags 0 and 1. When the value of the state is not directly observable, so that the values of state x_t must themselves be estimated from observations y_t, then we shall see in Section 2.6 that the form of $\mathcal{D}$ for the joint variables x and y immediately suggests a recursive evaluation of state estimates.

One may note, in passing, that the concept of state is elastic enough to be stretched to yield evaluations of considerable generality. The analogue of expression (2.2) if x obeys a linear autoregression of order p is obvious, and can be written as the asymptotic (large n) evaluation

$$\mathcal{D} \sim \text{const.} + \frac{n}{4\pi}\int_0^{2\pi}\left(\frac{\hat{f}(\omega)}{f(\omega)} + \log f(\omega)\right)d\omega. \qquad (2.3)$$

Here $f(\omega)$ is the spectral density of the process and $\hat{f}(\omega)$ is the periodogram evaluated from the sample indicated. This formula was demonstrated by Whittle (1951, 1953) using circulant approximations for the covariance matrix of the sample. However, the necessary and sufficient condition for its validity is simply that the x-process should have a stable autoregressive representation, not necessarily of finite order.

This example raises a number of points. For example, the role of the linear/Gaussian assumptions, and the question of state-observability, to which we shall return. However, one should also take a larger view. The point of collecting statistical data is to gain understanding, in the first instance, but then to achieve a basis for action, in the second. In the time series case this action can often be seen as conscious control of a dynamical system. The two aspects, of estimation and control, turn out often to be strangely dual. Between them they imply a stronger role for state structure and require a strengthening of the concept itself. These are the matters we shall discuss in this paper, initially for the well-worn LQG specifications (linear models, quadratic costs, Gaussian inputs). In Sections 2.7 and 2.8 we consider some relaxations of the LQG assumptions; the role of state structure in such cases is increased, if anything.

Some general background material on these matters can be found in Bertsekas (1987), Brockett (1970), Davis (1977, 1993).

2.2 Uncontrolled and controlled models

Let us consider a deterministic dynamic process in discrete time t with variable x, so that x_t is the value of x at time t. The process $\{x_t\}$ is said to have *state structure* if, for every t, the value of x_t is sufficient for prediction, in that it alone determines the path of the process from time t onwards. This implies then that the sequence $\{x_t\}$ obeys a forward recursion

$$x_{t+1} = f(x_t, t) \tag{2.4}$$

for some f. The stochastic analogue of this structure is that of a *Markov process*, for which $P(x_{t+1}|X_t) = P(x_{t+1}|x_t)$ for all t. Here X_t denotes the history $\{x_\tau; \tau \leq t\}$ of x up to time t and P denotes the conditional distribution indicated.

Consider now the case of a controlled process $\{x_t\}$ for which one wishes to optimise the value of the control variable u_t at each time t, taking into account its effect on the future course of the x-process. In such a case one might term a dynamic variable ξ an *operational state variable* if ξ_t is all that one needs to retain of observation history at time t in order to optimise control from that time. The variable must then have the properties that: (i) the optimal value of u_t is a function of ξ_t alone, and (ii) the sequence ξ_t is generated by a simple forward recursion, in that ξ_{t+1} is a function just of ξ_t and the observation gathered at time $t+1$.

This stronger demand will imply stronger structural conditions. In the deterministic case one conjectures that the condition

$$x_{t+1} = f(x_t, u_t, t) \tag{2.5}$$

would be the appropriate analogue of (2.4), if x itself is to remain a state variable. However, one must also set conditions on the control-optimisation criterion and on the nature of the observations. Feasibility of a control rule implies that the value chosen for u_t can be a function only of W_t, the observables which are available at time t. The following conditions imply that x is itself an operational state variable in the stochastic case:

Assumption 1: The Markov condition which is the stochastic analogue of (2.5): that is, $P(x_{t+1}|X_t, U_t) = P(x_{t+1}|x_t, u_t)$ for all relevant t. Here u_t must be regarded as a parameterising variable rather than a conditioning variable, but the structure of a temporal decision process in fact makes the usage consistent; see Whittle (1982, pp. 150–2). (That is, u_t is not initially defined as a random variable, but simply as a quantity to be prescribed which parameterises the conditional distribution $P(x_{t+1}|x_t)$. However, u_t will in general become a random variable once a control policy is specified, and the conditional distribution $P(x_{t+1}|x_t, u_t)$ would not in general be identical with the parameterised distribution just defined. However, the two expressions can in fact be identified if the control policy is realisable.)

Assumption 2: The current state value x_t is observable at the time t when u_t is to be determined, so that $W_t = (X_t, U_{t-1})$.

Assumption 3: The control rule is to be chosen to minimise $E\mathcal{C}(X, U)$, where X and U denote complete trajectories of the variables x and u and the cost function $\mathcal{C}$ has the additive form

$$\mathcal{C}(X, U) = \sum_{t<h} c(x_t, u_t, t) + K(x_h) = \sum_{t<h} c_t + K \tag{2.6}$$

say. Here h is the *horizon point*, the end of the planning period over which one is optimising. If one assumes h finite then one has a starting point for the backward determination of the optimal control rule.

Let $F(W_t)$ be the *value function*, the minimal expected future cost which can be incurred from time t conditional on the value of observables W_t at time t. Then F obeys the *dynamic programming equation*: the backward equation

$$F(W_t) = \inf_u \{c_t + E[F(W_{t+1})|W_t, u_t]\}. \tag{2.7}$$

Here the infimum is with respect to u_t, and the infimising value is the optimal value. Starting from $F(W_h) = K(x_h)$ one establishes by backward induction that both $F(W_t)$ and the optimal u_t are functions of x_t and t alone, whence it follows that x remains a state variable in this strengthened sense. The dynamic programming equation (2.7) then takes the reduced form

$$F(x,t) = \inf_u \{c(x,u,t) + E[F(x_{t+1}, t+1)|x_t = x, u_t = u]\}. \quad (2.8)$$

If one examines the above argument then one sees that the assumption (2.6) cannot be greatly weakened if one wishes to preserve the conclusion. If $\mathcal{C}_t$ is the cost incurred from time t onwards then the argument above will go through only if $\mathcal{C}_t$ obeys a *backward* recursion of the form

$$\mathcal{C}_t = g(x_t, u_t, t, \mathcal{C}_{t+1}).$$

Furthermore, the function g must be linear in its $\mathcal{C}$-argument if the process is stochastic, so that expectations are involved. The cost function (2.6) is then almost as general as is permissible.

If the current physical state x_t is imperfectly observed then the operational state variable for the control problem is the whole posterior distribution $P(x_t|W_t)$. This distinctly unwelcome ballooning of physical state to the much higher-dimensional 'information state' or 'hyperstate' has a resolution for the LQG processes to which we now turn.

2.3 LQG structure

Simplifications induced by state structure are quite distinct from the simplifications induced by LQG structure, although the two together produce the ultimate in tractability. The acronym LQG stands for Linear dynamic equations, Quadratic cost functions and Gaussian disturbances, but we can express the characterisation more pithily and exactly.

Let us suppose that the process variable x, observations y and control variable u are all vector-valued (i.e. take values in finite-dimensional Euclidean spaces). Denote the complete trajectories of these quantities over the relevant time period by X, Y and U. Denote the cost function whose expectation is to be minimised by $\mathcal{C}(X,U)$ and the probability density of X and Y for given U by $\exp[-\mathcal{D}(X,Y,U)]$. One may say that $\mathcal{D}$ measures the *discrepancy* of X and Y for given U; i.e their degree of improbability on a logarithmic scale. Then the problem has LQG structure if and only if $\mathcal{C}$ and $\mathcal{D}$ are quadratic functions of their arguments. (It is supposed that the density is relative to Lebesgue measure, with the implication that X and

Y are subject to no other constraints than the stochastic one implied by specification of $\mathcal{D}$.)

If there is no aspect of control in the problem then the discrepancy is a function of X and Y alone, and the LQG assumption implies that these variables are jointly normally distributed. At time t the set of observables is then just Y_t, and estimates of unobservables (X and future observations) are obtained by minimising $\mathcal{D}(X,Y)$ with respect to these quantities. These estimates, maximising the posterior density of the unobservables, have a dual and optimising property: they minimise the mean square estimation error. In the state-structured case (just that in which x_t and y_t jointly obey a first-order linear autoregression with Gaussian residuals uncorrelated over time) the estimate $\hat{x}_t$ of x_t based upon Y_t is generated recursively by the celebrated Kalman filter.

A controlled LQG process is special in that it obeys the *certainty equivalence principle* (CEP). In the state-structured case this principle is very clearly what it seems to claim. When current state is perfectly observed the optimal value of u_t is a function $u(x_t, t)$, linear in x_t. In the case of imperfect observation the optimal control is then just $u(\hat{x}_t, t)$ – the optimal perfect-observation rule with x_t replaced by its estimate.

In the general LQG case the principle would be applied in the following form. Consider the situation at time t, when the set of observables is $W_t = (Y_t, U_{t-1})$: that is, past and present observations and past control decisions. Determine an estimate of the state trajectory X for given U by minimising $\mathcal{D}(X, Y, U)$ with respect to all current unobservables. Replace X in $\mathcal{C}(X, U)$ by its estimate and minimise the resulting quantity with respect to all currently unmade decisions: $\{u_\tau; \tau \geq t\}$. Then the value of u_t determined in this way is optimal. This rather complicated-sounding procedure has a simple recursive realisation in the state-structured case; see the next section.

We shall see in Section 2.7 that the CEP has a version for processes with the so-called LEQG structure – more general than LQG structure. State structure turns out to have an added significance in this more general context. It not only gives the CEP the particular transparency which we have already noted in the LQG case, but also allows us to carry over the recursive algorithms of the LQG case.

2.4 An archetypal control problem

The simplest state-structured LQG problem is that for which x and u are vector-valued, x_t is observable at time t and the *plant equation* (or 'state equation'; see the next section) is a stochastic version of the linear relation

(2.5):
$$x_{t+1} = Ax_t + Bu_t + \epsilon_t. \tag{2.9}$$

Here A and B are matrices of appropriate dimension, which for simplicity we have supposed independent of t, and the disturbances ϵ_t are NID$(0, N)$. One supposes also that the instantaneous cost function $c(x, u, t)$ is quadratic in x and u. The choice of a fixed positive-definite quadratic form

$$c(x, u, t) = \tfrac{1}{2}(x_t'Rx_t + u_t'Qu_t + 2u_t'Sx_t) \tag{2.10}$$

is appropriate if one wishes to penalise deviations of x and u from zero. One finds from the dynamic programming equation (2.8) that the optimal control has the linear form $u_t = K_t x_t$ and the value function the quadratic form $F(x, t) = (1/2)x'\Pi_t x + \delta_t$, where

$$K_t = -(Q + B'\Pi_{t+1}B)^{-1}(S + B'\Pi_{t+1}A) \tag{2.11}$$

and the sequence of matrices Π_t is determined by the backward recursion

$$\Pi_t = R + A'\Pi_{t+1}A - (S' + A'\Pi_{t+1}B)(Q + B'\Pi_{t+1}B)^{-1}(S + B'\Pi_{t+1}A). \tag{2.12}$$

Relation (2.12) is the celebrated (although superficially unappealing) *matrix Riccati equation*. It occurs (as a forward equation for the covariance matrix V_t of estimation errors) when one considers state estimation – evidence of the duality between the optimisation of control and that of estimation (considered in the next section). This optimising control and the associated Riccati equation (2.12) were deduced by Kalman and other authors in the 1950s and are now standard. See, for example, Davis (1977, 1993), Whittle (1982, 1996) for details of the derivation.

Note that stationarity is not assumed, although, if a stabilising control exists and if the cost function positively penalises instability, then the matrices Π_t and K_t will become independent of t as the horizon becomes infinitely distant, and the controlled process will then converge to stationarity with increasing t.

2.5 State estimation and the Kalman filter

What turns out to be the most natural formulation of imperfect state observation is to assume that state x_t and observation y_t are generated by the pair of relations

$$x_{t+1} = Ax_t + \epsilon_t, \tag{2.13}$$
$$y_{t+1} = Cx_t + \eta_t, \tag{2.14}$$

where the stacked vector of plant noise ϵ and observation noise η is Gaussian and uncorrelated in time, with zero mean and covariance matrix

$$\operatorname{cov}\begin{bmatrix} \epsilon \\ \eta \end{bmatrix} = \begin{bmatrix} N & L \\ L' & M \end{bmatrix}. \tag{2.15}$$

Here y_t is the observation available at time t, so the assumption is that plant and observation variables jointly obey a first-order linear autoregression. The plant equation (2.13) is just a version of (2.9), but with the control term removed for simplicity.

It is perhaps at this point that one should clarify some of the differing conventions prevailing in stochastic control theory on the one hand, and in statistics (and perhaps econometrics) on the other. In control theory the system being controlled is the 'plant', and the variables describing it and the dynamic equations these obey are respectively the 'plant variables' and the 'plant equation'. These terms do not imply state structure, although in the present case we have such structure, so that the plant equation (2.13) is just what statisticians would term a 'state equation'. Equation (2.14) is the observation equation conventionally assumed in control theory; statisticians would usually regard y as an observation on *current* state, and so would replace x_t by x_{t+1} in (2.14). Prescription (2.14) has the advantage that the process $\{x_t, y_t\}$ is then Markov, so that observation of y alone corresponds to the idea of a partial observation of state. It is also this convention which leads to the complete duality of control and estimation. Treatments of the Kalman filter and related algorithms derived under the alternative hypothesis develop somewhat less naturally, although final results differ only slightly; see Durbin and Koopman (2001), pp. 67–8).

Kalman showed that the posterior distribution $P(x_t|W_t)$ is Gaussian with mean $\hat{x}_t$ and covariance matrix V_t, where these quantities obey the forward recursions

$$V_{t+1} = N + AV_t A' - (L + AV_t C')(M + CV_t C')^{-1}(L' + CV_t A') \tag{2.16}$$
$$\hat{x}_{t+1} = A\hat{x}_t + H_t(y_{t+1} - C\hat{x}_t), \tag{2.17}$$

where

$$H_t = (L + AV_t C')(M + CV_t C')^{-1}. \tag{2.18}$$

These relations demonstrate the complete duality of estimation and control optimisations. The forward Riccati equation (2.16) is the analogue of the backward equation (2.12), and the determination (2.17) of the matrix coefficient H is the analogue of the determination (2.11) of the control coefficient K.

2.6 Signal extraction and generating function methods

One would imagine that nothing new could be said about signal extraction and the deduction of the so-called 'smoothed' or 'retrospective' estimates of state in the time-homogeneous LQG model. For example, Durbin and Koopman (2001), pp. 70–80) have a very thorough treatment of the subject. However, there are points of some novelty to be made, mainly by taking advantage of the special form of the 'path integral' which the exponent $\mathcal{D}$ of the Gaussian likelihood presents.

We shall initially assume infinite realisations to past and future, with state and observation realisations denoted by X and Y. One can still write down formal probability densities, and the quadratic path integral which appears in the exponent of the joint density of X and Y can be written

$$\sum_t \begin{bmatrix} \epsilon_t \\ \eta_t \end{bmatrix}' \begin{bmatrix} N & L \\ L' & M \end{bmatrix}^{-1} \begin{bmatrix} \epsilon_t \\ \eta_t \end{bmatrix} = \sum_t (x_t' \Psi_j x_{t-j} - 2x_t' \Delta_j y_{t-j} + \cdots), \quad (2.19)$$

where the noise variables ϵ and η are to be expressed in terms of x and y by the equations (2.13) and (2.14), and the second expression picks out all terms in x. Note that $\Psi_{-j} = \Psi_j'$. The estimate $\tilde{X}$ of X given by the posterior mode is then determined by the system of equations

$$\Psi(\mathcal{T})\tilde{x}_t = \Delta(\mathcal{T})y_t, \quad (2.20)$$

where $\Psi(\mathcal{T}) = \sum_j \Psi_j \mathcal{T}^j$ etc., and $\mathcal{T}$ is the backward shift operator, with effect $\mathcal{T}x_t = x_{t-1}$. It is an immediate consequence of state structure that Ψ_j and Δ_j are zero for $|j| > 1$, so that the operators in (2.20) involve only unit lags into past and future. Suppose that the matrix of generating functions $\Psi(z)$ has the canonical factorisation

$$\Psi(z) = \psi(z)\psi(z^{-1})' = \psi(z)\overline{\psi(z)}, \quad (2.21)$$

say, where the canonical factor is such that the matrix $\psi(z)$ and its inverse can be validly expanded in nonnegative powers of z inside the unit circle. Then the system (2.20) has the valid inversion

$$\tilde{x}_t = \overline{\psi(\mathcal{T})}^{-1} \psi(\mathcal{T})^{-1} \Delta(\mathcal{T}) y_t. \quad (2.22)$$

Suppose that observations have only been taken up to time t, so that X must be estimated from Y_t. Denote the estimate $E(x_\tau | W_t)$ by $x_\tau^{(t)}$, so that

$\hat{x}_t = x_t^{(t)}$ and $\tilde{x}_t = x_t^{(\infty)}$. Then the equation system (2.20) would be modified to

$$\Psi(\mathcal{T})x_\tau^{(t)} = \Delta(\mathcal{T})y_\tau^{(t)}, \tag{2.23}$$

where the operator $\mathcal{T}$ now acts on the subscript τ. For $\tau > t$ we have $x_\tau^{(t)} = A^{\tau-t}\hat{x}_t$ and $y_\tau^{(t)} = CA^{\tau-t-1}\hat{x}_t$. By appeal to the canonical factorisation (2.21) we can economically deduce from (2.23) the Kalman recursion for $\hat{x}_t$ and similar recursive updatings for the retrospective estimates $x_\tau^{(t)}$ ($\tau < t$). (See Whittle 1996, pp. 361–3).)

The relation between canonical factorisation and solution of a Riccati equation is easily seen in the state-structured equilibrium case. We can look for the canonical factorisation in the form

$$\Psi(z) = (I - \Gamma z)V(I - \Gamma' z^{-1}), \tag{2.24}$$

where Γ is a stability matrix. (Recall that $\Psi(z)$ now contains only terms in z^j for $j = -1, 0, +1$.) Equating coefficients in (2.24) and eliminating Γ we find that V is determined by the archetypal Riccati equation:

$$V = \Psi_0 - \Psi_1 V^{-1} \Psi_{-1}. \tag{2.25}$$

The deduction of (2.20) by minimisation of (2.19) rather than by minimisation of a mean square error means that our Ψ is expressed in terms of information matrices (inverses of covariance matrices) rather than covariance matrices themselves. One can clear (2.19) of inverses by appeal to an identity of the form

$$\epsilon' N^{-1} \epsilon = \max_\lambda [2\lambda'\epsilon - \lambda' N \lambda]. \tag{2.26}$$

The extremising value of λ is determined by $N\lambda = \epsilon$. In case (2.19) we shall have a pair of auxiliary variables λ and μ which are related to the *estimated* noise variables by

$$\begin{bmatrix} N & L \\ L' & M \end{bmatrix} \begin{bmatrix} \lambda_t \\ \mu_t \end{bmatrix} = \begin{bmatrix} \epsilon_t \\ \eta_t \end{bmatrix}. \tag{2.27}$$

In terms of the expanded set of variables (x, λ, μ) we now obtain an expanded Ψ of the form

$$\Psi(z) = \begin{bmatrix} N & L & I - Az \\ L' & M & -Cz \\ I - A'z^{-1} & -C'z^{-1} & 0 \end{bmatrix} = \psi(z)\overline{\psi(z)}. \tag{2.28}$$

Although of greater dimension than the previous version its form is simple and revealing; exploitation of the canonical factorisation leads quickly to a range of familiar results.

This formalism has a control analogue. Consider the problem with cost function (2.10) and as plant equation the modified version

$$x_{t+1} = Ax_t + Bu_t + d_t + \epsilon_t \tag{2.29}$$

of (2.9), in which the sequence of disturbances has both random and known components, ϵ_t and d_t. We assume the current value of state to be observable. One can then write down a quadratic path integral analogous to (2.19) in variables x, u and λ, where λ is the Lagrange multiplier associated with the constraint represented by the plant equation (2.29). The analogue of expression (2.28) is now

$$\Phi(z) = \begin{bmatrix} R & S' & I - A'z^{-1} \\ S & Q & -B'z^{-1} \\ I - Az & -Bz & 0 \end{bmatrix}, \tag{2.30}$$

We seek now a canonical factorisation in the form

$$\Phi(z) = \phi(z^{-1})'\phi(z) = \overline{\phi(z)}\phi(z), \tag{2.31}$$

say, where the canonical factor is such that $\phi(z)$ and $\phi(z)^{-1}$ can be validly expanded in nonnegative powers of z inside the unit circle. The operational significance of this factorisation is most evident from the fact that the equation system analogous to (2.20) can be semi-inverted to a set of relations

$$\phi(\mathcal{T})\begin{bmatrix} x \\ u \\ \lambda \end{bmatrix}_t = \overline{\phi(\mathcal{T})}^{-1}\begin{bmatrix} 0 \\ 0 \\ d \end{bmatrix}_t \tag{2.32}$$

which determine the optimal control and the path of the optimal process. The term on the left in (2.32) is the *feed-back* term, giving the dependence of u_t upon past observations, and the term on the right is the *feed-forward* term, giving the dependence of u_t upon future disturbances.

As with most generating-function techniques, this approach is restricted to the equilibrium case (when Π_t has reached its infinite-horizon limit Π), but generalises immediately to higher-order recursions. It is then a technique which is not particularly linked to state structure. Nevertheless, the state-structured case provides the clearest example, and the reformulation of the unappealing Riccati equation (2.12) as the elegant canonical factorisation (2.31) is both satisfying and revealing.

2.7 Risk-sensitive criteria; the LEQG model

One would like to break away from the LQG model and obtain substantial results in more general cases. The LEQG model is a mild but significant move in this direction, with fascinating implications. The modification is to replace the quadratic cost function $\mathcal{C}$ by $\exp(-\theta\mathcal{C})$. The 'EQ' of 'LEQG' refers to the 'exponential of a quadratic' which now measures utility or disutility, according to the sign of θ. The policy is chosen to maximise or minimise the expectation of this quantity according as θ is positive or negative. The two cases correspond to those in which the optimiser is risk-seeking (optimistic) or risk-averse (pessimistic). This class of processes was first considered by Jacobson (1973, 1977), who showed how the determination (2.11), (2.12) of the optimal control rule generalises in the case of perfect state observation. The case of imperfect state observation resisted analysis until Whittle (1981) showed that a CEP held, although in a form which required a radically generalised understanding.

To summarise the approach, define the linear combination of cost and discrepancy

$$\mathcal{S} = \mathcal{C} + \theta^{-1}\mathcal{D} \qquad (2.33)$$

which we shall term *stress*. Then the *risk-sensitive CEP* states that: if one extremises stress with respect to all unobservables at time t and minimises it with respect to all present and future decisions, then the value of u_t thus determined is optimal. Here by 'extremise' we mean that one minimises or maximises according as θ is positive or negative.

This principle provides the essential lever on the problem, but one whose application is transparent and effective only if one makes the assumption of state structure. In this case the CEP also delivers a clear separation principle. The argument goes as follows.

Suppose that one is working at time t, and so seeking to optimise the immediate decision u_t. Suppose provisionally that x_t is known. Then this achieves a separation, in that past stress and future stress can be minimised/extremised separately. The calculations turn out to be just a modified form of those for the conventional risk-neutral case, with backward and forward Riccati equations associated with the optimisations of future control and estimation of the past respectively. These calculations are then coupled by imposition of the final extremisation of stress: that with respect to x_t. The extremising value is just the generalisation of the risk-neutral estimate $\hat{x}_t$, and the insertion of this estimate into the formula for u_t corresponds to the simple certainty equivalence principle of the risk-neutral case.

One cannot say that one has a separation of control and estimation, because control costs now affect estimates and noise statistics now affect control. However, one has separation in the sense that the state value x_t provides a pivot which separates past and future. For given x_t past and future optimisations are separated and of conventional form; these optimisations are then coupled by the determination and substitution of the stress-extremising value of x_t.

2.8 Non-LQG models and large deviations

Can one generalise any of this material to the radically non-LQG case? The subject of *nonlinear filtering* represents a sustained attempt to do so, in that it attempts to find general analogues of the Kalman filter (2.17) and of the signal extraction techniques of Section 2.6. The aspiration is frustrated at two levels. Firstly, the amenability of the linear/quadratic calculations of the LQG case simply does not transfer. Secondly, and more fundamentally, the Gaussian nature of the posterior distribution $P(x_t|W_t)$, parameterised by its first and second moments, scarcely ever transfers. There are very few cases for which this distribution has a finite number of parameters, whose updating achieves the updating of the distribution.

However, there is a class of cases for which one can make some degree of analytic headway. Consider a vector-valued continuous-time Markov process with infinitesimal generator Λ. Bartlett (1949) introduced the *derivate characteristic function* (abbreviated to DCF)

$$H(x, \alpha) = e^{-\alpha x} \Lambda e^{\alpha x}. \tag{2.34}$$

This is related to the characteristic function of the increment δx immediately after time t conditional on the value x of $x(t)$. The process with DCF $\kappa H(x, \kappa^{-1}\alpha)$ is one in which the increment is replaced by the average of κ independent realisations of the increment, although the transformation can be induced by other averaging operations over a system. As κ increases the process converges to its deterministic version, but just before that its stochastic behaviour begins to obey the stark but significant formalism of large-deviation theory. We shall sketch the application of this formalism for the evaluation of the posterior distribution.

Suppose that $x(t)$ and the integrated observation $s(t) = \int^t y(\tau) d\tau$ have joint DCF $\kappa H(x, \kappa^{-1}\alpha, \kappa^{-1}\beta)$. Then it was shown in Whittle (1991) that for large κ the posterior distribution of $x(t)$ has the evaluation

$$E\left[e^{\alpha x(t)}|W(t)\right] = \exp[\kappa \psi(\kappa^{-1}\alpha, t) + o(\kappa)] \tag{2.35}$$

where ψ is determined by the relations

$$\frac{\partial \psi(\alpha,t)}{\partial t} = \sigma(\alpha) - \sigma(0) \qquad (2.36)$$

$$\sigma(\alpha) = \inf_{\beta} \left[H\left(\frac{\partial H}{\partial \alpha}, \alpha, \beta\right) - \beta y \right]. \qquad (2.37)$$

The expression (2.35) for the characteristic function determines the distribution explicitly by the relation

$$P[x(t) \in A | W(t)] = \exp\{\kappa \sup_{x \in A} \inf_{\alpha} [\psi(\alpha,t) - \alpha x] + o(\kappa)\}. \qquad (2.38)$$

Relations (2.36) and (2.37) in principle determine the posterior distribution in a form stripped of asymptotic irrelevancies. Hijab (1984) developed a relation in a particular case (that in which the plant equation is a differential equation driven by low-power white noise) which he termed the 'wave equation of nonlinear filtering'. Relations (2.36) and (2.37) would seem much more deserving of this distinction. However, the reduction they achieve cannot solve the fundamental problem: that very seldom will the distribution have a finite parameterisation. They yield the known exact results in the LQG case. Surprisingly, they also yield the exact results in some others. Suppose, for example, that particles enter a chamber in a Poisson stream and leave it independently at a common fixed rate. The number $x(t)$ of particles in the chamber at time t is the state variable of a Markov process with a Poisson equilibrium distribution. Suppose that the only information on x comes from a sensor inserted in the chamber, on which particles present may register, independently and with a constant intensity. The total number of registrations up to time t is then just the integrated observation $s(t)$, whose history conditions the posterior distribution of $x(t)$. Introduction of the factor κ corresponds to multiplication of the input rate and division of variables by this factor, but relations (2.36) and (2.37) determine exact dynamic relations for the posterior distribution in any case.

3
An introduction to particle filters

Simon Maskell
Department of Engineering, University of Cambridge

Abstract

This paper introduces particle filtering to an audience who are more familiar with the Kalman filtering methodology. An example is used to draw comparisons and discuss differences between the two approaches and to motivate some avenues of current research.

3.1 Introduction

This paper introduces particle filtering to an audience that is unfamiliar with the literature. It compliments other introductions (Doucet, de Freitas and Gordon 2001) and tutorials (Arulampalam, Maskell, Gordon and Clapp 2002) with a contribution that appeals to intuition and documents an understanding that demystifies what can appear to be a much more complex subject than it truly is.

Particle filtering is a new statistical technique for sequentially updating estimates about a time evolving system as measurements are received. The approach has been developed in parallel by a number of different researchers and so is also known as: the CONDENSATION algorithm (Blake and Isard 1998), the bootstrap filter (Gordon, Salmond and Smith 1993), interacting particle approximations (Dan, Moral and Lyons 1999) and survival of the fittest (Kitagawa 1996). The approach opens the door to the analysis of time series using nonlinear non-Gaussian state space models. While linear Gaussian models can cope with a large variety of systems (e.g. Harvey (1989) and West and Harrison 1997)), nonlinear non-Gaussian models offer an even richer vocabulary with which to describe the evolution of a system and observations of this system.

Particle filtering is based on the idea that uncertainty over the value of a continuous random variable x can be represented using a probability density function, a pdf, $p(x)$. From this pdf, it is possible to deduce the estimates that are of interest; for example the mean, various quantiles and the covariance. It is rarely possible to uniquely deduce the pdf from such estimates. So, to use nonlinear, non-Gaussian state space models, we have to learn how to directly manipulate general pdfs. This paper therefore describes how to use such models to analyse time-series by considering how to manipulate pdfs.

Section 3.2 starts by describing the state space model and its generic solution. The way that the Kalman filter then implements this generic solution is described in Section 3.3 before Section 3.4 describes how the extended Kalman filter can tackle nonlinear problems by approximating the models and illustrates this using a simple example. Section 3.5 then describes particle filtering in the same context. Throughout, an example is used to aid the explanation. Finally, Section 3.6 illustrates the scope of the approach through reference to some specific examples of applications of particle filtering before Section 3.7 draws some conclusions.

3.2 State space model

3.2.1 Problem definition

At time t, we get some measurement, y_t, which is a function of the underlying state of the system, x_t. Sequential inference, or tracking, is the problem of describing the uncertainty over this underlying state of the system as a stream of measurements is received (Bar-Shalom and Li 1995). In the context of pdfs, at time t, this uncertainty is represented as $p(x_{1:t}|y_{1:t})$, where $x_{1:t} = \{x_1 \ldots x_t\}$ is the history of states and $y_{1:t} = \{y_1 \ldots y_t\}$ is the history of measurements. Hence $p(x_{1:t}|y_{1:t})$ is the uncertainty over the history of states that can be inferred from the history of measurements. From this, by the marginalisation theorem, one can obtain quantities that may be of more interest such as the pdf of the current state $p(x_t|y_{1:t})$ and the expected value of the current state $x_{t|t}$:

$$x_{t|t} = \int x_t p(x_{1:t}|y_{1:t}) \, \mathrm{d}x_t. \tag{3.1}$$

The problem is then to sequentially compute $p(x_{1:t}|y_{1:t})$.

3.2.2 Model definition

To be able to solve this problem, a model is required for the dynamics of the state and for the measurement process. Here, we use state space models, which are probabilistic and so represented using pdfs.

It is often assumed that the x_t process is Markov, so the state at a time step, x_{t-1}, is a sufficient statistic of the history of the process, $x_{1:t-1}$. Since the state captures all the information known about the system, the state at a time step is also assumed a sufficient statistic of the history of measurements, $y_{1:t-1}$. The current state, x_t, is therefore independent of the history of states and measurements if the previous state, x_{t-1}, is known:

$$p(x_t|x_{1:t-1}, y_{1:t-1}) = p(x_t|x_{t-1}, x_{1:t-2}, y_{1:t-1}) = p(x_t|x_{t-1}). \tag{3.2}$$

While in general, the measurement could be a function of the entire history of states, $x_{1:t}$, and the previous measurements, $y_{1:t-1}$, the case often encountered is that the measurement is independent of the history of states and the previous measurements:

$$p(y_t|x_{1:t}, y_{1:t-1}) = p(y_t|x_t, x_{1:t-1}, y_{1:t-1}) = p(y_t|x_t). \tag{3.3}$$

3.2.3 Prediction

To solve the tracking problem, each of the two components of the state space model is considered in turn. In the prediction step, the dynamic model is used to manipulate the pdf from the previous iteration via the multiplication theorem:

$$p(x_{1:t}|y_{1:t-1}) = p(x_t|x_{1:t-1}, y_{1:t-1}) p(x_{1:t-1}|y_{1:t-1}), \text{ using (3.2)}, \quad (3.4)$$
$$= p(x_t|x_{t-1}) p(x_{1:t-1}|y_{1:t-1}). \quad (3.5)$$

$p(x_{1:t}|y_{1:t-1})$ is then referred to as the predicted pdf.

In this paper, we start the recursion with a proper prior that integrates to unity, $p(x_0) = p(x_{1:0}) = p(x_{1:0}|y_{1:0})$ (that is, $p(x_{1:t-1}|y_{1:t-1})$ with $t = 1$). This prior is assumed to be known and able to be sampled from. This does exclude from the discussion the possibility of using truly diffuse priors (with infinite variance) as in Durbin and Koopman (2001), but does not exclude the possibility of using very large variances which may for practical purposes be sufficient.[1]

3.2.4 Update

Bayes theorem can then be used to manipulate this predicted pdf using the measurement model and the most recently received measurement:

$$p(x_{1:t}|y_{1:t}) = p(x_{1:t}|y_t, y_{1:t-1})$$
$$= \frac{p(y_t|x_{1:t}, y_{1:t-1}) p(x_{1:t}|y_{1:t-1})}{p(y_t|y_{1:t-1})}, \text{ using (3.3)}, \quad (3.6)$$
$$= \frac{p(y_t|x_t) p(x_{1:t}|y_{1:t-1})}{p(y_t|y_{1:t-1})}, \quad (3.7)$$

where the denominator is essentially just a normalising constant.

So, (3.5) and (3.7) form a solution to the tracking problem. The question that remains is how to carry out the required calculations in practice.

---

[1] It should be noted that the use of such priors will make pronounced any nonlinearity that is present; this in turn is likely to make it essential that care is taken in designing the filter. What nonlinearity means in this context and why this necessitates care in filter design is explained in the following sections.

3.3 Linear Gaussian filtering

These calculations are (relatively) simple when the models are linear and Gaussian, that is, one can express the models in the following way:

$$x_t = Ax_{t-1} + \omega_t^x, \quad y_t = Hx_t + \omega_t^y,$$
$$\omega_t^x \sim N(\omega_t^x; 0, Q), \quad \omega_t^y \sim N(\omega_t^y; 0, R).$$

Corresponding to this is the filtering density

$$p(x_{t-1}|y_{1:t-1}) = N(x_{t-1|t-1}, P_{t-1|t-1}),$$

where $P_{t|t}$ is the covariance associated with $x_{t|t}$.

If the models are of this form, it is possible to recursively calculate the parameters of $p(x_t|y_{1:t})$. The reason that this works is that the restriction on the model ensures that the resulting distribution is exactly Gaussian (Ho and Lee 1964). The problem then becomes one of deducing the parameters of $p(x_t|y_{1:t})$ (Kalman 1960). These parameters can be calculated as follows:

$$p(x_t|y_{1:t-1}) = N(x_{t|t-1}, P_{t|t-1}) \quad \text{and} \quad p(x_t|y_{1:t}) = N(x_{t|t}, P_{t|t}),$$

where

$$x_{t|t-1} = Ax_{t-1|t-1}, \tag{3.8}$$

$$P_{t|t-1} = AP_{t-1|t-1}A' + Q, \tag{3.9}$$

$$x_{t|t} = x_{t|t-1} + (P_{t|t-1}H')(HP_{t|t-1}H' + R)^{-1}(y_t - y_{t|t-1}),$$
$$P_{t|t} = P_{t|t-1} - (P_{t|t-1}H')(HP_{t|t-1}H' + R)^{-1}(P_{t|t-1}H')', \tag{3.10}$$

$$y_{t|t-1} = Hx_{t|t-1} \tag{3.11}$$

and (3.8)–(3.11) define the Kalman filter.

While algorithms exist for obtaining estimates of the entire path and so the parameters of $p(x_{1:t}|y_{1:t})$, these are not the focus of interest here and not discussed. The problem of obtaining the parameters of $p(x_\tau|y_{1:t})$ for $\tau < t$ is known as smoothing rather than filtering (e.g. Harvey (1989), West and Harrison 1997 and Durbin and Koopman (2001)).

3.4 Approximate linear Gaussian filtering

While there do exist a few special cases for which other analytic solutions exist (including when the state space is discrete, when the model is often called a hidden Markov model, e.g. Rabiner (1989a)), the family of possible models that one would like to use is far bigger than those that permit analytic solutions.

One can approximate the models as linear and Gaussian in locality of estimates and then use a Kalman filter with these approximating models (Bar-Shalom and Li 1995). This results in the *extended Kalman filter*, or EKF.

Example 3.1 Throughout this and subsequent sections, we will refer to a concrete example. This example consists of using (noisy) measurements of the position of some system to calculate estimates regarding the position and velocity of the system.

So, in our example, the system is taken to evolve according to:

$$x_t = Ax_{t-1} + \omega_t^x, \qquad (3.12)$$

where ω_t^x, the process noise, is a sample from a Gaussian distribution:

$$p(\omega_t^x) = N(\omega_t^x; 0, Q), \qquad (3.13)$$

where $N(x; m, C)$ is a Gaussian distribution for x parameterised a mean, m, and a covariance, C. Here

$$A = \begin{bmatrix} 1 & \Delta \\ 0 & 1 \end{bmatrix}, \quad Q = \begin{bmatrix} \frac{\Delta^3}{3} & \frac{\Delta^2}{2} \\ \frac{\Delta^2}{2} & \Delta \end{bmatrix} q, \qquad (3.14)$$

where q is a scalar quantity that defines the size of any deviations from straight line motion and Δ is the time between measurements. The probabilistic nature of the model dictates that while the predicted position at some point in the future takes account of the velocity at the present time, the uncertainty related to this estimate grows as the extrapolation extends further into the future. It is worth noting that this model is exactly equivalent to an integrated random walk evolving in continuous time and so Q has a determinant that is greater than zero; Q is positive definite. This means that even if we knew the position and velocity at time $t-1$ and the position at time t, there would still be uncertainty over the velocity at time t. As noted by a reviewer, this is related to the spline smoothing approach described in Durbin and Koopman (2001).

The definition of the model in terms of an equation for x_t in terms of x_{t-1} is equivalent to the following definition in terms of a distribution for x_t conditional on x_{t-1}:

$$p(x_t|x_{t-1}) = N(x_t; Ax_{t-1}, Q). \qquad (3.15)$$

Note that the process is Markov; the position and velocity at time t define the uncertainty over the position and velocity at all future times, $t' > t$.

Hence, further knowledge of the position and velocity at any previous time, $t'' < t$, has no effect on this uncertainty at the future time, t'.

In our example, we pick as simple a nonlinear measurement model as one can imagine:

$$y_t = (Fx_t)^2 + \omega_t^y, \tag{3.16}$$

where

$$p(\omega_t^y) = N(\omega_t^y; 0, R) \quad \text{and} \quad F = \begin{bmatrix} 1 & 0 \end{bmatrix}, \tag{3.17}$$

where R is a (scalar) variance defining the accuracy of the measurements.

Again, this is equivalent to a probabilistic description:

$$p(y_t|x_t) = N\left(y_t; (Fx_t)^2, R\right). \tag{3.18}$$

To be able to use an EKF for our example, we just need to linearise (3.18) about the prediction $x_{t|t-1}$:

$$p(y_t|x_t) \approx N\left(y_t; y_{t|t-1} + H\left(x_t - x_{t|t-1}\right), R\right),$$

where

$$y_{t|t-1} = \left(Fx_{t|t-1}\right)^2 \tag{3.19}$$

$$H = \begin{bmatrix} 2Fx_{t|t-1} & 0 \end{bmatrix}. \tag{3.20}$$

We can now use a Kalman filter with this approximating model; here (3.11) is simply replaced with (3.19) and the Kalman filter then operates as previously discussed.[2] $\Delta = 1$, $R = 1000$ and $q = 1$ and the filter is initialised with:

$$p(x_0) = N\left(x_0; x_0^{\text{true}} + \delta, C\right), \tag{3.21}$$

where

$$p(\delta) = N(\delta; 0, C), \quad \text{where} \quad C = \begin{bmatrix} 20 & 0 \\ 0 & 1 \end{bmatrix}. \tag{3.22}$$

We consider two cases. In the first case, we start with $x_0^{\text{true}} = \begin{bmatrix} 500 & 0 \end{bmatrix}'$. In the second, we start with $x_0^{\text{true}} = \begin{bmatrix} 5 & 0 \end{bmatrix}'$. Nine simulations and runs of the EKF result in the estimated position (and true position) shown in Figures 3.1 and 3.2 for the first and second cases respectively. The obvious question to ask is why the EKF, in a couple of examples seems to go drastically wrong

The answer is that this approximation to the models is good when the

---

[2] When the dynamics are nonlinear (3.8) will also need to be replaced with a linearisation based on $x_{t-1|t-1}$.

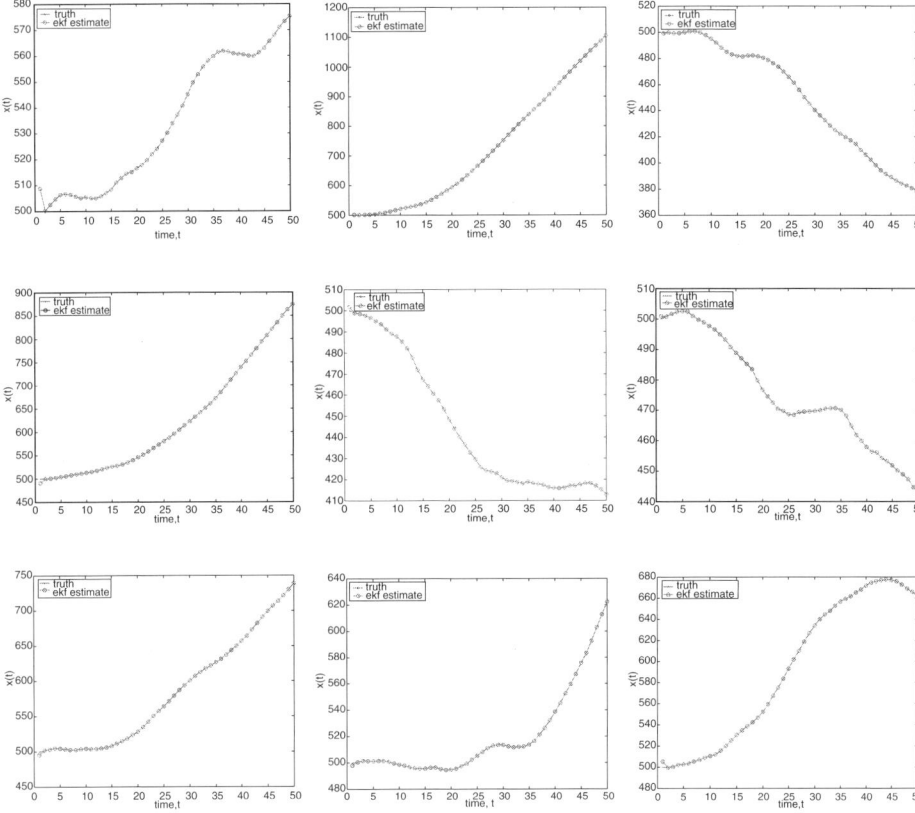

Fig. 3.1. Nine runs of the EKF for the model defined by (3.12)–(3.18) with $x_0 = [500, 0]'$.

nonlinearity and departure from Gaussian behavior is small. To consider if this is the case, it is instructive to consider what nonlinearity means. In this context what matters is how good an approximation would be achieved by using a local linearisation of the models. So, by looking at the expected departure from linear behavior, it is possible to get a handle on the amount of nonlinearity; if the variation in the linearity at different points in the posterior is large with respect to the linear coefficients then the nonlinearity is pronounced. So nonlinearity in this context is a function of both the nonlinear model and the diffuse nature of the pdf; an apparently less nonlinear model of two candidates can appear more nonlinear to the EKF if the distribution used with this less nonlinear model is more diffuse. In general, as the signal-to-noise drops, it will become increasingly likely that such scenarios are encountered since the support of the distribution will typically grow as the signal-to-noise falls; the underlying problem is the nonlinearity, not such things as the scale of the data.

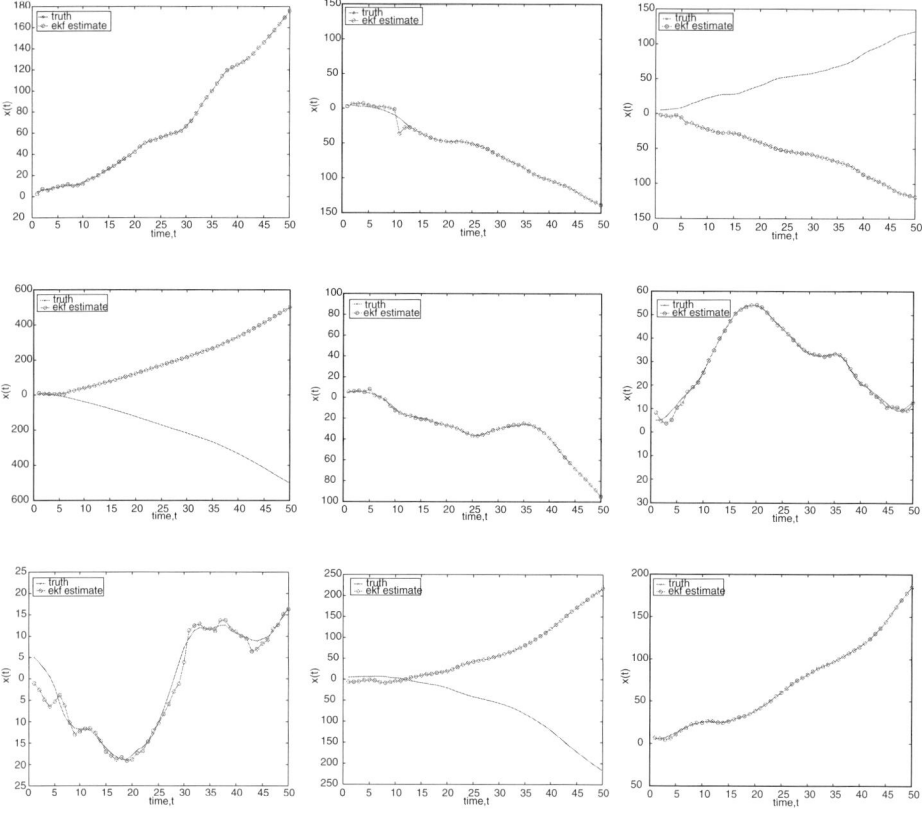

Fig. 3.2. Nine runs of the EKF for the model defined by (3.12)–(3.18) with $x_0 = [5, 0]'$.

In our example, the same change in the state will cause a much bigger (percentage) change in the linearisation when the position is small than when the position is large. So, the linear approximation is valid when the state is large and isn't valid when the state is small. This explains the observed behavior.

Another example of how to cater with mildly nonlinear non-Gaussian models is the unscented Kalman filter (Julier 1998), UKF. The UKF uses a deterministic sampling scheme to deduce the parameters of $p(x_t|y_{1:t})$ and matches the moments of the samples to those of the true distributions; this enables the UKF to capture the effects of higher order moments, not usually exploited by the EKF. The idea is that the posterior is well approximated by a Gaussian distribution, even though some of the models used and so intermediate distributions may be less well approximated as Gaussian.

This argument also helps explain why the iterated extended Kalman filter

(Jazwinski 1970), IEKF, often works well. The EKF (not IEKF) linearises once on the basis of the predicted estimate of the state, $x_{t|t-1}$. The uncertainty associated with this estimate is likely to be significantly larger than that associated with filtered estimate, $x_{t|t}$. So, by relinearising the measurement model using $x_{t|t}$, the parameters of the approximation to the models can be fine tuned. This can improve the accuracy of the parameters of the Gaussian approximation to $p(x_t|y_{1:t})$. Some further gains can be obtained by making this process iterate and using the estimates of $x_{t|t}$ from one iteration to relinearise the measurement model at the next iteration. One can also relinearise the dynamics using smoothed estimates of $x_{t-1|t}$ and then relinearise the measurement model using the prediction based on this revised linear approximation to the dynamics. In the extreme case, one can relinearise at all points in a time series by using a smoothed estimate of $x_{0|t}$ to relinearise the initial dynamic model and then relinearising at all subsequent points using the estimates that result; this is essentially the same as using expectation-maximisation (a form of hill-climbing) to learn the initial state of a system (Shumway and Stoffer 1982). It is worth stressing that, in this nonlinear non-Gaussian environment, learning the initial state in this way can affect the final state estimate; the final state estimate is a function of all the linearisations up to the final time so changing these linear approximations will change the final state estimate.

3.5 Particle filtering

It is possible to approximate $p(x_{1:t}|y_{1:t})$ directly. This is the approach adopted by the particle filter and is likely to be beneficial when the nonlinearity (as described above) and non-Gaussianity are pronounced.

Like the Kalman filter, the particle filter is a sequential algorithm. However, a better understanding of how a particle filter works in a sequential setting is made possible by considering the history of states; the particle filter is actually an approach to inferring the path through the state space over time which can be implemented by only storing quantities relating to the filtered time. So, to make it easy to see how the particle filter operates, the algorithm is described here from a path-based starting point.

The particle filter does not attempt to represent the distribution using an analytic form. Instead, the uncertainty (and so the distribution) is represented using the diversity of a set of N samples or particles. The ith of these particles consists of a hypothesised history of states of the target, $x_{1:t}^i$, and an associated weight, $\hat{w}_{1:t}^i$. Ideally one would want these samples to be samples from $p(x_{1:t}|y_{1:t})$. If this were the case, then

$$x^i_{1:t} \sim p(x_{1:t}|y_{1:t}) \quad \text{and} \quad \hat{w}^i_{1:t} = \frac{1}{N}. \tag{3.23}$$

The particles do represent the pdf and it is possible to think of the particles as samples from the pdf which can be used to estimate expectations with respect to the pdf. However, a minor point to note is that the particles do not really approximate the true pdf at any point x since the particle distribution is zero everywhere except under (the support of) the samples, where the delta function makes the distribution infinite:

$$p(x_{1:t}|y_{1:t}) \napprox \sum_{i=1}^{N} \hat{w}^i_{1:t} \delta(x^i_{1:t} - x_{1:t}), \tag{3.24}$$

where $\delta(x - x_0)$ is a delta function.

What makes these samples useful is the fact that they can be used to statistically estimate the expected value of functions of the state:

$$\int f(x_{1:t}) p(x_{1:t}|y_{1:t}) \, dx_{1:t} \approx \int f(x_{1:t}) \sum_{i=1}^{N} \hat{w}^i_{1:t} \delta(x^i_{1:t} - x_{1:t}) \, dx_{1:t}$$

$$= \sum_{i=1}^{N} \hat{w}^i_{1:t} f(x^i_{1:t}). \tag{3.25}$$

So, if we had an algorithm which could (somehow) sample N times from $p(x_{1:t}|y_{1:t})$, we would obtain $x^i_{1:t}$ for $i = 1, \ldots, N$. Then we could estimate quantities of interest such as $x_{t|t}$ and $P_{t|t}$ by just calculating the quantity from the weighted sample set using (3.25).

Unfortunately, it is unlikely that it will be possible to sample exactly from $p(x_{1:t}|y_{1:t})$; if one could sample exactly, it is probable that a particle filter would be unnecessary since $p(x_{1:t}|y_{1:t})$ would be likely to have an analytic form!

3.5.1 Importance sampling

Instead of simulating from $p(x_{1:t}|y_{1:t})$ directly we can choose a convenient *proposal distribution* or *importance function*, $q(x_{1:t}|y_{1:t})$, which is easy to sample from and evaluate. Samples are drawn from this proposal and the weights are modified using the principle of importance sampling. Where the proposal distribution proposes samples at which

$$q(x_{1:t}|y_{1:t}) > p(x_{1:t}|y_{1:t})$$

there are more samples than there should be. Hence these samples are given a reduced weight. Conversely, when samples are drawn in regions where

$$q(x_{1:t}|y_{1:t}) < p(x_{1:t}|y_{1:t}),$$

there are too few samples and the samples are given an increased weight to counter this effect. This can be formulated by looking at how to calculate expectations with respect to one distribution given samples from another distribution:

$$\int p(x_{1:t}|y_{1:t})f(x_{1:t})\mathrm{d}x_{1:t} = \int \frac{p(x_{1:t}|y_{1:t})}{q(x_{1:t}|y_{1:t})} f(x_{1:t})q(x_{1:t}|y_{1:t})\mathrm{d}x_{1:t} \quad (3.26)$$

$$\approx \int \frac{p(x_{1:t}|y_{1:t})}{q(x_{1:t}|y_{1:t})} f(x_{1:t}) \frac{1}{N} \sum_{i=1}^{N} \delta\left(x_{1:t} - x_{1:t}^i\right) \mathrm{d}x_{1:t}$$

$$\quad (3.27)$$

$$= \frac{1}{N} \sum_{i=1}^{N} \hat{w}_{1:t}^i f\left(x_{1:t}^i\right), \quad (3.28)$$

where

$$x_{1:t}^i \sim q(x_{1:t}|y_{1:t}) \quad \text{and} \quad \hat{w}_{1:t}^i = \frac{p(x_{1:t}^i|y_{1:t})}{q(x_{1:t}^i|y_{1:t})}. \quad (3.29)$$

So, the idea at the heart of the particle filter is to pick a convenient approximating proposal to sample particles from:

$$x_{1:t}^i \sim q(x_{1:t}|y_{1:t}) \approx p(x_{1:t}|y_{1:t}) = \frac{p(x_{1:t}, y_{1:t})}{p(y_{1:t})}. \quad (3.30)$$

The particle filter then defines an *unnormalised weight*:

$$\tilde{w}_{1:t}^i = \frac{p(x_{1:t}^i, y_{1:t})}{q(x_{1:t}^i|y_{1:t})}. \quad (3.31)$$

It is then possible to calculate an approximation to $p(y_{1:t})$,

$$p(y_{1:t}) = \int p(x_{1:t}, y_{1:t}) \, \mathrm{d}x_{1:t} \approx \sum_{i=1}^{N} \tilde{w}_{1:t}^i, \quad (3.32)$$

and so the *normalised weight* is

$$w_{1:t}^i = \frac{p(x_{1:t}^i|y_{1:t})}{q(x_{1:t}^i|y_{1:t})} = \frac{p(x_{1:t}^i, y_{1:t})}{q(x_{1:t}^i|y_{1:t})} \frac{1}{p(y_{1:t})} \quad (3.33)$$

$$= \frac{\tilde{w}_{1:t}^i}{\sum_{i=1}^{N} \tilde{w}_{1:t}^i}. \quad (3.34)$$

These normalised weights can then be used to estimate quantities of interest:

$$\int p(x_{1:t}|y_{1:t})f(x_{1:t})\mathrm{d}x_{1:t} \approx \sum_{i=1}^{N} w_{1:t}^{i} f\left(x_{1:t}^{i}\right). \qquad (3.35)$$

Such an algorithm would sample N times from $q(x_{1:t}|y_{1:t})$ to get $x_{1:t}^{i}$ for $i = 1, \ldots, N$. Each sample would have a weight $w_{1:t}^{i}$ as defined in (3.34). To calculate quantities of interest such as $x_{t|t}$ and $P_{t|t}$, one would then use (3.35). The author believes this is similar to the approach in Shephard and Pitt (1997) and Durbin and Koopman (1997), which use an (extended) Kalman smoother to obtain $q(x_{1:t}|y_{1:t})$.

Note that while both the denominator and numerator of (3.34) are unbiased estimates, if the normalised weight is used, since it is a ratio of estimates, the filter is biased.[3]

3.5.2 Sequential importance sampling

As described to this point, the particle filter proposes particles (consisting of the entire history of states) at each point in time with no reference to the previously sampled particles. This is clearly undesirable and not sequential. Indeed, to perform *sequential importance sampling*, SIS, the proposal is defined to be of the following particular form:[4]

$$q(x_{1:t}|y_{1:t}) \;=\; \underbrace{q(x_{1:t-1}|y_{1:t-1})}_{\text{Keep existing path}} \underbrace{q(x_t|x_{t-1}, y_{1:t})}_{\text{Extend path}},$$

which means that the particles at one iteration are generated by extending the existing history of the particle with a sample for the state at the current time using the information from the history (summarised by x_{t-1}^{i}), the measurement, y_t, and, potentially, the history of previous measurements, $y_{1:t-1}$. Hence, if $x_{1:t-1}$ is not of interest, we only need to store x_t^{i} and $w_{1:t}^{i}$ for each particle at each iteration; the particle filter is indeed a sequential algorithm.

---

[3] The formulations in Durbin and Koopman (1997) and Shephard and Pitt (1997) give rise to unbiased estimates, but a truly sequential unbiased particle filter appears to have eluded researchers up to the current time.

[4] This is slightly different to the usual formulation, for which $q(x_{1:t}|y_{1:t}) = q(x_{1:t-1}|y_{1:t-1})q(x_t|x_{t-1}, y_t)$, so as to accommodate all the proposals described in Section 3.5.7.

The unnormalised weight then takes on an intuitively appealing form:

$$\begin{aligned}
\tilde{w}^i_{1:t} &= \frac{p(x^i_{1:t}, y_t|y_{1:t-1})}{q(x^i_{1:t}|y_{1:t})} \\
&= \frac{p(x^i_{1:t-1}|y_{1:t-1})}{q(x^i_{1:t-1}|y_{1:t-1})} \frac{p(x^i_t, y_t|x^i_{t-1})}{q(x^i_t|x^i_{t-1}, y_{1:t})} \\
&= w^i_{1:t-1} \underbrace{\frac{p(y_t|x^i_t) p(x^i_t|x^i_{t-1})}{q(x^i_t|x^i_{t-1}, y_{1:t})}}_{\text{Incremental weight}}.
\end{aligned}$$

One can then approximate $p(y_t|y_{1:t-1})$ by:

$$p(y_t|y_{1:t-1}) = \int p(x_{1:t}, y_t|y_{1:t-1}) \mathrm{d}x_{1:t} \approx \sum_{i=1}^{N} \tilde{w}^i_{1:t}, \qquad (3.36)$$

which is an approximation[5] that can be used to estimate the likelihood of the data by substituting into

$$p(y_{1:t}) = \prod_{t'=1}^{t} p(y_{t'}|y_{1:t'-1}). \qquad (3.37)$$

Note that $p(y_1|y_{1:0})$ can be obtained from the first set of samples for which $i = 1$. The normalised weights can then be calculated:

$$w^i_{1:t} = \frac{p(x^i_{1:t}|y_{1:t})}{q(x^i_{1:t}|y_{1:t})} = \frac{p(x^i_{1:t}, y_t|y_{1:t-1})}{q(x^i_{1:t}|y_{1:t})} \frac{1}{p(y_t|y_{1:t-1})} \qquad (3.38)$$

$$= \frac{\tilde{w}^i_{1:t}}{\sum_{i=1}^{N} \tilde{w}^i_{1:t}}. \qquad (3.39)$$

So, to begin the algorithm samples N times from the prior, $p(x_0)$, and assigns each particle a weight of $1/N$. Then, at each iteration, an algorithm would now sample N times from $q(x_t|x_{t-1}, y_{1:t})$ to get x^i_t for $i = 1, \ldots, N$. Each sample would have a weight $w^i_{1:t}$ as defined in (3.39). To calculate quantities of interest such as $x_{t|t}$ and $P_{t|t}$, one would then use (3.35) as previously.

However, there is still a problem; whatever the choice of $q(x_t|x_{t-1}, y_{1:t})$, eventually one of the particles will have a weight close to unity while all the others will have a weight close to zero. To see why this might be problematic, consider what would happen to (3.35) in such a scenario; if the weights depart from uniformity and so some of the weights becomes much larger than

---

[5] Other approximations are proposed in Doucet, Godsill and Andrieu (2000) for example.

all the others, then these particles dominate any estimation. The other particles are then essentially wasted.

Another way to think about this is that the effective number of particles is reduced. This effective sample size can be approximated using the following statistic of the normalised weights:

$$N_{\text{eff}} \approx \frac{1}{\left(\sum_{i=1}^{N} {w_{1:t}^{i}}^{2}\right)}. \tag{3.40}$$

To appeal to intuition, consider two cases. When all the $w_{1:t}^{i} = 1/N$ for all i then $N_{\text{eff}} = N$ while when $w_{1:t}^{i} = 1$ for one value of i and $w_{1:t}^{i} = 0$ for all the others then $N_{\text{eff}} = 1$. So, the approximate effective sample size number is N when the weights are uniform and 1 in the extreme case of one of the particles having all the weight. So, by thresholding this effective sample size (perhaps with a threshold of $0.5N$), it is possible to spot when this problematic degeneracy is evident.

3.5.3 Resampling

The reason that this comes about is because the same number of samples is being used to conduct importance sampling on a pdf (of a history of states) that is increasing in dimensionality at every iteration. Eventually the samples will be insufficient in number. This is what is causing the aforementioned degeneracy phenomenon. In that it is the sequential setting that is of interest, a way to circumnavigate this problem is to intentionally move the degeneracy to somewhere that it doesn't matter; the path through the history of states.

So, to alleviate the effect of this degeneracy, a resampling stage is introduced. The resampling stage results in particles being (probabilistically) replicated and discarded. The particles are redistributed so as to spread the particles (and so the computation) evenly across the posterior. This redistribution introduces errors but makes the weights more uniform. After the resampling stage, $w_{1:t}^{i} = 1/N$ for all the particles. Particles that are replicated then share the same path and so the path pdf degenerates, but the filtered pdf is rejuvenated. To avoid any confusion, it should be stressed that the decision as to whether to resample is based on (3.40) being calculated at each iteration; hence since the statistic in (3.40) will typically fall below the aforementioned threshold several times, the resampling will typically take place a number of times.

There are a number of different resampling algorithms. The simplest to understand is *multinomial resampling* (e.g. Gordon, Salmond and Smith

(1993)). Each particle in the new particle set is a copy of the ith particle in the old particle set with probability w_t^i. So, the new generation is generated by sampling N times from the distribution formed by the weights on the old set of particles.

To understand the differences between this approach and other resampling algorithms, the new generation is defined to consist of N_i replicates of the ith particle in the old particle set. Different resampling algorithms are then the same with respect to $\mathbb{E}[N_i]$, the expected number of replicates, but differ with respect to $\mathbb{E}\left[(N_i - \mathbb{E}[N_i])^2\right]$, the variance in the number of replicates. This variance can be thought of as the size of an error; it is advantageous to introduce as few errors as possible and so sensible to use an algorithm that minimises this variance. While there are some candidate algorithms which don't guarantee constant N, these aren't discussed here.

3.5.3.1 Multinomial resampling

As previously stated, multinomial resampling is the simplest resampling algorithm. The new particle set is generated by drawing N independent samples from the weighted old particle set. This approach can be implemented in a time that scales linearly with the number of samples (e.g. Ripley 1987 and Carpenter, Clifford and Fearnhead 1999). The deficiency of this approach is that there is then a finite probability of any particle being replicated any number of times. This results in a large variance in the number of replicates.

3.5.3.2 Residual resampling

Residual resampling considers the expected number of replications of each particle and deterministically replicated some of the particles (e.g. Liu and Chen (1998)). Specifically, the ith particle is replicated $(\mathbb{E}[N_i])^-$ times, where A^- denotes the integer less than or equal to A. The remaining $N - \sum_{j=1}^{N}(\mathbb{E}[N_j])^-$ particles are then sampled according to the aforementioned multinomial scheme. This reduces $\mathbb{E}\left[(N_i - \mathbb{E}[N_i])^2\right]$ since the particles cannot be replicated less than $(\mathbb{E}[N_i])^-$ times.

3.5.3.3 Systematic resampling

Systematic resampling (e.g. Kitagawa (1996)) is the approach to be advocated. The new particle set is chosen so as to minimise $\mathbb{E}\left[(N_i - \mathbb{E}[N_i])^2\right]$; every particle is replicated either $(\mathbb{E}[N_i])^-$ or $(\mathbb{E}[N_i])^+$ times, where A^+ is the smallest integer greater than or equal to A. The approach works by forming the cumulative distribution function, cdf, over the N samples:

$$c_i = \begin{cases} 0 & i = 0, \\ c_{i-1} + w_{1:t}^i & i = 1, \ldots, N. \end{cases} \quad (3.41)$$

The new generation are then uniformly spaced in the cdf. A small initial value for the first new samples' cdf is then sampled from a uniform distribution:

$$\hat{c}_1 \sim U\left[\hat{c}_1; 0, \frac{1}{N}\right] \quad (3.42)$$

$$\hat{c}_i = \hat{c}_{i-1} + \frac{1}{N} \quad i = 2, \ldots, N, \quad (3.43)$$

where $U[x; a, b]$ is a uniform distribution over the region $a \leq x \leq b$.

The algorithm then iterates through both lists at the same time and so generates the new generation, $\hat{x}_{1:t}^i$ for $i = 1, \ldots, N$ from the old generation, $x_{1:t}^j$ for $j = 1, \ldots, N$. So, initially, $j = 1$ and $i = 1$. At each iteration the sample in the old set with the smallest value for c_j larger than $\hat{c}_i$ is replicated. So, if $c_j \geq \hat{c}_i$ then $\hat{x}_{1:t}^i = x_{1:t}^j$, while if $c_j < \hat{c}_i$ then j is incremented until $c_j \geq \hat{c}_i$ at which point $\hat{x}_{1:t}^i = x_{1:t}^j$. The process iterates incrementing i at each stage until the N samples comprising the new generation have been generated. The weights are (as stated previously) then all $1/N$.

So, at each iteration, as previously, an algorithm would now sample N times from $q(x_t|x_{t-1}, y_{1:t})$ to get x_t^i for $i = 1, \ldots, N$. Each sample would have a weight $w_{1:t}^i$ as defined in (3.39). To calculate quantities of interest such as $x_{t|t}$ and $P_{t|t}$, one would then use (3.35) as previously. However, crucially, (3.40) is now used to decide whether to perform the resampling which is implemented as just described. This gives rise to a particle filter algorithm, which can be implemented in a step-by-step fashion as follows:[6]

A step-by-step approach to implementing a particle filter

- Define the dynamics, $p(x_t|x_{t-1})$.
- Define the likelihood, $p(y_t|x_t)$.
- Define the initial prior, $p(x_0)$.

---

[6] Since the particle filter involves calculating the value of probability densities at specific points, to avoid errors resulting from rounding small values of probability densities to zero, the weights and probabilities should be manipulated in terms of logarithms. It is then useful to be able to calculate the logarithm of the sum of two probabilities by manipulating the logarithms of the probabilities as follows:

$$\log(a+b) = \log\left[a\left(1 + \frac{b}{a}\right)\right] = \log(a) + \log\{1 + \exp[\log(b) - \log(a)]\}, \quad (A)$$

where it should be ensured that $a \geq b$. Should $a < b$ then a and b should be interchanged in (A).

- Choose the number of samples, N.
- Choose the threshold on the effective sample size, N_T.
- Choose an appropriate proposal, $q(x_t|x_{t-1}, y_{1:t})$.
- Initialise the samples; $x_0^i \sim p(x_0)$ for $i = 1, \ldots, N$.
- Initialise the weights; $w_{1:0}^i = 1/N$ for $i = 1, \ldots, N$.
- FOR $t = 1, \ldots, T$.
  - Draw the samples:
  $$x_t^i \sim q(x_t|x_{t-1}^i, y_{1:t})$$
  for $i = 1, \ldots, N$.
  - Adjust the weights:
  $$\tilde{w}_{1:t}^i = w_{1:t-1}^i \frac{p(x_t^i|x_{t-1}^i) p(y_t|x_t^i)}{q(x_t^i|x_{t-1}^i, y_{1:t})}$$
  for $i = 1, \ldots, N$.
  - Normalise the weights:
  $$w_{1:t}^i = \frac{\tilde{w}_{1:t}^i}{\sum_{i'=1}^N \tilde{w}_{1:t}^{i'}}$$
  for $i = 1, \ldots, N$.
  - Calculate the approximate effective sample size:
  $$N_{eff} \approx \frac{1}{\left(\sum_{i=1}^N {w_{1:t}^i}^2\right)}.$$
  - IF $N_{eff} < N_T$ THEN resample;
    - Initialise the CDF; $c_0 = 0$.
    - Calculate the remainder of the CDF; $c_i = c_{i-1} + w_{1:t}^i$ for $i = 1, \ldots, N$.
    - Draw a starting point on the CDF; $\hat{c}_1 \sim U[\hat{c}_1; 0, 1/N]$.
    - Set up other points on the CDF; $\hat{c}_i = \hat{c}_{i-1} + 1/N$ for $i = 2, \ldots, N$.
    - Initialise pointer into existing samples; $j = 1$.
    - FOR $i = 1, \ldots, N$
      - WHILE $(c_j < \hat{c}_i)$, increment j; $j = j + 1$ END WHILE
      - Create new sample; $\hat{x}_{1:t}^i = x_{1:t}^j$
    - END FOR
    - Use new samples; $\hat{x}_{1:t}^i = x_{1:t}^i$ for $i = 1, \ldots, N$.
    - Set weights to be uniform; $w_{1:t}^i = 1/N$ for $i = 1, \ldots, N$.
  - END IF
- END FOR

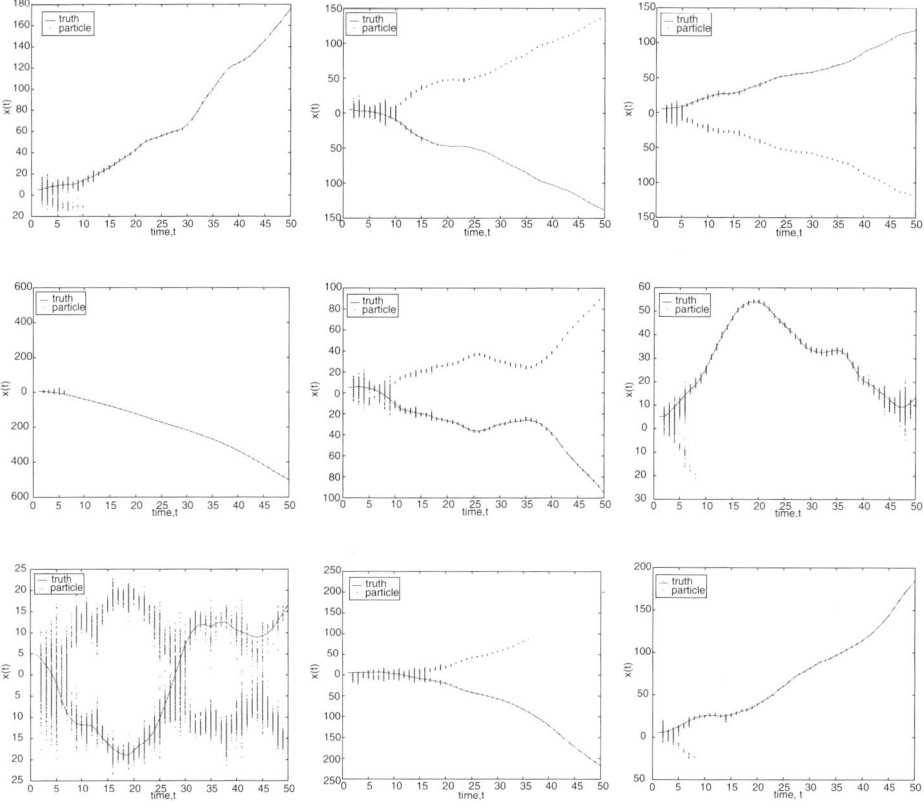

Fig. 3.3. Nine runs of the particle filter for the model defined by (3.12)–(3.18) with $x_0 = [5, 0]'$.

Example 3.2 So, we have all the ingredients with which to apply a particle filter to our example. We choose to use the dynamics as the proposal distribution:

$$q(x_t|x_{t-1}, y_{1:t}) = p(x_t|x_{t-1}).$$ (3.44)

We choose $N = 250$ particles and initialise the filter with N samples from the initial Gaussian distribution used for the EKF given in (3.21). We resample when N_{eff} falls below 125. We consider the second (difficult) case. We run the filter nine times as before. The particles' positions are shown with the true position in Figure 3.3.

The point is that the particle filter is able to model the multimodal distribution that results from the measurement model. Note that there could be occasions when the particle filter picks the wrong mode; persistently multimodal distributions do cause problems with the resampling step in particle filters and this is why so many particles are needed. It is also worth noting that the weights are not being displayed. As a result, the diversity of the particles is slightly misleading at some points when the particles appear to fan out; the particles on the edge of the fan typically have low weights.

As with the Kalman filter, schemes do exist to perform smoothing (Godsill, Doucet and West 2001), but these are not considered here.

3.5.4 Frequently encountered pitfalls

One of the first papers on the subject proposed sequential importance resampling, SIR (Gordon, Salmond and Smith 1993), which is a special case of SIS. SIR defines that the proposal is the dynamic model, or dynamic prior, $q(x_t|x_{t-1}, y_{1:t}) = p(x_t|x_{t-1})$, and proposes to use multinomial resampling at every t. Particle filters have been criticised on the basis of the degeneracy that results when using this filter with systems that mix slowly in comparison to the evolution of the posterior; this happens when conducting parameter estimation or when using very informative measurements. This is because of two effects: firstly, the errors introduced by the (unnecessary) resampling dominate; secondly, and more importantly, the dynamic prior is often an appalling choice of proposal.

The solution often proposed is to use millions of particles. However, a more interesting approach is to consider what the particles are all doing. To do this, the computational cost of particle filters needs to be discussed.

3.5.5 Integration methods

The computational cost of Monte Carlo integration is often (mis)quoted as being independent of the dimensionality of the integral. This isn't true. What is true is that the dependence of the computational cost on the number of samples is independent of the number of samples. The effect of doubling the number of samples is always to reduce the errors in by a factor of $\sqrt{2}$ since the samples are populating a probability mass that essentially doesn't care in what dimensional space it happens to reside:

$$\sigma \propto N^{-\frac{1}{2}}. \tag{3.45}$$

This contrasts with the computational cost of a grid based solution to calculating an integral for which the effect of doubling the number of samples is determined by the distance between the samples and so is heavily dependent on the dimensionality:

$$\sigma \propto N^{k-d}, \tag{3.46}$$

where k is defined by the integration method; $k = 1$, $k = 2$ and $k = 4$ for the Rectangle, Trapezium and Simpson's Rule respectively.

The problem with this argument is that there has been no mention of the constant of proportionality in (3.45). This constant of proportionality is not a function of dimensionality *per se*, since it is really a function of the peaked nature of the mismatch between the distribution from which the samples are drawn and the product of the function being integrated and the distribution. The flip-side is that the such mismatches are typically more peaked in higher dimensions since the product of a large number of flat scalar distributions will be very peaked in the higher dimensional space.

So, to bring this argument to a close, it is essential, if one wants to make particle filtering a viable option in real-time environments, that the particles inhabit as low a dimensional space as possible.

3.5.6 Ergodicity

To be able to understand the dimensionality of the space which the particles inhabit, it is necessary to understand the concept of ergodicity. Ergodicity of a system is a measure of how quickly mixing it is. To understand this concept, consider two examples. In the first case, parameter estimation is considered. Parameters remain constant as time evolves, so if the value of the parameter is known at a point in time, the uncertainty over the value does not change with time; the system does not forget. This is in contrast to dynamic systems for which a known state at some point in the past still results in uncertainty over the state at some point in the future. This differing rate with which the uncertainty grows with time is a measure of the history of the system. The faster the uncertainty grows, the shorter the memory.

Particle filters conduct importance sampling on the history of states. For the approach to be able to filter each measurement, at each time step, the particles need to have explored this history in terms of the uncertainty caused by the history on the prediction of the new measurement. The longer the memory of the system, the bigger the space that the particles must inhabit and so the more particles that will be required. Hence, it is advantageous to use systems with short memories.

3.5.7 Choice of proposal distribution

At first glance, the choice of system has been made at this point. However, if one considers the choice of proposal distribution as the choice of system which the particle filter simulates trajectories from, then different choices of proposal relate to different systems, different memories and so a potentially huge impact on the number of samples required. The closer that the proposal is to the posterior, the shorter the memory and so the fewer particles needed. The choice of proposal distribution is therefore a crucial step in the design of a particle filter.

The optimal choice of proposal (in terms of the variance of the importance weights when the incremental proposal is of the form, $q(x_t|x_{t-1}, y_t)$) is to use $q(x_t|x_{t-1}, y_{1:t}) = p(x_t|x_{t-1}, y_t)$. If the only departure from linear Gaussian models is a nonlinearity in the dynamic model then this pdf can be sampled from. This means that the model has the following form:

$$x_t = a(x_{t-1}) + \omega_t^x, \quad y_t = Hx_t + \omega_t^y, \qquad (3.47)$$

where $a(x)$ is a nonlinear function of x. As observed by a reviewer, this form of model is popular (e.g. Liu and Chen (1998) and Pitt and Shephard (1999b)); Pitt and Shephard (1999b) referred to such models as cases when it is possible to do *exact adaption*.

This proposal is optimal within a certain class of proposals in terms of minimising the variance of the importance weights (Doucet, Godsill and Andrieu 2000); so the sample used to extend the path of the particle turns out to have no effect on the weight update. This enables resampling to be carried out prior to the sampling, which can reduce the number of samples needed since the resampling ensures that the sampling is focused more efficiently. In some senses this is similar to the IEKF which readjusts its (linearisation) approximations in the light of the received measurements; here the samples are drawn from a distribution that has been readjusted to deter the samples from introducing degeneracy.

However, this optimal proposal cannot be used in the general case. Convenient alternatives include using the dynamics, $q(x_t|x_{t-1}, y_{1:t}) = p(x_t|x_{t-1})$. If one considers how close the posterior then is to the proposal, the difference will be small when the bulk of the information in the posterior comes from the history of the process rather than the measurement. Hence this proposal works well with diffuse likelihoods.

By a similar argument, if the bulk of the information in the posterior comes from the measurement rather than the history of the process then using what can be thought of as a global approximation to the likelihood function can work well (Doucet, Godsill and Andrieu 2000). In such cases, the proposal has some additional conditioning on the past measurements, but this is implemented in terms of an approximation to the particle cloud:

$$q(x_t|x_{t-1}, y_{1:t}) = p(x_t|y_{1:t}) = \frac{p(y_t|x_t, y_{1:t-1}) p(x_t|y_{1:t-1})}{p(y_t|y_{1:t-1})}$$

$$\propto p(y_t|x_t, y_{1:t-1}) \int p(x_t, x_{t-1}|y_{1:t-1}) \, \mathrm{d}x_{t-1}$$

$$\approx p(y_t|x_t, y_{1:t-1}) \int p(x_t|x_{t-1}) \hat{p}(x_{t-1}|y_{1:t-1}) \, \mathrm{d}x_{t-1},$$

where $\hat{p}(x_{t-1}|y_{1:t-1})$ is the Gaussian approximation to the particle cloud and the manipulations involved in the integral and multiplication use EKF or UKF based approximations.

An alternative that offers improved performance when there is a need to interpolate between the measurement and the individual particle's state is to form an approximation to the posterior, $p(x_t|x_{t-1}^i, y_t)$, using an EKF (Doucet, Godsill and Andrieu 2000) or UKF (van der Merwe, Doucet, de Freitas and Wan 2000) based approximation.

Example 3.3 Such an approach is advantageous when using a particle filter to tackle the first case considered previously (when $x_0^{\text{true}} = [500 \ 0]'$); while the (extended) Kalman filter is arguably more appropriate to use than a particle filter in this case, the EKF does provide a benchmark and the case does highlight the possible benefit that results from using alternative proposal distributions in particle filters. We choose to use an EKF-based approximation to the posterior, so the proposal takes on the following form:

$$q(x_t|x_{t-1}, y_{1:t}) = N(x_t; m_t, C_t),$$

where

$$C_t = \left(Q^{-1} + H'R^{-1}H\right)$$
$$m_t = x_{t-1} + C_t H' R^{-1} \left[y_t - (FAx_{t-1})^2\right].$$

We consider one long run with the same parameters as previously, but with a particle filter with just $N = 50$ particles and with resampling when N_{eff} falls below 15. The log squared estimation error for a Kalman filter, a particle filter with the dynamic prior as the proposal and a particle filter with this approximation to the posterior as the proposal are shown in Figure 3.4. Evidently, at various iterations, using an EKF-based approximation gives an improvement over the particle filter with the dynamic prior as a proposal in terms of this error.

As suggested by a reviewer, it is particularly informative to look in more

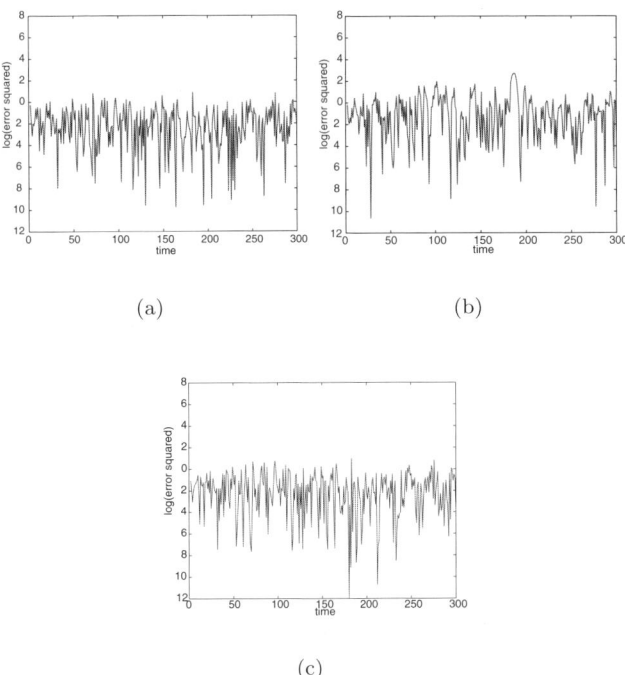

Fig. 3.4. A comparison of the errors for three filters for the first case: (a) EKF; (b) prior as proposal; (c) linearisation as proposal.

detail at the skewness of the particles' weights for both proposals. The effective sample sizes for the two proposals are given in Figure 3.5. It is very evident that with the prior as the proposal, the effective sample size is consistently very near its lower bound of 1. The particle filter that uses the approximation to the posterior as a proposal results in an effective sample size that is consistently higher. As a result, during the run, the particle filter that uses the prior as the proposal resampled at every time step whereas the particle filter with the more refined proposal resampled at 76 of the 300 iterations.

3.5.8 Jitter

An approach often advocated to rescue a particle filter from problems of degeneracy is to perturb the particles after resampling and so reintroduce diversity into the particle set. Since the resampling results in a number of replicants of some of the members of the previous particle set, it might be sensible to attempt to push the particles apart to ensure that they explore different parts of the posterior. This jittering has been proposed since the

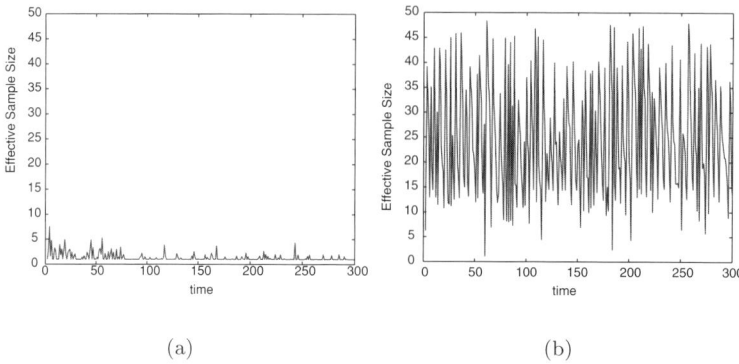

Fig. 3.5. A comparison of the effective sample size for a particle filter with a proposal that is either (a) the prior or (b) an approximation to the posterior based on linearised models.

outset of the recent research on particle filtering (Gordon, Salmond and Smith 1993). Some work has concentrated on formulating the approach as the sampling of the particles from a kernel density approximation to the posterior; this results in regularised particle filters (Musso, Oudjane and LeGland 2001). This enables, under some assumptions, the *optimal* kernel to be defined. However, a more rigorous approach is to accept any proposed jitter on the basis of a stochastic MCMC acceptance step (Gilks and Berzuini 2001); this ensures that the samples are actually samples from the posterior of interest and so can be thought of as statistically rigorous jitter.

Another approach used to recover from degeneracy is the auxiliary particle filter (Pitt and Shephard 1999b), often referred to as ASIR. This filter uses the same idea as mentioned previously in Section 3.5.7 within the context of the optimal proposal. However, since in the general case, the weights are not known prior to the samples being drawn, ASIR calculates the weights using samples from the proposal and then conducts resampling. The difference to (normal) SIR is that the samples used to extend the particle paths are then removed and new extensions to the paths sampled. The impact is that those paths that give rise to highly weighted samples are resampled and extended. A minor complication arises; if the new sample is $\hat{x}_t^i$ then the new weight needs to remove the effect on the weight of the old sample, x_t^i, and introduce the effect of the new sample:

$$\hat{x}_t^i \sim q(\hat{x}_t|x_{t-1}^i, y_{1:t}) \qquad (3.48)$$

$$\bar{w}_{1:t}^i = \frac{p(y_t|\hat{x}_t^i)p(\hat{x}_t^i|x_{t-1}^i)}{q(\hat{x}_t^i|x_{t-1}^i, y_{1:t})} \frac{q(x_t^i|x_{t-1}^i, y_{1:t})}{p(y_t|x_t^i)p(x_t^i|x_{t-1}^i)}. \qquad (3.49)$$

These weights then need to be normalised as previously.

3.5.9 Rao–Blackwellised particle filters

If the state space that the particles populate is smaller, then one can imagine that fewer particles will be necessary. So, if it is possible to restrict the particles to live in a subset of the problem, it may be that the computational requirements can be reduced. Another way of looking at the same thing is to notice that analytic integration introduces no errors whereas any approximation must, by its nature, introduce errors. So, if it is necessary to approximate, one wants to do so as little as possible.

Since

$$p(X,Y) = p(X)p(Y|X) = p(Y)p(X|Y), \qquad (3.50)$$

if one can calculate $p(X|Y)$ or $p(Y|X)$ analytically then there is no need to sample both X and Y.

This idea of mixing analytic integration with stochastic sampling is Rao–Blackwellisation. The benefit is sometimes difficult to quantify since, while one may need fewer particles, it may be that each Rao–Blackwellised particle is sufficiently more expensive than a vanilla particle that the overall computational cost is minimised by having more vanilla particles rather than fewer Rao–Blackwellised particles.[7]

On an intuitive level, this idea of Rao–Blackwellisation can be thought of in terms of a kernel density approximation to the posterior. If the kernels each only cover a small amount of the space, then one needs a large number of kernels and it might well be more expedient to populate the posterior with samples. Conversely, one may get a very close match to the posterior with only a few kernels.

The idea has been used implicitly since the outset of particle filtering when considering a likelihood that is a mixture distribution (Avitzour 1995).[8] Another example occurs when considering systems for which the measurement process is highly nonlinear while the dynamics are linear and Gaussian.

---
[7] The same argument also holds when considering using different proposal distributions.
[8] This has been explicitly posed as Rao–Blackwellising the discrete distribution over the index of the mixture component (Marrs, Maskell and Bar-Shalom (2002)).

Part of the state can then be Rao–Blackwellised. Each particle analytically integrates the distribution over part of the state using a Kalman filter (with correlated measurement and process noises) and the particles are only used to disambiguate the part of the state relating to the measurements (Maskell, Gordon, Rollason and Salmond 2002).

Example 3.4 In our example, tackling the problem using a filter that Rao–Blackwellises the velocity and only samples the position doesn't give much of an improvement since the kernels turn out to be small; for constant velocity models, the space of velocities that could give rise to the same sequence of positions is sufficiently small that integrating it out doesn't give much benefit. This hasn't been discussed here in detail since it would require a discussion of the aforementioned alternative derivations of Kalman filters with correlated noise structure and the benefit for the example is minimal.

As observed by a reviewer, were the parameters of the filter (e.g. R and q) treated as unknown and were it possible to choose prior distributions such that the distributions of these parameters conditional on the state sequence were analytically tractable, one could Rao–Blackwellise these parameters. This could be done if one were to consider a discrete distribution over some candidate values for the parameters. However, this has many undesirable properties, such as having to fix these candidate values at the start of the run. Again, the details of more complex approaches are beyond the scope of this paper.

However, from a pragmatic perspective, when considering the first case, the nonlinearity is sufficiently unpronounced that approximately integrating the whole state is appealing. The approximately Rao–Blackwellised particle filter that results is an EKF!

3.6 Some applications of particle filtering

This section gives an impression of the scope of the available literature. The details can be found in the references; here we simply illustrate the breadth of applications that have been considered by the approach by describing the models used.

3.6.1 Bearings-only target tracking model

Particle filtering was initially proposed as an alternative to the EKF for bearings-only tracking (Gordon, Salmond and Smith 1993). When viewing a target from a platform such as an aircraft, certain sensors (e.g. a camera) will give measurements of bearings with no associated estimates of range. The idea is to attempt to use the incoming bearings measurements to infer the two dimensional position. This is often made possible by the sensing platform outmanoeuvering the target and so moving relative to the target such that it is possible to triangulate over time. Typically the target itself is assumed to be moving according to a constant velocity process in two dimensional space:[9]

$$p(x_t|x_{t-1}) = N(x_t; Ax_{t-1}, Q), \qquad (3.51)$$
$$p(y_t|x_t) = N(y_t; f(x_t), R), \qquad (3.52)$$

where

$$A = \begin{bmatrix} 1 & \Delta & 0 & 0 \\ 0 & 1 & 0 & 0 \\ 0 & 0 & 1 & \Delta \\ 0 & 0 & 0 & 1 \end{bmatrix}, \quad Q = \begin{bmatrix} \frac{\Delta^3}{3} & \frac{\Delta^2}{2} & 0 & 0 \\ \frac{\Delta^2}{2} & \Delta & 0 & 0 \\ 0 & 0 & \frac{\Delta^3}{3} & \frac{\Delta^2}{2} \\ 0 & 0 & \frac{\Delta^2}{2} & \Delta \end{bmatrix} q, \qquad (3.53)$$

$$f(x_t) = \arctan \frac{x_t[1]}{x_t[3]}, \qquad (3.54)$$

where q, Δ and R are as defined previously and where $x[i]$ is the ith element of the vector x.

This is an application for which the nonlinearity can be too severe to use an EKF, and for which alternatives to the EKF (such as the particle filter) are often advocated. Since only bearing (and so no range) is observed, the uncertainty over position is typically large and the distribution therefore diffuse. Small changes in position at high range will have much less of an effect on the linear approximation than the same change at small range. So, the nonlinearity is likely to become pronounced and an EKF can then be outperformed by a particle filter. An alternative approach is to use a bank of EKFs, each of which initially considers the target to be within some portion of the ranges. Since, for each such range-parameterised EKF, the nonlinearity is less pronounced, one can then achieve similar performance to that possible with a particle filter at a fraction of the computational cost (Arulampalam and Ristic 2000).

---

[9] This two dimensional constant velocity model is similar to that considered in the example in this paper but is the result of two independent integrated diffusions rather than a single such process.

3.6.2 Simultaneous localisation and mapping

A paper (Montemerlo, Thrun, Koller and Wegbreit 2002) has used a Rao–Blackwellised particle filter to propose an elegant solution to the problem of simultaneous localisation and mapping (SLAM) for which a map of a robot's environment is constructed at the same time as the robot's location within the map. The robot observes the range and bearing of several landmarks which are assumed to be stationary. The particles then explore the uncertainty over the trajectory of the robot. For each particle, it transpires that the landmarks can then each be localised relative to the robot using a little Kalman filter for each landmark.

3.6.3 Navigation

Particle filters have also been used to process measurements from an inertial navigation system to localise a target within a known map using the knowledge that the vehicle is constrained to move on a road network (Gustafsson *et al.* 2002). The inertial navigation system provides measurements of the wheel speed (with assumed Gaussian noise) of the vehicle and the dynamic model is the aforementioned two dimensional constant velocity model. The constraints imposed by the road network (that the vehicle is on a road) are implemented as a non-linear Gaussian 'measurement' of the distance of the particle to the nearest road. The value of the 'measurement' is taken to be zero. This has the effect of concentrating the posterior's probability mass around the roads. Using this approach, the particle filter can deduce the position from the sequence and spacing of any changes in speed (a reduction in speed makes it more likely that the dynamics will result in a trajectory corresponding to a turn) experienced by the target.

3.6.4 Ellipsometry

A simple particle filter has been used to conduct on-line estimation of the alloy fraction of a semiconductor as a function of depth using nonlinear measurements of polarisation angle taken while the semiconductor is being grown (Marrs 2001). Gases passed over the surface of the semiconductor cause it to grow. The constituents of the gases determine the proportion of different alloys that form in the semiconductor, which in turn alter the angular difference in polarisation plane of incident and reflected polarised light. There is a need to adjust the relative concentrations of the gases so as to achieve the desired alloy fractions as a function of depth. Inference of the alloy fraction profile has previously necessitated the use of destructive analysis of the semiconductor samples at the end of the process. The use of a particle filter (which models the alloy content as evolving according to a random walk and makes use of the measurements of polarisation angle) makes it possible to monitor the semiconductor growth as the semiconductor is being grown and produces results that are in good agreement with the destructive approach.

3.6.5 Stochastic volatility model

If the volatility of a process is evolving according to a random walk, then the model is known as a stochastic volatility (SV) model (see Pitt and Shephard (1999b) and Kim, Shephard and Chib 1998 for papers on SV models and particle filters) for which:

$$x_t = \phi x_{t-1} + \omega_t^x, \quad y_t = \omega_t^y \beta \exp\left(\frac{x_t}{2}\right), \tag{3.55}$$

where ω_t^x and ω_t^y are noise sequences as previously and ϕ and β are model parameters which respectively govern the rate of decay of the volatility towards $x_t = 0$ and the volatility in this state.

3.6.6 Computer vision

Particle filters have been used extensively to track objects in video sequences, where the approach is often referred to as CONditional DENSity estimATION, CONDENSATION (Blake and Isard 1998). The states, which might represent the parameters of a hand in a sequence of images, are often modelled as evolving according to a diffuse random walk with the likelihood modelled as a complicated function of the image data.

3.6.7 Time-varying autoregressive model

A Rao–Blackwellised particle filter based smoother has been used to conduct audio signal enhancement (Fong, Godsill, Doucet and West 2002). The signal is modelled as an autoregressive process with time-varying parameters, $a_{t,i}$, which themselves evolve according to a random walk:

$$y_t = \sum_{i=1}^{p} a_{t,i} y_{t-i} + \omega_t^y. \tag{3.56}$$

A similar approach has also been proposed that allows the number of coefficients to change with time (Andrieu, Davy and Doucet 2002).

3.6.8 Jump Markov linear systems

A number of environments ranging from tracking manoeuvreing targets in clutter to code division multiple access (CDMA) for communications can be formulated as inference which could be conducted with a Kalman filter were a discrete systems history of states known. Such systems have dynamic and/or measurement models that are a mixture of Gaussians. Particle filters have been used to explore the uncertainty over the history of mixture components (Doucet, Gordon and Krishnamurthy 2001).

3.6.9 Out-of-sequence measurements

Out-of-sequence measurements are received in an order that is different to the order of the times to which it relates; one might as a result of communication or acoustic delays receive data in such a sequence as $\{1, 2, 4, 5, 6, 3, 8, 7\}$. Approaches to solving the filtering problem in this context have previously been proposed within a particle filter framework (Orton and Marrs 2001). The particle filter is appealing in this environment because, for each particle, such out-of-sequence measurements can be processed in the same way as in-sequence data; a state is sampled from a proposal and new weights calculated. Hence the samples at the most recent time remain fixed but the weights on these samples change to reflect the impact of the out-of-sequence measurement. Within an (extended) Kalman filter environment, the processing of such out-of-sequence measurements is more complicated since the Gaussian representing the uncertainty relating to the most recent time needs to reflect the impact of the out-of-sequence measurement.

3.7 Conclusions

Here the particle filter has been described within the context of a solution to the tracking problem. The fundamentals of representing uncertainty using a pdf have provided a starting point from which the discussion has progressed to considering methods for improving the computational cost of the particle filter; this is motivation for much of the current research on the subject. Examples of applications of the approach are given. The same example is used throughout so as to drawn attention to links with the Kalman filtering methodology and illustrate some of the ideas presented and described.

Acknowledgements

The author thanks the Royal Commission for the Exhibition of 1851 for his Industrial Fellowship and was also funded by the Ministry of Defence Corporate Research Programme. The author also gratefully acknowledges the constructive criticisms of the reviewers which strengthened this contribution significantly.

Part II

Testing

4

Frequency domain and wavelet-based estimation for long-memory signal plus noise models

Katsuto Tanaka

Hitotsubashi University, Tokyo

Abstract

This paper discusses the estimation problems associated with signal plus noise models, where the signal is assumed to follow a stationary or nonstationary long-memory process, whereas the noise is assumed to be an independent process. Moreover, the signal and noise are independent of each other. We take up various frequency domain and wavelet-based estimators, and examine finite sample properties of these estimators. It is found that frequency domain estimators perform better for stationary cases, whereas they tend to become worse for nonstationary cases. To overcome this deficiency we consider frequency domain estimators based on differenced series. It is also found that wavelet-based estimators perform well for both stationary and nonstationary cases.

4.1 Introduction

Let us consider the model

$$y_t = x_t + u_t, \qquad (t = 1, \ldots, T), \qquad (4.1)$$

where only y_t is observable, x_t is a stochastic signal and u_t is noise or a measurement error. We assume that x_t and u_t are independent of each other. Moreover, u_t is assumed to be normally independent and identically distributed with the mean 0 and variance $\rho\sigma^2$, which is abbreviated as NID(0, $\rho\sigma^2$) hereafter, where ρ is a nonnegative constant while σ^2 is a positive constant that is the variance of the innovation driving the signal x_t.

The signal process x_t is assumed to be of the form:

$$(1 - L)^d x_t = \varepsilon_t, \qquad (4.2)$$

where ε_t follows NID(0, σ^2), whereas the differencing parameter d is assumed to take any positive value, which yields a long-memory stationary or nonstationary process.

The model (4.2) is often referred to as the ARFIMA(0, d, 0) model. When $0 < d < 1/2$, x_t becomes stationary and is assumed to be generated from

$$x_t = (1 - L)^{-d} \varepsilon_t = \sum_{j=0}^{\infty} \frac{\Gamma(d+j)}{\Gamma(d)\,\Gamma(j+1)} \varepsilon_{t-j}.$$

When $d \geq 1/2$ so that $d = [d + 1/2] + \delta$, where $[a]$ is the integer part of a and δ is the remaining decimal part such that $-1/2 \leq \delta < 1/2$, $\{x_t\}$ is nonstationary and is assumed to be represented as the output of the filter $(1 - L)^\delta$ cascaded with the filter $(1 - L)^{[d+1/2]}$. Namely, the process $\{x_t\}$ is generated from $(1 - L)^{[d+1/2]} x_t = z_t$ with $(1 - L)^\delta z_t = \varepsilon_t$. Following this rule, we have, for example, $[d + 1/2] = 1$ and $\delta = -0.2$ for $d = 0.8$, while we have $[d + 1/2] = 2$ and $\delta = -0.5$ for $d = 1.5$. Note that it holds that

$$x_t = \begin{cases} O_p(1) & (0 < d < 1/2), \\ O_p\left(\sqrt{\log t}\right) & (d = 1/2), \\ O_p\left(t^{d-1/2}\right) & (1/2 < d). \end{cases}$$

The model (4.1) coupled with (4.2) is a special case of ARFIMA(0, d, ∞), where the MA coefficients decay slowly. Our main concern here is how to estimate the differencing parameter d, although the parameters ρ and σ^2 are also unknown and have to be estimated. The estimation problem for the case of no noise, that is, the case of $y_t = x_t$ has been extensively discussed both in the time and frequency domains. The existence of the noise term

$\{u_t\}$ makes the problem quite complicated, which will be recognised in the subsequent sections.

Let us put $y = (y_1, \ldots, y_T)' = x + u$, defining x and u similarly. We also put $\Omega(d, \rho) = E(y\,y')/\sigma^2 = (E(x\,x') + E(u\,u'))/\sigma^2 = \Phi(d) + \rho I_T$, where I_T is the identity matrix of $T \times T$. Then the log-likelihood for $y = (y_1, \ldots, y_T)'$ is given by

$$L(d, \rho, \sigma^2) = -\frac{T}{2}\log(2\pi\sigma^2) - \frac{1}{2}\log|\Phi(d) + \rho I_T| - \frac{1}{2\sigma^2} y'\,(\Phi(d) + \rho I_T)^{-1}\,y. \quad (4.3)$$

It turns out that the time domain method is much involved because the maximisation of (4.3) requires the inversion of the $T \times T$ matrix $\Omega(d, \rho) = \Phi(d) + \rho I_T$ and the computation of its determinant (e.g. Doornik and Ooms (2003))

Instead we consider the two alternative methods, which require neither the matrix inversion nor the computation of a determinant. One is the frequency domain method and the other the wavelet-based method. Section 4.2 discusses the former, while the latter is the subject of Sections 4.3 and 4.4. A brief introduction to wavelets is given in Section 4.3 and the estimation method based on the wavelet is described in Section 4.4. Section 4.5 reports simulation results, comparing the two methods. Section 4.6 gives concluding remarks together with a brief description of the testing problem associated with the present model.

4.2 Frequency domain method

Dividing this section into two subsections, we first discuss the case of no noise, that is, the case of $y_t = x_t$ in (4.1). Then we consider the general signal plus noise case.

4.2.1 Case of no noise

The estimation problem becomes much simpler if there is no noise, which we assume to be the case in this subsection. In this case the so-called GPH estimator suggested by Geweke and Porter-Hudak (1983) may be used. This is based on the expression of the spectrum $\sigma^2 f_x(\omega)$ of $\{x_t\}$, where

$$f_x(\omega) = \frac{1}{|1 - e^{-2\pi i \omega}|^{2d}} = \frac{1}{(4\sin^2 \pi\omega)^d} \quad (0 < \omega < 1/2), \quad (4.4)$$

which leads us to a frequency domain log-linear regression relation:

$$\log I_x(\omega_k) = \hat{a} - \hat{d} \log\left(4\sin^2 \pi\omega_k\right) + e_k, \quad \omega_k = k/T \quad (k=1,\ldots n < T/2). \tag{4.5}$$

Here $\hat{a}$ is the LSE of an intercept and $\hat{d}$ is the LSE of d, whereas $I_x(\omega)$ is the periodogram of $\{x_t\}$ defined by

$$I_x(\omega) = \frac{1}{T}\left|\sum_{t=1}^{T} x_t e^{-2\pi i\omega t}\right|^2.$$

The number n in (4.5) is chosen in such a way that $n = O(T^\alpha)$ with $0 < \alpha < 1$. Hurvich, Deo and Brodsky (1998) suggested using $n = O(T^{4/5})$ in terms of the mean square error (MSE) and also discussed the consistency of $\hat{d}$ in the case of $|d| < 1/2$. Note that the spectrum in (4.4) and the regression relation in (4.5) can be defined even for $d \geq 1/2$.

The frequency domain MLEs or the Whittle-type estimators of d and σ^2 can be obtained from the approximate log-likelihood

$$L_F(d,\sigma^2) = -\frac{T}{2}\log(2\pi\sigma^2) - \frac{1}{2}\sum_{k\neq 0}\log f_x(\omega_k) - \frac{1}{2\sigma^2}\sum_{k\neq 0}\frac{I_x(\omega_k)}{f_x(\omega_k)}. \tag{4.6}$$

Here and below k ($\neq 0$) runs over the range $-T/2 < k \leq [T/2]$ with $[x]$ being the integer part of x (Hannan 1973). This is a frequency domain version of the time domain likelihood given in (4.3) with ρ replaced by $\rho = 0$. Velasco and Robinson (2000) proved consistency of the MLE when $d < 1$.

4.2.2 Signal plus noise case

If there exists noise u_t so that $y_t = x_t + u_t$ as in (4.1), the log-linear relation in (4.5) breaks down. In fact, the spectrum of the process $\{y_t\}$ is given by $\sigma^2 f_y(\omega)$, where

$$f_y(\omega) = \frac{1}{|1 - e^{-2\pi i\omega}|^{2d}} + \rho = \left(4\sin^2 \pi\omega\right)^{-d} + \rho,$$

so that

$$\log\left(\sigma^2 f_y(\omega_k)\right) = \log \sigma^2 + \log\left[\left(4\sin^2 \pi\omega_k\right)^{-d} + \rho\right].$$

Because of the existence of the parameter ρ, this last relation leads us to an essentially nonlinear regression:

$$\log I_y(\omega_k) = \hat{a} + \log\left[\left(4\sin^2 \pi\omega_k\right)^{-\hat{d}} + \hat{\rho}\right] + e_k \quad (k=1,\ldots,n). \tag{4.7}$$

This nonlinear regression can still be estimated in spite of its complexity, which we show in the simulation study in Section 4.5.

The frequency domain MLEs of d, ρ and σ^2 can also be computed in the present case. We can consider the maximisation of

$$L_F(d, \rho, \sigma^2) = -\frac{T}{2}\log(2\pi\sigma^2) - \frac{1}{2}\sum_{k\neq 0}\log f_y(\omega_k) - \frac{1}{2\sigma^2}\sum_{k\neq 0}\frac{I_y(\omega_k)}{f_y(\omega_k)}, \quad (4.8)$$

which is quite similar to (4.6), but the present case contains an additional parameter ρ in the spectrum $f_y(\omega)$. The MLE of d will be inconsistent when $d \geq 1$. To overcome this difficulty we can consider differences of $\{y_t\}$. For example, if we consider $\{\triangle y_t\}$, the first differenced data of $\{y_t\}$, this has the spectrum

$$f_{\triangle y}(\omega) = \left(4\sin^2 \pi\omega\right)^{-d+1} + \rho\left(4\sin^2 \pi\omega\right).$$

Denoting the periodogram of $\{\triangle y_t\}$ by $I_{\triangle y}(\omega)$, we can also estimate the parameters by maximising (4.8) with $f_y(\omega_k)$ and $I_y(\omega_k)$ replaced by $f_{\triangle y}(\omega_k)$ and $I_{\triangle y}(\omega_k)$, respectively. A similar idea can be used to modify the nonlinear regression in (4.7). These estimators will also be examined in Section 4.5.

4.3 Wavelet transform

In this section we give a brief description of wavelets. The discrete wavelet transform (DWT) has been devised as a means of analysing discrete time series. However, it was the continuous wavelet transform (CWT) that was developed initially. Suppose that the continuous signal $\{x(t)\}$ satisfies

$$\|x\|^2 = \int_{-\infty}^{\infty} x^2(t)dt < \infty.$$

Let $\psi(t)$ be a wavelet function that plays an important role in the definition of the CWT, which satisfies

$$\int_{-\infty}^{\infty} \psi(t)dt = 0, \qquad \int_{-\infty}^{\infty} \psi^2(t)dt = 1. \quad (4.9)$$

The first condition implies that the wavelet is oscillating around zero, whereas the second condition means that its energy is finite and normalised to unity.

The wavelet is a generic function, unlike trigonometric or exponential functions. The function $\psi(t)$ becomes a wavelet so long as the conditions in (4.9) are satisfied. The simplest wavelet function, called the Haar wavelet, is given by

$$\psi_H(t) = \begin{cases} -1/\sqrt{2} & (0 \leq t < 1), \\ 1/\sqrt{2} & (1 \leq t < 2), \\ 0 & (\text{otherwise}). \end{cases}$$

The CWT of $x(t)$ is defined by

$$C_{a,b}(\psi, x) = \frac{1}{\sqrt{a}} \int_{-\infty}^{\infty} \psi\left(\frac{t-b}{a}\right) x(t) dt = \int_{-\infty}^{\infty} \tilde{\psi}_{a,b}(t) x(t) dt, \qquad (4.10)$$

where

$$\tilde{\psi}_{a,b}(t) = \frac{1}{\sqrt{a}} \psi\left(\frac{t-b}{a}\right) \qquad (0 < a, \ -\infty < b < \infty).$$

The function $\tilde{\psi}_{a,b}(t)$ is obtained by shifting $\psi(t)$ by b to the right and by dilating $\psi(t)$ a times. Note that $\tilde{\psi}_{a,b}(t)$ has the properties described in (4.9). In fact, we have

$$\int_{-\infty}^{\infty} \tilde{\psi}_{a,b}^2(t) dt = \frac{1}{a} \int_{-\infty}^{\infty} \psi^2\left(\frac{t-b}{a}\right) dt = \int_{-\infty}^{\infty} \psi^2(t) dt = 1.$$

If the Haar wavelet is used, the CWT takes the following form:

$$\begin{aligned} C_{a,b}(\psi_H, x) &= \frac{1}{\sqrt{a}} \int_{-\infty}^{\infty} \psi_H\left(\frac{t-b}{a}\right) x(t) dt \\ &= \frac{1}{\sqrt{2a}} \left[\int_{a+b}^{2a+b} x(t) dt - \int_{b}^{a+b} x(t) dt \right]. \end{aligned}$$

It is seen that the resulting CWT is essentially the difference between the averages of $x(t)$ over the adjacent intervals $[b, a+b]$ and $[a+b, 2a+b]$ of width a. We call b the shift parameter and a the dilation parameter. The use of the two parameters makes the wavelet transform localised in time and scale, whereas the Fourier transform has only one parameter λ as in

$$\hat{x}(\lambda) = \int_{-\infty}^{\infty} x(t) e^{-2\pi i \lambda t} dt,$$

which is not localised in time in any meaningful sense.

We next move on to the description of the DWT corresponding to the CWT (see, for details, Percival and Walden (2000)). Let $\{X_t\}$ be a discrete time series and suppose that $\mathbf{X} = (\mathbf{X}_1, \ldots, \mathbf{X}_T)'$ is observed, where the sample size T is assumed to be $T = 2^J$ with J being a positive integer.

Let us denote the DWT of $\mathbf{X}$ by $\mathbf{W}$, which is given by

$$\mathbf{W} = \mathcal{W}\mathbf{X} \quad \Leftrightarrow \quad \begin{pmatrix} \mathbf{W}_1 \\ \mathbf{W}_2 \\ \vdots \\ \mathbf{W}_J \\ \mathbf{V}_J \end{pmatrix} = \begin{pmatrix} \mathcal{W}_1 \\ \mathcal{W}_2 \\ \vdots \\ \mathcal{W}_J \\ \mathcal{V}_J \end{pmatrix} \mathbf{X}, \quad (4.11)$$

where $\mathbf{W}_j$ is a vector of $T_j = T/2^j$ components and is called the jth level wavelet, whereas $\mathbf{V}_J$ is a scalar and is called the Jth level scaling coefficient. The matrix $\mathcal{W}$ is the orthonormal matrix called the wavelet transform matrix with $\mathcal{W}_j$ being a $T_j \times T$ matrix and $\mathcal{V}_J$ being a row vector of T components. Note that $\mathcal{W}_j \mathcal{W}'_k = I_{T_j}$ for $j = k$ and 0 for $j \ne k$. Note also that $\mathcal{V}_J \mathcal{V}'_J = 1$ and that $\mathcal{V}_J$ is orthogonal to every $\mathcal{W}_j$.

Define the $T_j \times 1$ vectors

$$\mathbf{W}_j = (W_{j,1}, \ldots, W_{j,t}, \ldots, W_{j,T_j})', \quad j = 1, \ldots, J.$$

Each element $W_{j,t}$ of $\mathbf{W}_j$ is a discrete version of the CWT $C_{a,b}(\psi, x)$ in (4.10), where the shift parameter b is replaced by t and the dilation parameter a by j. In terms of frequency or scale, $\mathbf{W}_1$ gives the highest resolution because j or a is the smallest so that the average is taken for the shortest period. As the level j becomes large, $\mathbf{W}_j$ gives a lower resolution. On the other hand, $\mathbf{V}_J$, the last component of $\mathbf{W}$, gives the lowest resolution. In fact, $\mathbf{V}_J = \sqrt{T}\bar{X}$.

The computation of the DWT can be accomplished most efficiently by the pyramid algorithm (Percival and Walden 2000). More specifically, we first prepare the wavelet filter $\{h_l : l = 0, 1, \ldots, M-1\}$ which serves as a high-pass filter, where M, the width of the filter, is an even integer. The wavelet filter $\{h_l\}$ is required to satisfy

$$\sum_{l=0}^{M-1} h_l = 0, \quad \sum_{l=0}^{M-1} h_l^2 = 1, \quad \sum_{l=0}^{M-1} h_l h_{l+2n} = 0 \quad (n: \text{any nonzero integer}).$$
(4.12)

Note that $\{h_l\}$ corresponds to the wavelet function $\psi(t)$ in the CWT. One such example of $\{h_l\}$ is the Haar wavelet filter, where $h_0 = 1/\sqrt{2}$, $h_1 = -1/\sqrt{2}$ with $M = 2$. Then each component of the first level wavelet $\mathbf{W}_1$ can be computed following

$$W_{1,t} = \sum_{l=0}^{M-1} h_l X_{2t-l \bmod T} = \sum_{l=1}^{T} h_{2t-l \bmod T} X_l.$$

For the Haar wavelet filter, this gives us, for example,

$$W_{1,1} = h_0 X_2 + h_1 X_1 = (X_2 - X_1)/\sqrt{2}.$$

For various even numbers of width M, Daubechies 1992 suggested a useful class of wavelet filters, all of which yield a DWT in accordance with the notion of differences of adjacent averages. More specifically, the Daubechies wavelet filters are interpreted as a convolution of two cascaded filters; one is a differencing filter of width $M/2$, and the other a moving average filter resulting in a low-pass filter. The Haar wavelet filter is the simplest filter that belongs to the class of the Daubechies wavelet filters.

To compute wavelet coefficients of higher levels, we also need a filter $\{g_l : l = 0, 1, \ldots, M-1\}$ which serves as a low-pass filter. This is obtained as the quadruture mirror filter (QMF) of $\{h_l\}$ determined by

$$g_l = (-1)^{l+1} h_{M-1-l} \Leftrightarrow h_l = (-1)^l g_{M-1-l} \quad (l = 0, 1, \ldots, M-1). \tag{4.13}$$

The filter $\{g_l\}$ is called the scaling filter. It holds that

$$\sum_{l=0}^{M-1} g_l = \sqrt{2}, \quad \sum_{l=0}^{M-1} g_l^2 = 1, \quad \sum_{l=0}^{M-1} g_l g_{l+2n} = 0 \quad (n\text{: any nonzero integer}). \tag{4.14}$$

It also holds that the scaling and wavelet filters and any of their even shifts are orthogonal to each other, that is,

$$\sum_{l=0}^{M-1} g_l h_{l+2n} = 0 \quad (n\text{: any integer}). \tag{4.15}$$

Then, at the jth stage of the pyramid algorithm, we have the jth level wavelet and scaling coefficients given by, respectively,

$$W_{j,t} = \sum_{l=0}^{M-1} h_l V_{j-1, 2t-l \bmod T_{j-1}}, \quad V_{j,t} = \sum_{l=0}^{M-1} g_l V_{j-1, 2t-l \bmod T_{j-1}}, \tag{4.16}$$

for $t = 1, \ldots, T_j$, where $V_{0,t} = X_t$. Defining $\mathbf{V}_0 = X$, these may be rewritten as

$$\mathbf{W}_j = \mathcal{B}_j \mathbf{V}_{j-1} = \mathcal{B}_j \mathcal{A}_{j-1} \cdots \mathcal{A}_1 \mathbf{X} = \mathcal{W}_j \mathbf{X}, \tag{4.17}$$

$$\mathbf{V}_j = \mathcal{A}_j \mathbf{V}_{j-1} = \mathcal{A}_j \mathcal{A}_{j-1} \cdots \mathcal{A}_1 \mathbf{X} = \mathcal{V}_j \mathbf{X}, \tag{4.18}$$

where $\mathcal{A}_j$ and $\mathcal{B}_j$ are $T_j \times T_{j-1}$ matrices. The matrix $\mathcal{A}_j$ is composed of $\{g_l\}$ only, whereas $\mathcal{B}_j$ is composed of $\{h_l\}$ only. Executing the computation recursively from $j = 1$ to $j = J$, we arrive at **W** in (4.11). An important point of the pyramid algorithm is that the jth level wavelet $\mathbf{W_j}$ is computed, not from $\mathbf{W}_{j-1}$, but from the $(j-1)$st scaling vector $\mathbf{V}_{j-1}$, as is recognised in (4.17). This indirect computation, however, requires only $O(T)$ multiplications, whereas brute force computation of the product in $\mathbf{W} = \mathcal{W}\mathbf{X}$ requires $O(T^2)$ multiplications (Percival and Walden 2000, p. 68).

A salient feature of the DWT applied to long-memory processes is the stationarity property, which means that when M, the width of the Daubechies wavelet filter, is at least as large as $2d$ with d being the differencing parameter, then the DWT $\{W_{j,t}\}$ becomes stationary for each level j (Percival and Walden 2000, p. 304). This justifies the definition of the jth level wavelet variance given by $\sigma_j^2 = V(W_{j,t})$ for each j.

Another feature of the DWT applied to long-memory processes is the approximate decorrelation property, which means that $\{W_{j,t}\}$ is approximately uncorrelated for each j and is also approximately uncorrelated with $\{W_{k,u}\}$ when $j \neq k$. The approximate decorrelation property of the DWT makes the covariance matrix of a long-memory process almost diagonal, as in the Fourier transform.

The above two properties of the DWT will be effectively used in the estimation of signal plus noise models, which will be discussed in the next section.

The DWT discussed so far, however, has some limitations, among which are:

(a) The sample size T must be a power of 2.
(b) The number of components of the jth level wavelet $\mathbf{W}_j$ decreases by a factor of 2 for each increasing level j so that wavelet coefficients for higher levels cannot be used for inference purposes. In fact $\mathbf{W}_J$ becomes a scalar.
(c) The transforms are not associated with zero phase filters, thus making it difficult to line up features with the original time series meaningfully.
(d) The wavelet coefficients are not circularly shift invariant, that is, circularly shifting the time series by some amount will not circularly shift the coefficients by the same amount.

These deficiencies can be overcome by using the maximal overlap DWT (MODWT), which is a highly redundant nonorthogonal transform yielding the jth level wavelet of T components for each j. Thus some computational price has to be paid, although its computational burden is the same as the widely used fast Fourier transform. See Percival and Walden (2000) for details. Transforms that are essentially the same as the MODWT have been discussed in the wavelet literature under the names 'nondecimated DWT', 'translation invariant DWT', 'time invariant DWT' and so on. The MODWT, however, does not enjoy the decorrelation property, although it has the stationarity property. Thus the MODWT is not very useful so far as our purpose is concerned.

4.4 Wavelet-based estimation

In this section we discuss the wavelet-based estimation. We mainly discuss the signal plus noise case. The case of no noise is briefly described as a by-product.

Let us return to the following model:

$$y_t = x_t + u_t, \qquad (1 - L)^d x_t = \varepsilon_t. \tag{4.19}$$

Our main concern here is to estimate the differencing parameter d, for which we consider two types of wavelet-based estimators. One is the LSE and the other the MLE.

We first consider the LSE. Let $\mathbf{W}_y$ be the wavelet vector obtained from $y = (y_1, \ldots, y_T)'$. We also define $\mathbf{W}_x$ and $\mathbf{W}_u$ similarly. Then we have $\mathbf{W}_y = \mathbf{W}_x + \mathbf{W}_u$. In terms of the wavelet variance defined in the last section, it holds that

$$\sigma_{jy}^2 = \sigma_{jx}^2 + \sigma_{ju}^2 = \sigma^2 \left[\int_{-1/2}^{1/2} f_{jx}(\omega) d\omega + \rho \right] = \sigma^2 \left[\int_{-1/2}^{1/2} \mathcal{H}_j(\omega) f_x(\omega) d\omega + \rho \right], \tag{4.20}$$

where σ_{jy}^2 is the jth level wavelet variance associated with $\{y_t\}$, and σ_{jx}^2 and σ_{ju}^2 are defined similarly, whereas $\sigma^2 f_{jx}(\omega)$ is the spectrum of the jth level wavelet associated with $\{x_t\}$, which can be expressed as the product of the squared gain $\mathcal{H}_j(\omega)$ associated with the jth level wavelet filter and the spectrum $\sigma^2 f_x(\omega)$ of $\{x_t\}$. Note that $\sigma_{ju}^2 = \rho \sigma^2$ because $\{u_t\}$ is an NID$(0, \rho \sigma^2)$ sequence.

Then, utilising the fact that the jth level wavelet filter is a band-pass filter with nominal pass-band given by $[-1/2^j, -1/2^{j+1}] \cup [1/2^{j+1}, 1/2^j]$, we can make the crude approximation (McCoy and Walden 1996)

$$\mathcal{H}_j(\omega) \approx \begin{cases} 2^j & (1/2^{j+1} \leq |\omega| \leq 1/2^j), \\ 0 & (\text{otherwise}). \end{cases} \tag{4.21}$$

Substituting (4.21) into (4.20) and noting that $f_x(\omega) = 1/(4 \sin^2 \pi \omega)^d \approx$

$1/(4\pi^2\omega^2)^d$ for ω close to 0, it follows that, when j is not small,

$$\sigma_{jy}^2 \approx 2^{j+1} \int_{1/2^{j+1}}^{1/2^j} \frac{\sigma^2}{(4\sin^2 \pi\omega)^d}\, d\omega + \rho\sigma^2$$

$$\approx 2^{j+1} \int_{1/2^{j+1}}^{1/2^j} \frac{\sigma^2}{(4\pi^2\omega^2)^d}\, d\omega + \rho\sigma^2$$

$$= a + b \times 4^{jd}, \tag{4.22}$$

where a and b are positive constants.

Thus, denoting by $\hat\sigma_{jy}^2$ the sample variance of $W_{j,1},\ldots,W_{j,T_j}$, we have the following nonlinear regression relation:

$$\hat\sigma_{jy}^2 = \hat a + \hat b \times 4^{j\hat d} + e_{1j}, \qquad (j=j_0,\ldots,J_0), \tag{4.23}$$

where e_{1j} is the residual. Here it is required that $j_0 > 1$ and $J_0 < J$. The former restriction comes from the approximation $\sin x \approx x$ used in (4.22), whereas the latter comes from the condition for $\hat\sigma_{jy}^2$ to be computed. In fact, only one $W_{j,t}$ is available when $j = J$. These restrictions reduce the effective number of data points used to compute the regression relation (4.23). As an example, suppose that $T = 512 = 2^9$ so that $J = 9$. Then we may put $j_0 = 2$ at least and $J_0 = 6$ at most. Note that $J_0 = 7$ gives us only four data points to estimate the wavelet variance, which is too few, whereas $J_0 = 6$ gives us eight data points. Thus the three parameters in the nonlinear regression relation (4.23) must be estimated from five pairs of data when $j_0 = 2$ and $J_0 = 6$. It is found that T is required to be much larger to obtain satisfactory results.

In the regression relation (4.23), the wavelet variance σ_{jy}^2 is estimated by using the DWT. We can also use the MODWT to estimate σ_{jy}^2. It is found that the use of the MODWT gives little improvement if the noise term $\{u_t\}$ is present.

The estimation problem becomes much simpler if there is no noise, that is, if $y_t = x_t$ in (4.19). In that case, we have the log-linear relation

$$\log \hat\sigma_{jx}^2 = \hat a + \hat d \log 4^j + e_{2j} \qquad (j=j_0,\ldots,J_0), \tag{4.24}$$

which is discussed in Jensen 1999, where the DWT is used to compute $\hat\sigma_{jx}^2$. This is a wavelet version of the GPH estimator in the frequency domain given in (4.5). It is found that reasonably good results can be obtained even when the sample size T is not large. We shall examine the performance of the LSE in the next section using the DWT and MODWT to estimate σ_{jx}^2.

Finally we discuss the wavelet-based MLE, which is derived from the approximate decorrelation property of the wavelet $\{W_{j,t}\}$ between times and levels. Denoting by $\mathbf{W}_y = \mathcal{W} y$ the wavelet vector associated with $y = (y_1, \cdots, y_T)'$, it holds approximately that $\mathbf{W}_y \sim N(0, \sigma^2 \Sigma_W)$, where

$$\Sigma_W = \text{diag}\left(h_1 I_{T_1}, h_2 I_{T_2}, \ldots, h_J I_{T_J}, h_{J+1} I_{T_{J+1}}\right), \tag{4.25}$$

with $T_j = T/2^j$ $(j = 1, \ldots, J)$, $T_{J+1} = 1$ and

$$h_j = 2^{j+1} \int_{1/2^{j+1}}^{1/2^j} \left(4 \sin^2 \pi \omega\right)^{-d} d\omega + \rho, \qquad (j = 1, \ldots, J+1). \tag{4.26}$$

Then we derive from (4.3) the approximate wavelet-based log-likelihood as

$$L_W(d, \rho, \sigma^2) = -\frac{T}{2} \log(2\pi\sigma^2) - \frac{1}{2} \sum_{j=1}^{J+1} T_j \log h_j - \frac{1}{2\sigma^2} \sum_{j=1}^{J+1} \frac{\mathbf{W}_j' \mathbf{W}_j}{h_j}, \tag{4.27}$$

where $\mathbf{W}_j$ is the jth level wavelet for $j = 1, \ldots, J$, whereas $\mathbf{W_{J+1}}$ is the scaling coefficient $\mathbf{V}_J$ defined in (4.11).

The wavelet-based MLE of d for the case of $\rho = 0$ was discussed in McCoy and Walden (1996) and Jensen (2000), and proved to be well behaved. We shall examine the performance of the MLE of d for the case of $\rho > 0$ in the next section.

4.5 Some simulations

In this section we examine the performance of the frequency domain and wavelet-based estimators we have discussed so far. For this purpose we conduct some simulations. The simulations presented here are illustrative only, and we do not attempt the very large task of a definitive simulation study.

We consider two types of models; one is the model without noise, that is,

$$\text{Model A:} \qquad (1-L)^d x_t = \varepsilon_t, \qquad (t = 1, \ldots, T), \tag{4.28}$$

and the other is the signal plus noise model

$$\text{Model B:} \qquad y_t = x_t + u_t, \qquad (1-L)^d x_t = \varepsilon_t, \qquad (t = 1, \ldots, T), \tag{4.29}$$

where $\{u_t\} \sim \text{NID}(0, \rho\sigma^2)$ and $\{\varepsilon_t\} \sim \text{NID}(0, \sigma^2)$. Here $\{u_t\}$ and $\{\varepsilon_t\}$ are independent of each other. Throughout the simulations we put $\sigma^2 = 1$, although it is also estimated. We also fix T at $512 = 2^9$ so that $J = 9$. We use the S-Plus function *arima.fracdiff* to generate x_t and *nlmin* or *nlminb* to estimate the parameters. The DWT and MODWT are computed via the Haar wavelet filter. The simulation results reported below are based on 1000 replications.

For each model we concentrate on the estimation of d and examine the following estimators:

Model A

> GPH: The frequency domain LSE based on (4.5) with n fixed at $n = 150 = T^{0.80}$.
>
> F-ML: The frequency domain MLE derived from (4.6).
>
> DWT: The wavelet-based LSE based on (4.24) using the DWT.
>
> MODWT: The wavelet-based LSE based on (4.24) using the MODWT.
>
> W-ML: The wavelet-based MLE derived from (4.27) with $\rho = 0$.

Model B

> GPH: The frequency domain nonlinear estimator based on (4.7) with $n = 150$.
>
> D-GPH: The frequency domain nonlinear estimator based on (4.7) with the first differenced data and $n = 150$.
>
> F-ML: The frequency domain MLE derived from (4.8).
>
> D-F-ML: The frequency domain MLE derived from (4.8) with the first differenced data.
>
> W-ML: The wavelet-based MLE derived from (4.27) with unknown ρ.

Note that the estimators D-GPH and D-F-ML estimate the value of $d - 1$ in the process $\triangle y_t = z_t + \triangle u_t$ with $(1-L)^{d-1} z_t = \varepsilon_t$. As was described before, the first differenced process has the spectrum given by

$$f_{\triangle y}(\omega) = \left(4\sin^2 \pi\omega\right)^{-d+1} + \rho\left(4\sin^2 \pi\omega\right).$$

Note also that, to obtain D-GPH and D-F-ML, the regression relation in (4.7) and the likelihood in (4.8) are changed accordingly by substituting $f_{\triangle y}(\omega)$ into $f_y(\omega)$.

Table 4.1 is concerned with Model A and reports estimation results on d for various values of d. We may conclude that

(a) The F-ML estimator performs best when $d \leq 1$, but it deteriorates when d is greater than 1.

(b) The GPH estimator performs better than the DWT estimator and performs as well as the MODWT estimator when $d \leq 1$. The simulation results reported in Jensen 1999, where the GPH and DWT estimators are compared, show evident superiority of the latter, which

Table 4.1. *Model A: case of no noise*

	GPH	F-ML	DWT	MODWT	W-ML
$d = 0.1$					
Mean	0.091	0.092	0.072	0.086	0.083
Var	0.0036	0.0014	0.0073	0.0029	0.0014
$d = 0.3$					
Mean	0.289	0.292	0.263	0.277	0.283
Var	0.0037	0.0014	0.0076	0.0034	0.0015
$d = 0.45$					
Mean	0.439	0.443	0.411	0.427	0.452
Var	0.0038	0.0014	0.0083	0.0037	0.0021
$d = 0.6$					
Mean	0.591	0.595	0.560	0.582	0.596
Var	0.0038	0.0014	0.0088	0.0043	0.0021
$d = 0.8$					
Mean	0.794	0.798	0.759	0.788	0.803
Var	0.0039	0.0015	0.0090	0.0048	0.0023
$d = 1$					
Mean	0.985	0.990	0.955	0.955	1.024
Var	0.0037	0.0014	0.0092	0.0036	0.0028
$d = 1.2$					
Mean	1.124	1.135	1.138	1.032	1.257
Var	0.0067	0.0035	0.0082	0.0049	0.0038
$d = 1.4$					
Mean	1.177	1.202	1.291	1.019	1.496
Var	0.0204	0.0145	0.0064	0.0071	0.0053
$d = 1.6$					
Mean	1.183	1.214	1.397	1.021	1.612
Var	0.0364	0.0300	0.0041	0.0088	0.0020
$d = 1.8$					
Mean	1.168	1.202	1.460	1.024	1.652
Var	0.0424	0.0391	0.0019	0.0097	0.0005

is contrary to the present results. When d is large, these estimators become worse in that the downward bias becomes serious.

(c) The DWT estimator performs worst when $d \leq 1$, but it performs better than the GPH, F-ML and MODWT estimators as d becomes large. However, it tends to be biased downward as d approaches 2.

(d) The W-ML estimator performs reasonably well for all values of d examined here. In particular it is best for $d > 1$, although it tends to give upward bias for some values of $d > 1$. However, this is not serious, whereas the other estimators give serious downward bias when d is greater than 1.

Table 4.2 reports simulation results for Model B, where only the estimation results on d are shown. The parameter value of ρ was chosen so that the signal to noise ratio (SNR) is between 5 and 10, where the SNR is defined as

$$\mathrm{SNR} = V(x_T)/V(u_T) = \begin{cases} \Gamma(1-2d)/\left(\rho\Gamma^2(1-d)\right) & (d < 1/2), \\ T^{2d-1}/\left(\rho\left(2d-1\right)\Gamma^2(d)\right) & (d > 1/2). \end{cases}$$

The SNR is reported in Table 4.2 together with SNR1=$V(\triangle x_T)/V(\triangle u_T)$, which is the SNR of the first differenced process.

It may be concluded from Table 4.2 that:

(a) The performance of all of the estimators for Model B deteriorates to a large extent as compared with Model A because of the inclusion of the unknown parameter ρ.
(b) The F-ML estimator performs well for $d \leq 1$ as in Model A, but it becomes quite bad and tends to become worse than the GPH estimator as d approaches 2.
(c) The W-ML estimator is relatively stable and performs best when $d > 1/2$.
(d) When $d > 1$, the D-GPH and D-F-ML estimators based on the first differenced data improve the performance of the corresponding GPH and F-ML to some extent, but they are not as good as W-ML. This is because differencing makes the SNR smaller and this causes the identification of the signal to be difficult. For example, when $d = 1$, the SNR of the original series is $T/\rho = 512/100 = 5.12$, while that of the first differenced series is $V(\varepsilon_t)/V(\triangle u_t) = 1/200 = 0.005$.
(e) All of the estimators of ρ, though not shown here, are quite unreliable. This is particularly true when the SNR is small. This seems a main reason for the poor performance of the estimators of d in the present model.

4.6 Concluding remarks

We considered the estimation problems associated with long-memory processes, where we allowed for nonstationary as well as stationary processes. It

Table 4.2. *Model B: signal plus noise case*

	GPH	D-GPH	F-ML	D-F-ML	W-ML
\multicolumn{6}{c}{$d = 0.1$, $\rho = 0.1$ (SNR=10.19, SNR1=9.06)}					
Mean	0.161	0.923	0.158	0.719	0.218
Var	0.2829	0.2302	0.0257	0.1006	0.1183
\multicolumn{6}{c}{$d = 0.3$, $\rho = 0.16$ (SNR=8.23, SNR1=4.70)}					
Mean	0.340	0.555	0.310	0.465	0.342
Var	0.0180	0.0574	0.0063	0.0187	0.0117
\multicolumn{6}{c}{$d = 0.45$, $\rho = 0.5$ (SNR=7.28, SNR1=1.32)}					
Mean	0.451	0.570	0.432	0.521	0.527
Var	0.0159	0.0264	0.0085	0.0127	0.0132
\multicolumn{6}{c}{$d = 0.6$, $\rho = 1.5$ (SNR=5.23, SNR1=0.39)}					
Mean	0.574	0.660	0.568	0.629	0.622
Var	0.0219	0.0274	0.0132	0.0136	0.0131
\multicolumn{6}{c}{$d = 0.8$, $\rho = 10$ (SNR=5.19, SNR1=0.053)}					
Mean	0.750	0.846	0.748	0.814	0.790
Var	0.0454	0.0530	0.0278	0.0259	0.0218
\multicolumn{6}{c}{$d = 1$, $\rho = 100$ (SNR=5.12, SNR1=0.005)}					
Mean	0.923	1.075	0.895	1.070	0.952
Var	0.1340	0.1363	0.0773	0.4700	0.0425
\multicolumn{6}{c}{$d = 1.2$, $\rho = 1000$ (SNR=5.26, SNR1=0.00055)}					
Mean	1.050	1.250	1.112	1.199	1.122
Var	0.4479	0.2186	0.1086	0.0622	0.0544
\multicolumn{6}{c}{$d = 1.4$, $\rho = 10000$ (SNR=5.31, SNR1=0.00010)}					
Mean	1.170	1.507	1.229	1.502	1.582
Var	1.751	0.143	0.583	0.076	0.030
\multicolumn{6}{c}{$d = 1.6$, $\rho = 100000$ (SNR=5.20, SNR1=0.000039)}					
Mean	1.190	1.759	1.180	1.731	1.795
Var	1.072	0.635	2.304	0.315	0.060
\multicolumn{6}{c}{$d = 1.8$, $\rho = 900000$ (SNR=5.45, SNR1=0.000029)}					
Mean	1.598	2.040	1.368	2.006	1.976
Var	1.345	0.305	2.713	0.333	0.069

was recognised that, if the long-memory process is contaminated by noise, the computational burden for estimating parameters increases to a large extent.

For two types of models, that is, the model of pure long-memory signal and the model of long-memory signal plus noise, we examined various frequency domain and wavelet-based estimators of the differencing parameter d. It was found that, for both types of models, the frequency domain ML estimator performs better for stationary cases, whereas the wavelet-based ML estimator performs better for nonstationary cases. The performance of the estimators, however, becomes worse when the model is corrupted by noise. It may be possible to use some other estimation methods for this case. For example, the EM algorithm may be applicable. In fact, such an attempt was made by Deriche and Tewfik (1993), although it is not based on wavelets and the model was estimated under simplified conditions. The estimation method based on the EM algorithm combined with wavelets may be worth pursuing.

We can also consider the problem of testing if noise really exists. The problem may be formulated from the model (4.1) coupled with (4.2) as H_0: $\rho = 0$ against H_1: $\rho > 0$. This was dealt with by Tanaka 2002 in the time domain. It was found that the power of the suggested test is not necessarily monotonic. This is especially true if d is small. It is of some interest to see if the difficulty can be overcome by the frequency domain or the wavelet-based approach.

Acknowledgments

I am grateful to Andrew Harvey and two anonymous referees for useful comments on the earlier version of the present paper.

5

A goodness-of-fit test for AR(1) models and power against state space alternatives

T.W. Anderson
Departments of Statistics and Economics, Stanford University

Michael A. Stephens
Department of Statistics and Actuarial Science, Simon Fraser University

Abstract

A test that a time series is an autoregressive model of order 1 is discussed. The test is of Cramér–von Mises type, based on comparing the standardised spectral distribution with its sample estimate. A power study suggests that the statistic does well in detecting the AR(1) model against the alternative of a state space model. In order to introduce the Cramér–von Mises statistic, a survey of goodness-of-fit statistics is given for testing distributions, a field to which Professor Durbin has made major contributions.

5.1 Introduction: EDF statistics

5.1.1 Test for a given probability distribution

In this section we describe statistics based on the empirical distribution function (EDF), also known as the sample distribution function (Durbin 1973) and its use in testing goodness-of-fit for a probability distribution.

Let $x_1, \ldots, x_n$ be n observations (we shall assume a random sample) from a continuous distribution $F(x)$. The EDF $F_n(x)$ is defined by

$$F_n(x) = \{\# \text{ of } x_i \leq x\}/n, \quad -\infty \leq x \leq \infty. \tag{5.1}$$

Suppose the null hypothesis is that $F(x)$ is a given (continuous) distribution $F_0(x)$, which for the moment we consider to be completely specified – all parameters are known. Several well-known test statistics are based on the discrepancy between $F_n(x)$ and $F_0(x)$. The oldest is the Kolmogorov–Smirnov statistic defined by $D_n = \sup|F_n(x) - F_0(x)|$. The Cramér–von Mises statistic is

$$W_n^2 = n \int_{-\infty}^{\infty} [F_n(x) - F_0(x)]^2 \mathrm{d}F_0(x). \tag{5.2}$$

Two closely related statistics are Watson's U_n^2 defined by

$$U_n^2 = n \int_{-\infty}^{\infty} \left[F_n(x) - F_0(x) - \int_{-\infty}^{\infty} \{F_n(x) - F_0(x)\} \mathrm{d}F_0(x) \right]^2 \mathrm{d}F_0(x) \tag{5.3}$$

and the Anderson–Darling statistic

$$A_n^2 = n \int_{-\infty}^{\infty} [F_n(x) - F_0(x)]^2 \psi(x) \mathrm{d}F_0(x), \tag{5.4}$$

where the weight function $\psi(x) = [F_0(x)\{(1 - F_0(x))\}]^{-1}$; see Anderson and Darling (1952).

Watson (1961) introduced U_n^2 to be used for observations on a circle since its computed value does not depend on the origin used to calculate $F_n(x)$. The Anderson–Darling statistic gives more weight to the observations in the tails of the distribution, and in many test situations it is more powerful than W_n^2. The Kolmogorov–Smirnov D_n, although perhaps better known than the Cramér–von Mises family, is in general not as powerful as a test statistic.

5.1.2 Test for uniformity

Suppose the null hypothesis is true, that is, $F(x) = F_0(x)$. It is well-known that the monotonic transformation $z = F_0(x)$ (the probability integral transformation, PIT) gives a random variable z which is uniformly distributed on $[0, 1]$. When the tested distribution $F_0(x)$ is completely specified, the PIT can be made and the $z_i = F_0(x_i)$, $i = 1, \ldots, n$, can be tested as observations from the uniform distribution on $[0, 1]$. This situation is called Case 0. Then suppose $F_n(z)$ ($0 \leq z \leq 1$) is the EDF of the z-values. The Cramér–von Mises definitions above simplify to

$$W_n^2 = n \int_0^1 (F_n(z) - z)^2 dz, \tag{5.5}$$

$$U_n^2 = n \int_0^1 \left[(F_n(z) - z) - \int_0^1 \{F_n(z) - z\} dz \right]^2 dz \tag{5.6}$$

and

$$A_n^2 = n \int_0^1 (F_n(z) - z)^2 \{z(1-z)\}^{-1} dz. \tag{5.7}$$

Suppose $\bar{z}$ is the mean, and $z_{(1)} < z_{(2)} < \cdots < z_{(n)}$ are the order statistics of the z-set. The statistics have simple computing formulae:

$$W_n^2 = \sum_{i=1}^n \{z_{(i)} - (2i-1)/(2n)\}^2 + 1/(12n), \tag{5.8}$$

$$U_n^2 = W_n^2 - n(\bar{z} - 0.5)^2, \tag{5.9}$$

$$A_n^2 = -n - (1/n) \sum_{i=1}^n (2i-1)[\log\{z_{(i)}\} + \log\{1 - z_{(n+1-i)}\}]. \tag{5.10}$$

For Case 0 a great deal is known about the distribution theory of these statistics – in particular the finite-sample and asymptotic distributions can be found. Stephens (1970) showed how suitable modifications of the statistics can be made and then referred only to the asymptotic points to make the test of fit.

Some years ago, Durbin (1973) wrote an excellent monograph on these statistics. This monograph includes distribution theory when parameters must be estimated, and gives many new results on calculating the distributions, both finite-sample and asymptotic.

5.1.3 Tests when parameters must be estimated

Suppose now, returning to tests for distributions, that parameters in the tested distribution are unknown and must be estimated from the given sample. Further, let these be estimated efficiently. The PIT can still be made, using these estimates of the unknown parameters, and the statistics calculated from the formulae above, although now the z-values will not be uniformly distributed. The distribution theory, especially for finite samples, is largely unknown, although Durbin (1975) provided the distribution of the Kolmogorov–Smirnov statistic for testing the exponential distribution with unknown scale.

However, asymptotic distributions of the Cramér–von Mises statistics can be found. The calculation follows a procedure described in Durbin (1973) and in Stephens (1976). It is based on the fact that $y_n(z) = \sqrt{n}\{F_n(z) - z\}$, $0 \leq z \leq 1$, tends to a Gaussian process $y(z)$ as $n \to \infty$. Since, in the time series application which we discuss below, the test statistic is similar to W_n^2, only the asymptotic theory for this statistic will be given.

Firstly, the asymptotic distribution of W_n^2 will be that of $W^2 = \int_0^1 y^2(z) dz$. The mean of $y(z)$ is zero; suppose the covariance function $E\{y(s)y(t)\}, 0 \leq s, t \leq 1$ is $\rho(s,t)$. In Case 0, when all the parameters are known, this covariance is

$$\rho_0(s,t) = \min(s,t) - st. \tag{5.11}$$

More generally, when parameters are estimated efficiently, the covariance will take the form $\rho(s,t) = \rho_0(s,t) - G(s,t)$ where $G(s,t)$ depends on the true parameter values. However, it will not depend on location and scale parameters but will depend on shape parameters.

The limiting distribution of W_n^2 is then

$$W^2 = \sum_{i=1}^{\infty} \omega_i X_i^2, \tag{5.12}$$

where the X_i are independent standard normal random variables and the weights ω_i are the solutions of the integral equation

$$\omega g(s) = \int_0^1 \rho(s,t) \, g(t) \, dt. \tag{5.13}$$

Once the weights are known, the percentage points of the distributions can be calculated by Imhof's method. EDF tests for many distributions when parameters are unknown, and tables with which to make the tests, are given by Stephens (1986).

Tests of Cramér–von Mises type have appeared in many other areas of statistics, for example in tests of independence and in tests for multivariate normality; other tests for time series, in the same family, are given in Chapter 6.

5.2 Applications to time series analysis

5.2.1 The EDF and the periodogram

Durbin (1969) has also discussed tests based on the cumulated periodogram calculated from a time series $x_1, x_2, \ldots, x_T$. Durbin defined the periodogram as follows. Let $T = 2m$, and define $p_{i1} = \sqrt{(2/T)} \sum_{t=1}^{T} x_t \sin(2\pi it/T)$ and $p_{i2} = \sqrt{(2/T)} \sum_{t=1}^{T} x_t \cos(2\pi it/T)$: then the periodogram coordinates are $p_i = p_{i1}^2 + p_{i2}^2, i = 1, 2, \ldots, m$. On the null hypothesis that the x-series consists of independent Gaussian variables, the p_i will be iid exponential variables. Suppose $P_r = \sum_{1}^{r} p_i$ is the cumulated periodogram, and define $y_i = P_i/P_m$; let $n = m - 1$. On the null hypothesis the $y_i, i = 1, \ldots, n$ will then be distributed as uniform order statistics from $U[0, 1]$. It is customary to plot y_i on the y-axis against i/n on the x-axis, so that essentially this plot behaves like the EDF turned through 90 degrees. Then EDF statistics could be used to test the null hypothesis; however, it is customary to use statistics $C^+ = \max(y_i - i/n), C^- = \max(i/n - y_i)$ and $C = \max(C^+, C^-)$. Although not the same, these are clearly related to the Kolmogorov–Smirnov statistic.

5.2.2 A test for the time series model AR(1)

We now develop a Cramér–von Mises statistic to test the hypothesis that an observed time series is a realisation of some AR(1) process. The basic time series model is a stationary stochastic process $\{y_t\}$, $t = \ldots, -1, 0, 1, \ldots$, with $E(y_t) = \mu$, autocovariance function $E(y_t - \mu)(y_{t+h} - \mu) = \sigma(h)$ and autocorrelation function $\rho_h = \sigma(h)/\sigma(0)$, $h = \ldots, -1, 0, 1, \ldots$. The AR(1) model under consideration is

$$y_t = \rho y_{t-1} + u_t, \tag{5.14}$$

where $-1 < \rho < 1$, and the u_t are uncorrelated with mean 0 and variance σ^2. For this model $\rho_h = \rho_{-h} = \rho^h$, $h = 0, 1, \ldots$.

The test will be developed for the general time series model, and then particularised to testing the AR(1) model. The work comes from Anderson, Lockhart and Stephens (1997); the theory used is based on Anderson

(1997). Although we give some preliminary results below, the work is still in progress.

For the general model, the standardised spectral density is

$$f(\lambda) = \frac{1}{2\pi} \sum_{h=-\infty}^{\infty} \rho_h \cos \lambda h, \quad -\pi \leq \lambda \leq \pi. \tag{5.15}$$

Note that the coefficients of the trigonometric functions are the autocorrelations, not the autocovariances; the standardised spectral density $f(\lambda)$ is the conventional spectral density divided by the variance $\sigma(0)$. Since $f(\lambda) = f(-\lambda)$, we define the standardised spectral distribution as

$$F(\lambda) = 2 \int_0^\lambda f(v) \mathrm{d}v = \frac{1}{\pi}\left(\lambda + 2\sum_{h=1}^{\infty} \rho_h \frac{\sin \lambda h}{h}\right). \tag{5.16}$$

The standardised spectral distribution has the properties (nonnegative increments) of a probability distribution on $[0, \pi]$, and $F(\pi) = 1$. Note that we have used the symbol $F(.)$ to describe a probability distribution in Section 5.1, as is standard, but this should cause no confusion with its use to denote the standardised spectral distribution throughout this section.

The test is based on a comparison of $F(\lambda)$ with an estimate $F_T(\lambda)$, obtained as follows. Suppose observations are taken in time, say $y_1, y_2, \ldots, y_T$. For a given h, when μ is known, the unbiased estimate c_h of the autocovariance is defined by

$$c_h = c_{-h} = \frac{1}{T-h} \sum_{t=1}^{T-h} (y_t - \mu)(y_{t+h} - \mu), \quad h = 0, 1, \ldots, T-1. \tag{5.17}$$

When μ is not known, μ in the above expression is replaced by $\bar{y}$, the mean of the T observations. The asymptotic theory which follows in this paper will be the same whether μ or $\bar{y}$ is used in (5.17).

The sample autocorrelation sequence is then defined as $r_h = c_h/c_0$, $h = -(T-1), \ldots, T-1$. Suppose $\hat{f}(\lambda)$ is $f(\lambda)$ with ρ_h replaced by r_h. The standardised sample spectral distribution function is then

$$F_T(\lambda) = 2 \int_0^\lambda \hat{f}(v) \mathrm{d}v = \frac{1}{\pi}\left(\lambda + 2\sum_{h=1}^{T-1} r_h \frac{\sin \lambda h}{h}\right). \tag{5.18}$$

The similarities between the empirical and theoretical standardised spectral distributions and empirical and probability distributions suggest that we can employ EDF statistics, used above for probability distributions, to test hypotheses about the standardised spectral distribution. They will be functionals of the process $\sqrt{T}\left[F_T(\lambda) - F(\lambda)\right]$, as a stochastic process over $[0, \pi]$. This process can be transformed to a process on the unit interval by the monotonic transformation $u = G(\lambda)/G(\pi)$, equivalently $\lambda = G^{-1}[G(\pi)u] = \lambda(u)$, say, where

$$G(\lambda) = 2 \int_0^\lambda f^2(v) \mathrm{d}v. \tag{5.19}$$

Then the process is transformed to

$$Y_T(u) = \sqrt{T} \left\{ F_T[\lambda(u)] - F[\lambda(u)] \right\}. \tag{5.20}$$

Let $Y_T^*(u) = Y_T(u)/\left[2\sqrt{\pi G(\pi)}\right]$. For testing a given series with spectral distribution $F(\lambda)$, we propose the Cramér–von Mises statistic

$$W_T^2 = \int_0^1 [Y_T^*(u)]^2 \mathrm{d}u, \tag{5.21}$$

This statistic has the computing formula

$$W_T^2 = \frac{T}{4\pi^4 G^2(\pi)} \sum_{h=1}^{\infty} \left\{ \sum_{g=1}^{T-1} \frac{(r_g - \rho_g)(\rho_{h+g} - \rho_{h-g})}{g} \right\}^2. \tag{5.22}$$

Anderson (1993) showed that the process $Y_T^*(u)$ converges weakly to the Gaussian process $Y^*(u)$ with mean 0 and covariance function $\rho(u, v) = \min(u, v) - uv + q(u)q(v)$, where $q(u) = u - F[\lambda(u)]$. Then, as in Section 5.1.3, the limiting distribution of W_T^2 is given by (5.12) with weights ω_i the solutions of (5.13) with u, v replacing s, t. Anderson and Stephens (1993) showed how to compute the ω_is and from them the significance points, and gave tables of asymptotic significance points of W^2 for both AR(1) and MA(1) models. Tables of points for finite T, for both models, when the first-order autocorrelation ρ is known, are given by Anderson, Lockhart and Stephens (1995).

5.2.3 Test for AR(1) when ρ is unknown

However, in most applications, the parameter ρ in an AR(1) model will not be known. The test will now be given for this case. Only the practical details will be given; the theory is in Anderson, Lockhart and Stephens (1997). For the AR(1) model it is necessary only to estimate $\rho_1 = \rho$ by $r_1 = c_1/c_0$, using (5.17), since $\rho_h = \rho^h$ and so is estimated by r_1^h. The computing formula for

W_T^2 is then

$$W_T^2 = \frac{T(1-r_1^2)^2}{\pi^2(1+r_1^2)^2} \sum_{h=1}^{H} \left[\sum_{g=2}^{T-1} \frac{(r_g - r_1^g)(r_1^{h+g} - r_1^{|h-g|})}{g} \right]^2. \quad (5.23)$$

Here, H is theoretically infinite, but in practice, it must be chosen so that the term in the sum is negligibly small for $h > H$. An equivalent formula for W_T^2, which avoids the problem of determining H, is

$$W_T^2 = \frac{T(1-r_1^2)^2}{\pi^2(1+r_1^2)^2} \sum_{g_1=2}^{T-1} \sum_{g_2=2}^{T-1} \left(\frac{r_{g_1} - r_1^{g_1}}{g_1} \right) \left(\frac{r_{g_2} - r_1^{g_2}}{g_2} \right)$$

$$\times \left\{ \frac{1+r_1^2}{1-r_1^2} \left(r_1^{|g_1-g_2|} - r_1^{g_1+g_2} \right) + |g_1 - g_2| r_1^{|g_1-g_2|} - (g_1+g_2) r_1^{g_1+g_2} \right\}. \quad (5.24)$$

The null hypothesis of an AR(1) process will be rejected at significance level α if the value of W_T^2 exceeds the appropriate significance point corresponding to r_1, T and α. The significance points of the test statistics depend on the unknown autocorrelation parameter ρ; since this will be estimated by r_1, tables of such points have been constructed which are conditional on the value of r_1. The tables will not be given here; they may be found in Anderson, Lockhart and Stephens (2003), which is an update of the test for AR(1) given earlier in Anderson, Lockhart and Stephens (1997).

5.2.4 Power against state space alternatives

Since the focus of this volume is on state space models, we give some results on power for this particular alternative, and only for W_T^2. However, the Cramér–von Mises test is an omnibus test, and more extensive results, for power against other AR and MA models, and using other statistics for comparison, are given by Anderson, Lockhart and Stephens (2003).

The state space model is characterised by the fact that the model is a stochastic process but each observation is made with an independent error. Such models are sometimes referred to as models with unobserved components.

We consider as the basic process only the AR(1) process. Thus we first suppose values y_t^* generated by

$$y_t^* = \rho_1 y_{t-1}^* + u_t, \quad (5.25)$$

and then the observed data are

$$y_t = y_t^* + \epsilon_t, \quad (5.26)$$

Table 5.1. *Power levels for the test for AR(1), ρ and μ unknown. The alternative is the state space model, with ρ_1 and σ as shown. The sample size is 100, and the test size is $\alpha = 0.05$. Monte Carlo sample size is 5000.*

ρ_1	σ					
	0.5	1.0	1.5	2.0	3.0	4.0
−0.8	0.104	0.355	0.503	0.469	0.277	0.158
−0.6	0.073	0.140	0.140	0.116	0.071	0.057
−0.4	0.056	0.061	0.056	0.054	0.053	0.051
0.4	0.046	0.053	0.050	0.046	0.046	0.047
0.6	0.047	0.096	0.103	0.087	0.057	0.048
0.8	0.059	0.260	0.391	0.361	0.211	0.119

where u_t and ϵ_t are sets of independent normals with mean zero, and with standard deviations 1 and σ respectively. Hopefully, the test will judge when the error is detected as modifying the underlying AR(1) model; that is, the test is that (5.25) is correct, together with $\sigma = 0$ in (5.26).

Tables 5.1 and 5.2 give the powers of the test statistic for various values of ρ_1 and σ in (5.25) and (5.26). They are based on 5000 samples of size T, and the tests were made at levels $\alpha = 0.05$ and $\alpha = 0.10$. The tables show that W_T^2 has some power, although small, when the basic AR(1) process is weak (that is, when ρ is close to 0). The power increases, and then decreases, as the standard deviation σ of ϵ_t increases. When σ is small, one expects low power, since the state space modification will not be great; when σ is large, the error process dominates the background process, and the model begins to resemble white noise. Since this is a special case of the AR(1) process, with $\rho = 0$, the power diminishes towards α. As one would expect, the powers increase as the ρ_1 of the AR(1) process increases in absolute value. Of course, one cannot expect large powers for relatively small sample sizes in a test against the state space model where there are two random noise elements, but the work reported in Anderson, Lockhart and Stephens (2003b) suggests that the W_T^2 statistic gives good results against some rivals. However, more extensive studies are being made.

In Anderson, Lockhart and Stephens (2004) the statistic W_T^2 is defined using the more conventional biased estimates of the covariances, given by $(T-h)/T$ times the c_h used above, and percentage points are given for this definition of W_T^2. It would be interesting to compare the powers using the two statistics to make the test.

Table 5.2. *Power levels for the test for AR(1), ρ and μ unknown. The parameters are as for Table 5.1, but with test size $\alpha = 0.10$.*

ρ_1	σ					
	0.5	1.0	1.5	2.0	3.0	4.0
-0.8	0.168	0.465	0.600	0.570	0.371	0.238
-0.6	0.125	0.210	0.211	0.176	0.132	0.110
-0.4	0.105	0.115	0.112	0.104	0.101	0.100
0.4	0.101	0.101	0.103	0.097	0.097	0.101
0.6	0.100	0.158	0.171	0.146	0.111	0.102
0.8	0.111	0.366	0.507	0.464	0.295	0.186

Acknowledgment

The research in this report was partly supported by the Natural Sciences and Engineering Research Council of Canada. The authors are grateful for comments of referees, and for personal communications from Andrew Harvey, which have drawn our attention to further statistics to be included in the power comparisons.

6
Tests for cycles
Andrew C. Harvey
Faculty of Economics and Politics, University of Cambridge

Abstract

This article presents a model-based approach to the investigation of cyclical properties of a time series. It is shown how models for stochastic cycles, both stationary and nonstationary, may be set up and how deterministic cycles emerge as a special case. The Lagrange multiplier principle is used to formulate a test of the null of a deterministic cycle against the alternative of a stochastic cycle and a test of a nonstationary against a stationary cycle. Similar ideas are used to set up a test against the presence of any kind of cycle. All the test statistics have asymptotic distributions that belong to the Cramér–von Mises family under the null. A Wald test against a deterministic cycles is also described as is a test of the null hypothesis that the series contains a permanent cycle. The modelling framework may be extended to include other components, such as trends, and explanatory variables. Finally it is argued that cycles are best detected by fitting models rather than by examining the periodogram.

State Space and Unobserved Component Models: Theory and Applications, eds. Andrew C. Harvey, Siem Jan Koopman and Neil Shephard. Published by Cambridge University Press. © Cambridge University Press 2004

6.1 Introduction

The traditional paradigm of spectral analysis is one in which deterministic cycles are embedded in a stationary indeterministic process. By allowing the initial conditions to be random, these deterministic cycles can be set up so as to be stationary, as in the Wold decomposition. Deterministic cycles are often identified from large periodogram ordinates. Tests of significance of these ordinates may then be carried out, a famous early example being that of Fisher (1929). However, Priestley (1981, Chapter 8) cautioned against identifying cycles from the periodogram since this is normally computed at only $T/2$ discrete points and these points may not correspond to the frequencies in the series.

In many situations, restricting attention to deterministic cycles is unsatisfactory. However, it is not difficult to set up models which produce cycles that evolve over time. If these cycles are nonstationary, they are persistent in that the effects of shocks are permanent and forecasts project the most recent cyclical pattern indefinitely into the future. Cycles may also be stationary, but near persistent in that predictions retain their pattern for a long time before eventually dying down to zero. Both kinds of cycles have applications in a wide range of subjects.

This paper shows how models for stationary and nonstationary cycles may be formulated and estimated within an unobserved components (UC) framework. Deterministic cycles emerge as special cases and it is possible to test the null hypothesis that a cycle is deterministic against the alternative that it evolves over time. A test of whether a cycle is nonstationary against the alternative that it is stationary can also be set up. Tests against the presence of any trend of cycle, be it deterministic or stochastic, can be constructed. In each of these cases it is possible to set up a test based on the Lagrange multiplier principle and to use test statistics which have asymptotic distributions under the null hypothesis that belong to the class of Cramér–von Mises distributions. The Cramér–von Mises distribution was originally developed for goodness-of-fit tests; see Chapter 5.

Traditional spectral analysis is useful for an initial exploration of the data. However, it has limitations and if not used with care it can be misleading. A model-based approach is more flexible and more reliable. It encompasses nonstationary cycles and it allows other components, such as trends and seasonals, to be handled at the same time as the properties of cycles are being investigated. However, not all parts of a model need be fully specified since there is considerable scope for the use of semi-nonparametric methods in carrying out tests.

The plan of the paper is as follows. Section 6.2 explores the properties of the stochastic cycle model, shows that it can be stationary or nonstationary, indeterministic or deterministic, and relates it to the Wold decomposition. It is shown how the model may be handled by state space algorithms and how it may be generalised to include several cycles. This stochastic cycle model lays the foundation for the tests described in the sections that follow.

Section 6.3 describes how to test for a deterministic cycle in a stationary time series. Section 6.4 explains how to test against a nonstationary stochastic cycle, the null hypothesis being that the cycle is deterministic. Section 6.5 describes a test against the presence of any kind of permanent cycle, be it deterministic or stochastic; an application is a test for trading day effects. Section 6.6 describes a test of Bierens (2001) which reverses the null and alternative hypotheses of Section 6.2, taking a permanent cycle to be the null. Section 6.7 gives the test against a stationary cycle when the null is of a deterministic, but stationary cycle; this was originally proposed in Harvey and Streibel (1998b). Section 6.8 shows how complex unit root tests can be formulated in such a way that the test statistics have Cramér–von Mises distributions under the null hypothesis of nonstationarity. Some extensions of the tests and models are laid out in Section 6.9, while Section 6.10 investigates the important practical problem of detecting a cycle with unknown period. It is argued that fitting UC models is a more effective way of locating cycles than is the periodogram. Section 6.11 concludes.

There are many links with Jim Durbin's work in the paper. He is best known in economics for his work on serial correlation and he has written a great deal on the frequency domain. For example, his 1969 paper on the cumulative periodogram test combines the two themes, while Durbin (1970) is a seminal work on the application of what is effectively the Lagrange multiplier principle to testing for serial correlation. Jim has also worked with the class of Cramér–von Mises distributions, primarily in the context of boundary crossing problems.

6.2 Stochastic and deterministic cycles

The starting point is

$$y_t = \psi_t + \varepsilon_t, \quad t = 1, \ldots, T, \qquad (6.1)$$

where ε_t is a white noise term with variance σ_ε^2. The *fixed deterministic cycle* model is

$$y_t = \alpha \cos \lambda_c t + \beta \sin \lambda_c t + \varepsilon_t, \quad 0 < \lambda_c \leq \pi, \quad t = 1, \ldots, T, \quad (6.2)$$

where λ_c is the frequency in radians and α and β are parameters. The period of the cycle is $2\pi/\lambda_c$. Given λ_c, α and β are simply estimated by ordinary least squares (OLS).

The cycle can be built up recursively by

$$\begin{bmatrix} \psi_t \\ \psi_t^* \end{bmatrix} = \begin{bmatrix} \cos \lambda_c & \sin \lambda_c \\ -\sin \lambda_c & \cos \lambda_c \end{bmatrix} \begin{bmatrix} \psi_{t-1} \\ \psi_{t-1}^* \end{bmatrix}, \quad t = 1, \ldots, T, \quad (6.3)$$

where $\psi_0 = \alpha$ and $\psi_0^* = \beta$. Now suppose two mutually uncorrelated white noise disturbances, κ_t and κ_t^*, with zero means and common variance σ_κ^2 are added to the right hand side of (6.3), and a parameter ρ is introduced so that

$$\begin{bmatrix} \psi_t \\ \psi_t^* \end{bmatrix} = \rho \begin{bmatrix} \cos \lambda_c & \sin \lambda_c \\ -\sin \lambda_c & \cos \lambda_c \end{bmatrix} \begin{bmatrix} \psi_{t-1} \\ \psi_{t-1}^* \end{bmatrix} + \begin{bmatrix} \kappa_t \\ \kappa_t^* \end{bmatrix}, \quad t = 1, \ldots, T, \quad (6.4)$$

with $0 \leq \rho \leq 1$. This is the *stochastic cycle* model.

6.2.1 Nonstationary cycle

If $\rho = 1$ in (6.4) the cycle has complex unit roots, and can be made stationary by the operator $1 - 2\cos \lambda_c L + L^2$. The initial values, ψ_0 and ψ_0^*, are fixed, but the model could be modified to let them have diffuse priors. The cycle evolves over time and forecasts project the latest cyclical pattern indefinitely into the future.

6.2.2 Stationary cycle

If the initial conditions are fixed, the stochastic cycle is asymptotically stationary if $\rho < 1$. Exact stationarity requires that the distribution of the initial vector $(\psi_0, \psi_0^*)'$ be the same as the unconditional distribution of $(\psi_t, \psi_t^*)'$ for all t. If the initial conditions are such that the vector $(\psi_0, \psi_0^*)'$ has zero mean and covariance matrix $\sigma_\psi^2 I$, it is straightforward to show that the process $\psi_t, t = 0, \ldots, T$ is stationary and indeterministic with zero mean and variance $\sigma_\psi^2 = \sigma_\kappa^2/(1-\rho^2)$.

The autocorrelation function (ACF) of ψ_t is

$$\rho_\psi(\tau) = \rho^\tau \cos \lambda_c \tau, \quad \tau = 0, 1, 2, \ldots. \quad (6.5)$$

The spectrum,

$$f(\lambda) = \frac{\sigma_\psi^2}{2\pi} \frac{(1+\rho^2 - 2\rho\cos\lambda_c \cos\lambda)(1-\rho^2)}{1+\rho^4 + 4\rho^2 \cos^2\lambda_c - 4\rho(1+\rho^2)\cos\lambda_c \cos\lambda + 2\rho^2 \cos 2\lambda},$$

displays a peak, centred around λ_c, which becomes sharper as ρ moves closer to one. Figure 6.1 shows the spectral densities for $\rho = 0.7$ and 0.9.

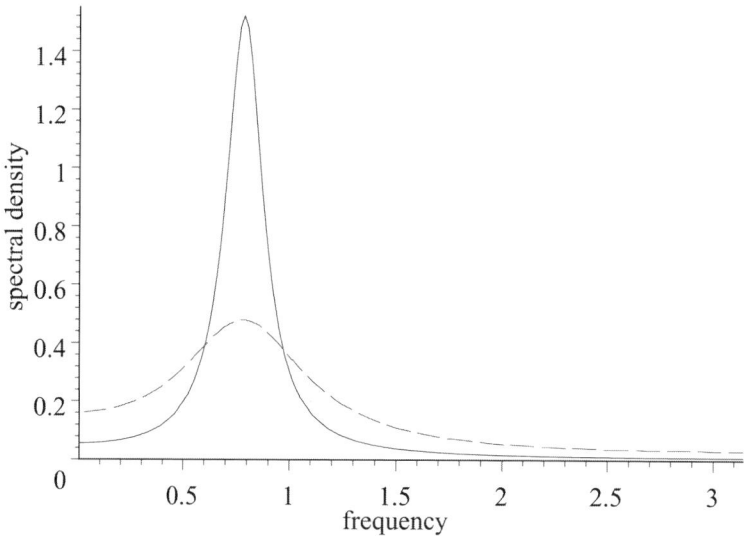

Fig. 6.1. Spectra for cycles with $\lambda_c = \pi/4$ and $\rho = 0.7$ (broken line) and 0.9.

6.2.3 Stationary deterministic cycle

If σ_κ^2 is fixed, setting ρ equal to 1 in (6.4) generates a nonstationary cycle with complex unit roots. If, on the other hand, σ_ψ^2 is taken to be a fixed parameter and σ_κ^2 is defined by $\sigma_\kappa^2 = (1-\rho^2)\sigma_\psi^2$, then $\sigma_\kappa^2 \to 0$ as $\rho \to 1$ and a stationary deterministic cycle results when $\rho = 1$. Specifically

$$\psi_t = \psi_0 \cos\lambda_c t + \psi_0^* \sin\lambda_c t, \quad t = 1,\ldots,T, \qquad (6.6)$$

where ψ_0 and ψ_0^* are mutually uncorrelated random variables with mean zero and variance σ_ψ^2. The expression for the ACF, (6.5), remains valid for $\rho = 1$, but the process now has a purely discrete or line spectrum; see Priestley (1981, p. 230).

The model in which ψ_t is generated by (6.4), with initial conditions such that $(\psi_0, \psi_0^*)'$ has mean zero and covariance matrix $\sigma_\psi^2 I$, and in which the parameters are $0 < \lambda_c \leq \pi$, $0 \leq \rho \leq 1$ and $\sigma_\psi^2 \geq 0$, is called the *stationary cycle* model. The *stationary deterministic cycle* model results when $\rho = 1$.

If noise is added to the stochastic cycle, the variance of y_t decomposes as

$$\sigma_y^2 = \sigma_\psi^2 + \sigma_\varepsilon^2 \tag{6.7}$$

while its ACF is

$$\rho(\tau) = (\sigma_\psi^2 / \{\sigma_\psi^2 + \sigma_\varepsilon^2\}) \rho^\tau \cos \lambda_c \tau, \quad \tau \geq 1, \quad 0 \leq \rho \leq 1.$$

When ρ is 1, the frequency domain properties of the model are displayed by the spectral distribution function, which shows a jump at λ_c.

The Wold decomposition views deterministic and indeterministic parts of a series as completely separate entities, with cycles belonging to the former group. The model which has just been set up has a deterministic cycle as a special case of a stochastic cycle. Correspondingly, the sudden jump in the spectral distribution function is a limiting case of a spectral distribution function that is continuous for ρ strictly less than 1.

6.2.4 Estimation

The stochastic cycle plus noise model can be put in state space form by letting $(\psi_t, \psi_t^*)'$ be the state and defining the measurement equation

$$y_t = \begin{bmatrix} 1 & 0 \end{bmatrix} \begin{bmatrix} \psi_t \\ \psi_t^* \end{bmatrix} + \varepsilon_t, \quad t = 1, \ldots, T$$

to accompany the transition equation in (6.4). The initial conditions on $(\psi_t, \psi_t^*)'$ can be either a diffuse or a proper prior depending on whether σ_ψ^2 or σ_κ^2 is taken to be fixed. Fixing σ_κ^2 fails to exploit the proper prior when the cycle is stationary but it allows the nonstationary cycle to emerge as a special case. Given the hyperparameters λ_c, ρ and q, where $q = \sigma_\psi^2 / \sigma_\varepsilon^2$ or $\sigma_\kappa^2 / \sigma_\varepsilon^2$, the Kalman filter can be applied. The cyclical component may be extracted from the noise by a smoothing algorithm and forecasts of future observations, together with their root mean square errors, can be made.

In the *Gaussian* stochastic cycle plus noise model, all disturbances, ε_t, κ_t and $\kappa_t^*, t = 1, \ldots, T$, together with ψ_0 and ψ_0^*, are assumed to be normally distributed. The disturbances driving the cycle are usually uncorrelated with ε_t, though this is not necessary for identifiability. The model is still identifiable if ε_t is allowed to be correlated with κ_t and κ_t^*, but the effect of such correlation is to make the signal extraction filters asymmetric. The model may be estimated by maximum likelihood and this is implemented in

Table 6.1. *Estimated cycles for deep sea cores*

cycle	period (years)	ρ	σ_ψ^2
1	102 000	0.998	1.200
2	40 800	0.999	0.140
3	22 800	0.983	0.109

the STAMP package of Koopman, Harvey, Doornik and Shephard (2000). It is not necessary to assume that λ_c is such that T/P is an integer. When $\rho < 1$ the standard regularity conditions hold for applying asymptotic theory to obtain the properties of the estimators. Even if the model is set up so that it is still stationary when $\rho = 1$, the fact that it is not ergodic means standard asymptotic theory cannot be used to make inferences about ρ.

6.2.5 Several cycles

The model can be extended so as to include more than one cycle. Thus

$$y_t = \psi_{1t} + \cdots + \psi_{nt} + \varepsilon_t, \quad t = 1, \ldots, T, \tag{6.8}$$

where n is the number of cycles. Estimation using the state space form is straightforward. An example is provided by the study of Newton, North and Crowley (1991) on global ice volume as estimated by oxygen-18 levels from 220 observations on deep sea cores recorded at intervals of 2000 years. Newton, North and Crowley (1991) set up a long autoregression to capture cycles, but an alternative is to proceed as suggested in the STAMP manual – Koopman, Harvey, Doornik and Shephard (2000, p. 57) – and fit three stochastic cycles by searching over all frequencies to find the best fit. The results are as shown in Table 6.1. All three cycles are very close to being nonstationary. The irregular disappears as σ_ε^2 is estimated to be zero. The first order residual serial correlation is rather high at 0.48, but the others are quite small and $R^2 = 0.983$.

6.3 Test against a deterministic cycle

A test against a deterministic cycle is easily constructed by working with the fixed cycle model of (6.2). Let $Z_t = (\cos \lambda_c t, \sin \lambda_c t)$ and $\delta = (\alpha, \beta)'$ so that

$$y_t = \alpha \cos \lambda_c t + \beta \sin \lambda_c t + \varepsilon_t = Z_t \delta + \varepsilon_t, \quad t = 1, \ldots, T.$$

The frequency, λ_c, is assumed known. The Wald statistic for testing the null hypothesis that $\delta = 0$ is

$$W = \widehat{\delta}' \left(\widehat{\sigma}_e^{-2} \sum_{t=1}^{T} Z_t Z_t' \right) \widehat{\delta}, \qquad (6.9)$$

where $\widehat{\delta}$ is the OLS estimator of δ and $\widehat{\sigma}_e^2 = T^{-1} \sum_{t=1}^{T} e_t^2$, where e_t is the tth OLS residual. The asymptotic distribution of W is χ_2^2 under the null hypothesis that $\delta = 0$; in a Gaussian model W can be transformed to a variable having an F-distribution under the null. If λ_c is such that T/P is an integer, then W is the standardised periodogram ordinate, that is, $W = p(\lambda_c)/\widehat{\sigma}_e^2$, where

$$p(\lambda_c) = \frac{2}{T} \left[\left(\sum_{t=1}^{T} y_t \cos \lambda_c t \right)^2 + \left(\sum_{t=1}^{T} y_t \sin \lambda_c t \right)^2 \right] = \frac{T(\widehat{\alpha}^2 + \widehat{\beta}^2)}{2}.$$

The Wald test can be carried out by fitting a parametric unobserved components model that specifies a serially correlated component as well as a deterministic cycle. Alternatively a nonparametric test[1] can be set up by replacing $\widehat{\sigma}_e^2$ in (6.9) by an estimate of (2π times) the spectrum at λ_c computed from the spectral generating function

$$\widehat{g}(\lambda_c; m) = \sum_{\tau=-m}^{m} w(\tau, m) \widehat{\gamma}_e(\tau) \cos \lambda_c \tau, \qquad (6.10)$$

where $w(\tau, m)$ is a weighting function or kernel, such as the Bartlett window,

$$w(\tau, m) = 1 - |\tau|/(m+1),$$

and

$$\widehat{\gamma}_e(j) = T^{-1} \sum_{t=j+1}^{T} e_t e_{t-j}$$

is the sample autocovariance of the OLS residuals at lag j. Alternative options for the kernel $w(.,.)$, such as the Parzen or quadratic spectral (QS) windows, may be found in Andrews (1991, p. 821). Andrews also discussed conditions required on the lag truncation parameter to ensure consistency of the estimated variance of $\widehat{\delta}$. The spectrum could also be estimated by directly weighting the periodogram ordinates, though this is rarely considered in the econometric literature.

---

[1] The OLS estimator is asymptotically efficient. Indeed if T/P is an integer and the stationary component is circular, OLS and GLS are the same.

The above modification allows for serial correlation from a covariance stationary process. Conditional heteroskedasticity may be present;[2] see the conditions in Andrews (1991, p. 823). The test can be extended to deal with unconditional heteroskedasticity as well as serial correlation as allowed in the conditions stated in Andrews (1991, p. 839). Thus

$$W^* = T\widehat{\delta}'[Q^{-1}\widehat{\Omega}(m)Q^{-1}]^{-1}\widehat{\delta},$$

where

$$Q = T^{-1}\sum_{t=1}^{T} Z_t Z_t'$$

and $\widehat{\Omega}$ is a nonparametric estimator of the 'long run variance' of the vector $Z_t\varepsilon_t$, that is,

$$\widehat{\Omega}(m) = \sum_{\tau=-m}^{m} w(\tau, m)\widehat{\Gamma}(\tau), \qquad (6.11)$$

where $w(\tau, m)$ is a kernel as in (6.10) and

$$\widehat{\Gamma}(\tau) = T^{-1} \sum_{t=\tau+1}^{T} Z_t e_t e_{t-\tau} Z_{t-\tau}'$$

is the sample autocovariance matrix at lag τ formed from $Z_t e_t$.

One might correct for serial correlation by fitting lagged dependent variables, but this is not very satisfactory as there is likely to be confounding with the cycles. Indeed in the context of seasonality, Canova and Hansen (1995, p. 239) explicitly point out that two or more lags may absorb some pairs of complex roots.

6.4 Test against a nonstationary cycle

Following the work on seasonality testing by Canova and Hansen (1995), it can be shown that the locally best invariant (LBI) test of the null hypothesis of a fixed deterministic cycle against the alternative of a nonstationary stochastic cycle at a given frequency, λ_c, that is, $H_0: \sigma_\kappa^2 = 0$ against $H_1: \sigma_\kappa^2 > 0$, is to reject large values of

$$\kappa_c = 2T^{-2}\widehat{\sigma}_e^{-2} \sum_{t=1}^{T} \left[\left(\sum_{i=1}^{t} e_i \cos \lambda_c i\right)^2 + \left(\sum_{i=1}^{t} e_i \sin \lambda_c i\right)^2\right], \qquad (6.12)$$

---

[2] In which case the OLS estimators are no longer efficient.

where e_t is the tth OLS residual from regressing on the pair of trigonometric terms at λ_c and $\widehat{\sigma}_e^2$ is the sample variance of the OLS residuals, as in (6.9). In deriving the test from the LBI principle, one initially obtains the summations running in reverse, that is, from $t = i$ to T, but, as a consequence of fitting the deterministic cycle by OLS,

$$\sum_{t=1}^{T} e_t \cos \lambda_c t = \sum_{t=1}^{T} e_t \sin \lambda_c t = 0, \qquad (6.13)$$

from which it follows that the two statistics are identical. Asymptotically, κ has the Cramér–von Mises distribution with two degrees of freedom under the null hypothesis.

The test procedures may be extended to construct joint tests at two frequencies, λ_1 and λ_2, simply by adding up the individual test statistics. The asymptotic distribution is then Cramér–von Mises distribution with four degrees of freedom. The seasonality test statistic is $CvM(s-1)$, where s is the number of seasons in the year.

Canova and Hansen (1995) introduced a modification to the variance estimator in the denominator of (6.12) which allows the test to be valid when the noise, within which the cycle is embedded, is generated by any indeterministic stationary process. The statistic is

$$\kappa_c(m) = T^{-2} \text{trace} \left(\widehat{\Omega}(m) \right)^{-1} \sum_{t=1}^{T} S_t S_t' \qquad (6.14)$$

$$= T^{-2} \left(\widehat{\Omega}(m) \right)^{-1} \sum_{t=1}^{T} S_t' \left(\widehat{\Omega}(m) \right)^{-1} S_t, \qquad (6.15)$$

where $S_t = \sum_{i=1}^{t} Z_i e_i$. Some patterns of serial correlation can be dealt with by the inclusion of the lagged dependent variable, y_{t-1}, as a regressor. This will leave the asymptotic distribution of the test statistic unchanged under the null and will have very little effect on the power of the test as long as λ_c is not close to zero or π; see Canova and Hansen (1995, pp. 239, 247). As noted in the previous section, more lags could absorb the cycle.

If there is no need to guard against unconditional heteroskedasticity, the test statistic can be set up as

$$\kappa_c(m) = \frac{2 \sum_{t=1}^{T} \left[\left(\sum_{i=1}^{t} e_i \cos \lambda_c i \right)^2 + \left(\sum_{i=1}^{t} e_i \sin \lambda_c i \right)^2 \right]}{T^2 \widehat{g}(\lambda_c; m)}, \qquad 0 < \lambda_c < \pi, \qquad (6.16)$$

where $\widehat{g}(\lambda_j; m)$ is as defined in (6.10). In the context of seasonality testing, Busetti and Harvey (2003) called this the *spectral nonparametric test statistic*.

An alternative strategy is to estimate a fully specified model under the alternative hypothesis and use the estimates of the nuisance parameters to construct an LBI test from the innovations obtained under the null. Busetti and Harvey (2003) compared this parametric option with the nonparametric tests in the context of tests at the seasonal frequencies. Their results are based on a model that has a random walk component added to the seasonal and irregular. The parametric test is somewhat oversized at frequency π but this is not the case at $\lambda_c = \pi/2$, which is more relevant here. The parametric test is clearly more powerful than the nonparametric one.

6.5 Test against a permanent cycle

The test against a nonstationary cyclical component takes the null hypothesis to be a model in which the cycle is deterministic. Sometimes we may wish to test whether there is any cycle at all at a given frequency. We then want power against a deterministic cycle as well as a stochastic one.

If a test statistic is constructed as in (6.12), but without fitting a deterministic cycle at λ_c, it will be LBI against a nonstationary stochastic cycle with zero initial conditions.[3] Under the null hypothesis, the asymptotic distribution of this statistic, denoted κ_c^0, will be a function of Brownian motions rather than Brownian bridges. It is still of the Cramér–von Mises family, but denoted CvM_0. The distribution is unaffected by the inclusion of a constant or a constant and a time trend ; see Busetti and Harvey (2003). Again parametric and nonparametric forms of the test may be constructed.

An example of the above tests arises in the detection of trading day effects. Cleveland and Devlin (1980) showed that peaks at certain frequencies in the estimated spectra of monthly time series indicate the presence of trading day effects. Specifically there is a peak at a frequency of $0.348 \times 2\pi$ radians, with the possibility of subsidiary peaks at $0.432 \times 2\pi$ and $0.304 \times 2\pi$ radians. Busetti and Harvey (2003) suggested testing for trading day effects by carrying out a κ_c^0 test at the relevant frequency or a joint test at all three frequencies.

An alternative approach to the κ_c^0 test is simply to use the Wald test against a deterministic cycle as described in Section 6.3. The evidence in Busetti and Harvey (2003) in the seasonal case suggests this is almost as effective as the κ_c^0 test.

A variant of the Wald test was proposed by Bierens (2001, p. 976). Rather than dealing with serial correlation by the correction based on (6.10), he

---

[3] There is now an issue of whether summations should be forwards or backwards since (4.2) no longer holds; see the discussion in Busetti and Harvey (2003).

suggested fitting a pth order autoregressive model

$$\phi(L)y_t = \mu + \varepsilon_t,$$

where

$$\phi(L) = 1 - \phi_1 L - \cdots - \phi_p L^p.$$

He proposed then testing the stationary autoregressive hypothesis against a pair of complex unit roots based on the statistic

$$\widehat{\sigma}^{-2} p(\lambda_c) \left|\phi(\mathrm{e}^{-\mathrm{i}\lambda_c})\right|^2.$$

This reduces to the Wald statistic of (6.9) when there is no serial correlation. Under the null hypothesis of stationarity this is asymptotically χ_2^2 while under the alternative it goes to infinity. The test extends to treating several pairs of complex roots. The problem with this approach is that the autoregression is likely to be confounded with the complex roots in the cycle.

6.6 Permanent cycle as the null hypothesis

Bierens (2001) proposed a test of the null hypothesis of a nonstationary cycle based on the standardised periodogram, namely

$$\xi_c = \frac{p(\lambda_c)}{T\widehat{\sigma}_y^2} = \frac{2}{T^2 \widehat{\sigma}_y^2}\left[\left(\sum_{t=1}^T y_t \cos\lambda_c t\right)^2 + \left(\sum_{t=1}^T y_t \sin\lambda_c t\right)^2\right] < c,$$

where $c = 0.14$ at the 5% level from the asymptotic distribution. The theory is based on a second-order autoregressive process with complex roots, but it generalises to a model in which the disturbance follows an indeterministic process and it could equally be based on the components model (6.1) with ε_t an indeterministic process. The test is nonparametric because, under the null hypothesis, $\widehat{\sigma}_y^2/T$ has a limiting distribution and ξ_c has a limiting distribution that does not depend on the process generating the stationary part of the model.

A slightly different test is obtained if a deterministic cycle is removed in calculating the denominator, that is, $\widehat{\sigma}_y^2$ is replaced by $\widehat{\sigma}_e^2$. This statistic, denoted ξ_c^*, is related to the simple Wald test statistic, (6.9), in that it is W/T. The limiting distribution of $\widehat{\sigma}_e^2/T$ is now a function of demeaned Brownian motion so the asymptotic distribution of ξ_c^* differs slightly from that of ξ_c.

What if the cycle is deterministic? Then

$$p\lim \xi_c = \frac{\alpha^2 + \beta^2}{2\sigma_\varepsilon^2 + \alpha^2 + \beta^2}.$$

If we remove the cycle in computing the denominator, then

$$p\lim \xi_c^* = \frac{\alpha^2 + \beta^2}{2\sigma_\varepsilon^2}.$$

In both cases the null is unlikely to be rejected unless the amplitude of the deterministic cycle is small relative to σ_ε^2. For this reason ξ_c and ξ_c^* are more appropriately classified as tests of the null of a permanent cycle rather than as tests of the null of a nonstationary cycle.

Bierens assumed that λ_c is known for the purposes of asymptotic theory. In practice he estimated it by the largest periodogram ordinate. For the reasons given in Priestley (1981, Ch. 8), this is unsatisfactory. (Indeed Bierens admitted to a 'pretesting type of sin' in selecting λ_c in his application.) However, estimating λ_c by fitting a nonstationary model would detract from the nonparametric nature of the test.

6.7 Testing against a stationary stochastic cycle

Suppose that we wish to test for the presence of an indeterministic stationary cycle at a particular frequency, $0 < \lambda_c < \pi$. The model contains a white noise component and may also contain deterministic components including a fixed cycle at frequency λ_c. The Lagrange multiplier test statistic of the null hypothesis that ρ is zero in (6.4) is Tr_1^2, where r_1 is the first order sample autocorrelation; see Harvey (1989, pp. 244 5). Furthermore, King and Evans (1988) showed that a test of the form $r_1 \cos \lambda_c$ is LBI for testing $H_0: \rho = 0$ against $H_1: \rho > 0$. If the period, P, is known to be greater than 4, $\cos \lambda_c$ is positive and a one sided test against positive serial correlation can be carried out using r_1 or the Durbin–Watson d-statistic.[4]

The above test is designed to have high power in the region of $\rho = 0$. If we want high power in the region of $\rho = 1$ we can exploit the fact that a deterministic cycle can be nested within a stationary cycle as in Section 6.2.3. An LBI test of $H_0: \rho = 1$ against $H_a: \rho < 1$ can be constructed. Harvey and Streibel (1998b) showed that this is exactly the same as the κ_c test against a nonstationary stochastic cycle. They compared this test with

---

[4] The cumulative periodogram test of Durbin (1969) offers a more general alternative.

point optimal tests against specific values of ρ. Unlike the LBI test, point optimal tests depend on q.

The test proposed by Harvey and Streibel (1998b) is a radical departure from what is normally done where the question asked is whether the series consists of a deterministic cycle plus some arbitrary indeterministic component. Instead the question is whether a deterministic cycle can be distinguished from an indeterministic process capable of producing pseudocyclical behaviour.

6.8 Nonstationary versus stationary cycles: test for complex unit roots

Within the framework of (6.4), with σ_κ^2 fixed and the initial conditions, ψ_0 and ψ_0^*, treated as fixed, a parametric test of the null hypothesis that the component at a particular frequency is nonstationary against the alternative that it is stationary, that is, $H_0: \rho = 1$ against $H_1: \rho < 1$, can be constructed from the null hypothesis innovations, $\widetilde{\nu}_t$, as

$$\kappa_c^\dagger = 2T^{-2} \sum_{i=1}^{T} \left[\left(\sum_{t=1}^{i} \widetilde{\nu}_t \cos \lambda_c t \right)^2 + \left(\sum_{t=1}^{i} \widetilde{\nu}_t \sin \lambda_c t \right)^2 \right] < c. \qquad (6.17)$$

The test is suggested by the Lagrange multiplier (LM) principle, as outlined in Harvey (2004), though it is only, strictly speaking, an LM test if cyclical slopes, as described in Section 6.9.2, are included.[5] Under the null hypothesis the asymptotic distribution is $CvM_0(2)$ since if the nonstationary seasonal operator $1 - 2\cos\lambda_c L + L^2$ were to be applied it would remove the corresponding deterministic cycle. The lower tail of the distribution is now appropriate. An alternative approach would be to set up a test using an autoregressive model, along the lines of the seasonal unit root tests proposed by Rodrigues (2002).

A Wald test based on an autoregressive model could be carried out as in Ahtola and Tiao (1987). This is a generalisation of the augmented Dickey–Fuller test. It is based on the result that, in a nonstationary model, the limiting distribution of $T(\widehat{\phi}_2 + 1)$, where $\widehat{\phi}_2$ is the estimator of the second AR parameter, is independent of λ_c. However, tests based on autoregressions may perform poorly in situations where a UC model is appropriate since

---

[5] As in the unit root test there is an issue of consistency of the test; see Harvey (2004). This is because there may be a finite probability that the variance of the seasonal disturbance is zero under the alternative.

the reduced form contains an MA component; the simulation evidence in Hylleberg (1995) illustrates this point for seasonal unit roots.

The Bierens test is unlikely to be very useful against a strongly stationary process since it cannot compete with a complex unit root test such as (6.17); see the evidence in Busetti and Harvey (2002) for frequency zero. Thus Bierens' example of testing against a stationary business cycle is unconvincing.

6.9 Extensions

6.9.1 Trends, seasonals and explanatory variables

A stochastic trend component can be added to a structural time series model; a stochastic seasonal may also be included. In a nonparametric framework differences must be taken. A seasonal difference is needed to deal with a random walk and a nonstationary stochastic seasonal. Note that differencing will slightly shift the peak in the spectrum of a stationary indeterministic cycle.

Time trends and seasonal dummies can be added to the model without affecting the asymptotic distribution of κ_c; the proof follows along the lines of Busetti and Harvey (2003). The asymptotic distribution remains unaffected if the regressors are stochastic provided they do not contain complex unit roots at λ_c.

6.9.2 Higher order cycles

The model with cyclical slopes is

$$\begin{bmatrix} \psi_t \\ \psi_t^* \end{bmatrix} = \begin{bmatrix} \cos\lambda_c & \sin\lambda_c \\ -\sin\lambda_c & \cos\lambda_c \end{bmatrix} \begin{bmatrix} \psi_{t-1} \\ \psi_{t-1}^* \end{bmatrix} + \begin{bmatrix} \psi_{\beta,t} \\ \psi_{\beta,t}^* \end{bmatrix} + \begin{bmatrix} \kappa_t \\ \kappa_t^* \end{bmatrix}, \quad (6.18)$$

$$\begin{bmatrix} \psi_{\beta,t} \\ \psi_{\beta,t}^* \end{bmatrix} = \begin{bmatrix} \cos\lambda_c & \sin\lambda_c \\ -\sin\lambda_c & \cos\lambda_c \end{bmatrix} \begin{bmatrix} \psi_{\beta,t-1} \\ \psi_{\beta,t-1}^* \end{bmatrix},$$

where κ_t and κ_t^* are as in the first order cycle, (6.4), and ρ and λ_c satisfy the same conditions. This model can be rewritten as

$$y_t = \psi_t + \psi_{\beta,0} t \cos\lambda_c t + \psi_{\beta,0}^* t \sin\lambda_c t + \varepsilon_t, \quad t = 1, \ldots, T,$$

where ψ_t is as in (6.4).

Following a similar line of reasoning as with the trend, a smoother cycle can be obtained by shifting the disturbance to the equations for the slopes.

Tests for cycles

If a damping factor, ρ, is introduced into the model, we obtain what Harvey and Trimbur (2003) called a double, or *second order*, stochastic cycle:

$$\begin{bmatrix} \psi_t \\ \psi_t^* \end{bmatrix} = \rho \begin{bmatrix} \cos\lambda_c & \sin\lambda_c \\ -\sin\lambda_c & \cos\lambda_c \end{bmatrix} \begin{bmatrix} \psi_{t-1} \\ \psi_{t-1}^* \end{bmatrix} + \begin{bmatrix} \psi_{\beta,t-1} \\ \psi_{\beta,t-1}^* \end{bmatrix},$$

$$\begin{bmatrix} \psi_{\beta,t} \\ \psi_{\beta,t}^* \end{bmatrix} = \rho \begin{bmatrix} \cos\lambda_c & \sin\lambda_c \\ -\sin\lambda_c & \cos\lambda_c \end{bmatrix} \begin{bmatrix} \psi_{\beta,t-1} \\ \psi_{\beta,t-1}^* \end{bmatrix} + \begin{bmatrix} \kappa_t \\ \kappa_t^* \end{bmatrix},$$

(6.19)

where κ_t and κ_t^* are as in the first order cycle, (6.4), and ρ and λ_c satisfy the same conditions. Harvey and Trimbur (2003) derived expressions for the ACF and spectrum and fitted the model to US macroeconomic time series.

6.10 Unknown period

When λ_c is not known *a priori*, it may be taken to be located at the largest periodogram ordinate or it may be estimated within an unobserved components model by maximum likelihood using a search procedure.[6] The latter approach is preferable since there is no reason for T/P to be an integer. Despite all the warnings in Priestley (1981, Chapter 8), it is interesting that Bierens (2001, pp. 974–5) identified six cycles in a US unemployment series from the six largest periodogram ordinates[7] and then found them to be statistically significant. This is somewhat reminiscent of the work in the 1920s where large numbers of (spurious) cycles were discovered in the Beveridge wheat price index.

The data on the numbers of various furs traded by the Hudson Bay Company in Canada from 1848 to 1909 have been used as a testbed for many time series techniques. We consider the (logarithm of the) mink series and focus on the period, 1854–1907 because the observations in the two or three years at the beginning and the end seem abnormally low. The largest periodogram ordinate is at 7.71 years, but there is another large one at 9.00 years. The message is not changed significantly by smoothing the periodogram. Estimating the nonstationary stochastic cycle model gives a period of 9.55; all the diagnostics are satisfactory. Estimating the unrestricted stochastic cycle model gives a period of 9.56 and $\widetilde{\rho} = 0.983$ so the results are very similar. Figure 6.2 shows a graph of the smoothed cycle.

Fitting an AR(2) model with a constant gives regression coefficients of 0.63 and 0.27 which imply $\widehat{\rho} = 0.52$ and a period of 6.83. The Ahtola–Tiao

---

[6] The effect that estimating λ_c has on the tests is something that requires investigation.
[7] Not one of the higher frequencies is a harmonic of the two lowest frequencies which correspond to periods of 117 and 50 months.

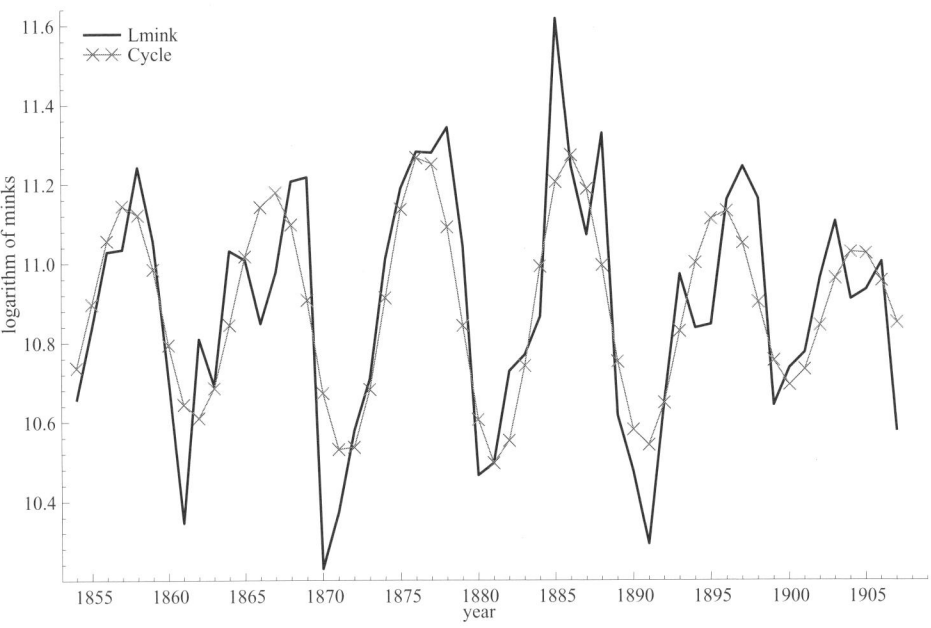

Fig. 6.2. Smoothed cycle in mink series.

test statistic, $T(\widehat{\phi}_2 + 1) = T(\widehat{\rho}^2 + 1)$, is -68.58 and so the null of a pair of complex unit roots is clearly rejected since the 5% critical value is -2.03. However, the Box–Ljung Q-statistic with six residual autocorrelations has a probability value, based on χ_4^2, of 0.038. Applying the procedure suggested by Ahtola and Tiao (1987, p. 9) based on an $AR(4)$ model, for which the Q-statistic, now with two degrees of freedom, has a probability value of 0.145, gives $\rho = 0.77$ and a period of 7.93. The unit root test statistic is now -20.2 so again the unit root is rejected. However, the fact that the implied period is only 7.93 casts doubt on the ability of the autoregressive model to adequately capture the cycle.

The global ice volume example in Section 6.2.5 illustrated how several cycles of unknown period could be extracted using a UC model. In this case the periods of the cycles identified by examining the periodogram or by extracting roots from a long autoregression – Newton, North and Crowley (1991) used 36 lags – were similar to those obtained by fitting the unobserved components model.

6.11 Conclusion

The important practical message of this paper is that tests concerning cycles need to be understood in the context of models. While the periodogram interpretation of various tests is useful, the periodogram should only be used to give a rough indication of the location and statistical significance of cycles. The modelling framework allows various tests to be carried out on the type of cycle that might be present. Standard regularity conditions do not apply in many cases but the Lagrange multiplier principle and the Cramér–von Mises distribution provide a systematic basis for the form of many of the test statistics and their asymptotic distributions.

Acknowledgements

I am grateful to Fabio Busetti, Paulo Rodrigues, Robert Taylor and an anonymous referee for helpful comments on the topics covered in this paper.

Part III

Bayesian inference and bootstrap

7
Efficient Bayesian parameter estimation

Sylvia Frühwirth-Schnatter
Department of Applied Statistics (IFAS), Johannes Kepler University

Abstract

The paper contributes to practical Bayesian estimation of state space models using Markov chain Monte Carlo (MCMC) and data augmentation methods. The main topic is the relation between parameterisation and computational efficiency of the resulting MCMC sampler. Standard data augmentation based directly on the dynamic process of the model, also called centred parameterisation, is computationally efficient as long as the variability of the dynamic process is important to explain the variability of the marginal process. This centred parameterisation, however, is shown to be computationally inefficient for models where the unobserved component is hardly changing over time. For such a state space model various reparameterisation techniques are considered, among them noncentring the location as in Pitt and Shephard (1999a) and noncentring the variance as suggested in Meng and van Dyk (1998) for random effects models. A new reparameterisation technique based on the standardised disturbances is suggested that is computationally efficient over a wide range of overidentified models.

7.1 Introduction

State space models are a rather intuitive tool for modelling empirical time series through hidden processes. The state space model is usually formulated in a natural way, where the model parameters governing the dynamic part of the process appear only in the transition equation. Consider, as a typical example, the following time varying parameter model:

$$\beta_t = \phi \beta_{t-1} + (1-\phi)\mu + w_t, \qquad w_t \sim N\left(0, \sigma_w^2\right), \qquad (7.1)$$

$$y_t = Z_t \beta_t + \varepsilon_t, \qquad \varepsilon_t \sim N\left(0, \sigma_\varepsilon^2\right), \qquad (7.2)$$

$t = 1, \ldots, N$, where β_0 is an unknown starting value and ε_t and w_t are independent processes. Model (7.1) and (7.2) is a state space model with state process β_t. Among the unknown model parameters θ, where $\theta = (\mu, \phi, \sigma_w^2, \sigma_\varepsilon^2)$, all parameters that determine the stochastic law for the hidden process, namely (μ, ϕ, σ_w^2), appear only in the transition equation (7.1).

For unknown model parameters θ, Bayesian inference usually relies on data augmentation by choosing missing data $\tilde{X}$ and sampling from the joint posterior $\pi(\tilde{X}, \theta | y)$, where $y = (y_1, \ldots, y_N)$ is the observed time series. A natural candidate for the missing data is the state process: $\tilde{X} = (\beta_0, \ldots, \beta_N)$. Whereas the joint posterior $\pi(\tilde{X}, \theta | y)$ is rather complex, the conditional posterior densities $\pi(\tilde{X} | \theta, y)$ and $\pi(\theta | \tilde{X}, y)$ are often of closed form, especially for a Gaussian state space model. It has been recognised by various authors (Carlin, Polson and Stoffer 1992a, Carter and Kohn 1994, Frühwirth-Schnatter 1994c) that due to this specific structure MCMC estimation based on data augmentation could be implemented for joint estimation of $\tilde{X}$ and θ through a two-block sampler where one samples the missing data conditional on the model parameters and the model parameters conditional on the missing data.

As an example, consider again model (7.1) and (7.2). Conditional on a known value of θ, the model is a Gaussian linear state space model and the moments of the joint posterior $\pi(\tilde{X} | \theta, y)$ may be obtained with the Kalman filter and the associated smoothing formulae. Conditional on a known state process $\tilde{X}$, the model factors in two independent components, the autoregressive state equation (7.1) and the regression model (7.2) for which parameter estimation is pretty standard. Therefore a straightforward MCMC sampler for this model consists of two main blocks:

(a) Sample a path of $\tilde{X}$ conditional on known model parameters

$$\theta = (\mu, \sigma_w^2, \sigma_\varepsilon^2, \phi)$$

using one of the multimove samplers discussed in Carter and Kohn (1994),

Frühwirth-Schnatter (1994c), de Jong and Shephard (1995) or Durbin and Koopman (2002).

(b) Sample the model parameters $(\mu, \sigma_w^2, \sigma_\varepsilon^2, \phi)$ conditional on $\tilde{X}$. Conditional on knowing $\tilde{X}$, the parameters μ, ϕ and σ_w^2 appearing in the state equation (7.1) are independent from the parameter σ_ε^2 appearing in the observation equation (7.2).

Bayesian parameter estimation for state space models using an MCMC sampler, which is alternating between the updates of $\tilde{X}$ and θ, is very convenient and often leads to sensible results. Under certain circumstances, however, it turns out that the convergence properties of this sampler may be extremely poor leading to a painfully slow and computationally inefficient estimation method. For the time varying parameter model (7.1) and (7.2), for instance, this sampler will be mixing poorly, if β_t is changing only slowly over time (σ_w^2 close to 0), see also Figure 7.1. The same problem is also encountered in many other state space models of practical relevance for financial econometrics such as dynamic factor models or stochastic volatility models, see e.g. Kim, Shephard and Chib (1998), Frühwirth-Schnatter and Geyer (1996), Shephard (1996), and Roberts, Papaspiliopoulos and Dellaportas (2004).

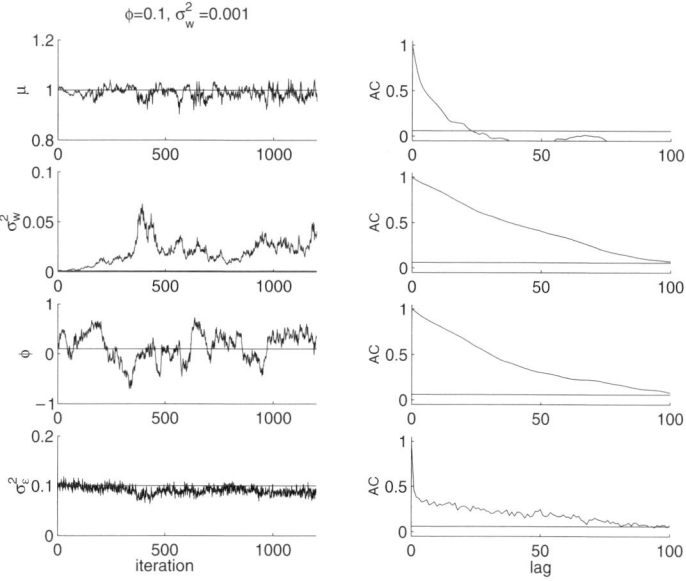

Fig. 7.1. Data simulated from model (7.1) and (7.2) with $Z_t = 1$, $\phi = 0.1$ and $\sigma_w^2 = 0.001$; MCMC draws with empirical autocorrelation obtained for μ, σ_w^2, σ_ε^2 and ϕ for parameterisation (7.1) and (7.2).

Somewhat surprisingly, considerable improvement is often possible by simple reparameterisation techniques. The relation between the convergence properties of the alternating MCMC sampler and the parameterisation is best understood for the random-effects model which is that special case of model (7.1) and (7.2) where $\phi = 0$. Gelfand, Sahu and Carlin (1995) discuss the effect of recentring the location μ of β_t, whereas Meng and van Dyk (1998) discussed the effect of recentring the scale σ_w of β_t. It has been recognised, that there exists a whole continuum of reparameterisations both of the locations (Papaspiliopoulos, Roberts and Sköld 2003) and the scale (Meng and van Dyk 1998). Far less is known about optimal parameterisation for time series models where $\phi \neq 0$. Recentring the location of a time series has been studied by Pitt and Shephard (1999a). Reparameterisation techniques for stochastic volatility models are discussed in Shephard (1996), Roberts, Papaspiliopoulos and Dellaportas (2004) and Frühwirth-Schnatter and Sögner (2002).

This contribution reviews some of the existing reparameterisation techniques and introduces some new techniques, especially designed for time series models. The discussion will be confined entirely to the time varying parameter model (7.1) and (7.2). Section 7.2 reviews reparametrisation techniques for an unknown location parameter μ for the case where $\sigma_w^2, \sigma_\varepsilon^2$ and ϕ are known. Section 7.3 explores the case where additionally the variance parameters σ_w^2 and σ_ε^2 are unknown, whereas ϕ is still known. A new parameterisation for time series models is suggested that is based on recentring the scale of the hidden process. Section 7.4 considers the most interesting case where all parameters are unknown. Here an MCMC sampler is introduced that is based on the noncentred disturbances of the hidden process. Section 7.5 presents more details on implementing MCMC estimation for all parameterisations. Section 7.6 concludes by discussing the results of this paper in the light of more general state space models.

7.2 Reparameterisation of the location

This section reviews the reparametrisation technique for an unknown location parameter μ of the time varying parameter model (7.1) and (7.2) for the case where all other model parameters, namely $\sigma_w^2, \sigma_\varepsilon^2$ and ϕ are known. The material is based on Gelfand, Sahu and Carlin (1995), Papaspiliopoulos, Roberts and Sköld (2003) and Pitt and Shephard (1999a).

7.2.1 Noncentring the location for the random-effects model

7.2.1.1 Recentring of the location

The relation between the convergence properties of the alternating MCMC sampler reviewed in Section 7.1 and the parameterisation chosen for the location is best understood for a random-effects model under the assumption of known variances σ_w^2 and σ_ε^2. A random-effects model is that special case of model (7.1) and (7.2) where $\phi = 0$:

$$\beta_t = \mu + w_t, \qquad w_t \sim N\left(0, \sigma_w^2\right), \tag{7.3}$$
$$y_t = Z_t \beta_t + \varepsilon_t, \qquad \varepsilon_t \sim N\left(0, \sigma_\varepsilon^2\right). \tag{7.4}$$

A very clear discussion of how to parameterise the location for a random-effects models appears in Gelfand, Sahu and Carlin (1995). They called parameterisation (7.3) and (7.4) the centred parameterisation, as the random effect β_t is centred around the *a priori* expected value: $E(\beta_t) = \mu$. An alternative parameterisation is based on the transformation $\tilde{\beta}_t = \beta_t - \mu$:

$$\tilde{\beta}_t = w_t, \qquad w_t \sim N\left(0, \sigma_w^2\right), \tag{7.5}$$
$$y_t = Z_t \mu + Z_t \tilde{\beta}_t + \varepsilon_t, \qquad \varepsilon_t \sim N\left(0, \sigma_\varepsilon^2\right). \tag{7.6}$$

Gelfand, Sahu and Carlin (1995) called parameterisation (7.5) and (7.6) the noncentred parameterisation, as the corresponding random effect $\tilde{\beta}_t$ is no longer centred around the *a priori* expected value, but around 0: $E(\tilde{\beta}_t) = 0$. An interesting effect of this reparameterisation has been to move the location parameter μ from the transition equation (7.3) to the observation equation (7.6) where it now appears as a fixed effect.

7.2.1.2 To centre or not to centre the location?

For known variances σ_w^2 and σ_ε^2, Gelfand, Sahu and Carlin (1995) gave general conditions, under which the centred parameterisation yields MCMC algorithms with faster convergence than the noncentred one. In general, the noncentred parameterisation should be preferred, if only a small fraction of the (marginal) variability $V(y_t)$ of y_t is caused by heterogeneity in β_t, i.e. $V(y_t|\beta_t)$ is only slightly smaller than $V(y_t)$. For a random design Z_t this is equivalent to the condition that the coefficient of determination defined by

$$D = 1 - \frac{V(y_t|\beta_t)}{V(y_t)} = \frac{E(Z_t^2)\sigma_w^2}{\sigma_\varepsilon^2 + E(Z_t^2)\sigma_w^2} = \frac{E(Z_t^2)\sigma_w^2/\sigma_\varepsilon^2}{1 + E(Z_t^2)\sigma_w^2/\sigma_\varepsilon^2} \tag{7.7}$$

is close to 0; $\sigma_w^2/\sigma_\varepsilon^2$ is known as the signal-to-noise ratio.

Roberts and Sahu (1997) showed that for a model with known variances the coefficient D determines the convergence rate of the alternating MCMC sampler under the two different parameterisation. Whereas the convergence rate is $1 - D$ for the centred parameterisation, it is equal to D under the noncentred parameterisation. Therefore, for the random effects model the parameterisation that is centred in the location is preferred if D is bigger than $1/2$, whereas the noncentred parameterization is preferred if D is smaller than $1/2$.

An important (and disturbing) conclusion from Roberts and Sahu (1997) is the following. If the signal-to-noise ratio $\sigma_w^2/\sigma_\varepsilon^2$ goes to infinity, the noncentred parameterisation will not converge geometrically, whereas the centred parameterisation will not converge geometrically if the signal-to-noise ratio $\sigma_w^2/\sigma_\varepsilon^2$ goes to 0. Each of the two parameterisations may be arbitrarily bad under unfortunate parameter constellations.

7.2.2 Noncentring the location for a time series model

7.2.2.1 Noncentring the location

Far less is known about recentring the location of a time series models where $\phi \neq 0$, a notable exception being Pitt and Shephard (1999a). Following Gelfand, Sahu and Carlin (1995), Pitt and Shephard (1999a) called parameterisation (7.1) and (7.2) the centred one also for a time series model, as the long-run mean of β_t is equal to μ. As an alternative, Pitt and Shephard (1999a) considered a model that is noncentred in the location:

$$\tilde{\beta}_t = \phi \tilde{\beta}_{t-1} + w_t, \qquad w_t \sim N\left(0, \sigma_w^2\right), \tag{7.8}$$

$$y_t = Z_t \mu + Z_t \tilde{\beta}_t + \varepsilon_t, \qquad \varepsilon_t \sim N\left(0, \sigma_\varepsilon^2\right). \tag{7.9}$$

For $\phi \to 0$, this corresponds to the noncentred random-effects model (7.3) and (7.4).

7.2.2.2 To centre or not to centre the location?

For a model where $Z_t = 1$ and the variances are assumed to be known, the following results are proven in Pitt and Shephard (1999a, p. 71). For $\phi \to 1$, the convergence rate of the centred parameterisation goes to 0, whereas the convergence rate of the noncentred parameterisation goes to 1. Thus for the limiting random walk model the noncentred parameterisation does not converge geometrically regardless of the signal-to-noise ratio $\sigma_w^2/\sigma_\varepsilon^2$.

For $\phi < 1$, the variances matter, and the centred parameterisation is better than the noncentred parameterisation if

$$\frac{\sigma_w^2}{(1-\phi)^2} > \sigma_\varepsilon^2. \tag{7.10}$$

For $\phi = 0$, this result coincides with the previous result, that the centred parameterisation is better if $\sigma_w^2 > \sigma_\varepsilon^2$. Also for $\phi > 0$, the centred parameterisation is better in the case of high heterogeneity, whereas the noncentred parameterisation is better in the case of low heterogeneity. With ϕ approaching 1, however, the range of variance parameters σ_w^2 for which the noncentred parameterisation is better than the centred one becomes smaller and may be just a very small region close to 0 for highly persistent processes.

7.2.2.3 Illustration for simulated data

For illustration, we simulated time series of length $N = 500$ from a model where $\mu = 1$ and $\sigma_\varepsilon^2 = 0.1$ under three different levels of heterogeneity: a low level of heterogeneity with $\sigma_w^2 = 0.001$, a medium level of heterogeneity with $\sigma_w^2 = 0.05$ and a high level of heterogeneity with $\sigma_w^2 = 1$. Here, Z_t takes randomly the values $-1, 0, 1$, a choice that guarantees identifiability also for ϕ approaching 0.

MCMC sampling of $\tilde{X}$ and μ, with all other model parameters fixed, is carried out under the two parameterisations for all settings of heterogeneity for $\phi = 0.1$ and $\phi = 0.95$. Figure 7.2 illustrates the following results already obtained in Section 7.2.2.2 for a time series model, where the variances are known:

- For time series with low autocorrelation ($\phi = 0.1$ in the example) we have a result similar to that of the random-effects model, namely that the centred parameterisation is better than the noncentred one for a high level of heterogeneity.
- For highly persistent time series ($\phi = 0.95$ in the example), the centred parameterisation is more efficient even for a low level of heterogeneity.

7.2.3 Partially noncentring of the location

Papaspiliopoulos, Roberts and Sköld (2003) showed for a random-effects model with known variances that there always exists a reparameterisation which has the optimal convergence rate 0. They call this parameterisation partially noncentred. The partially noncentred parameterisation is defined as a continuum between the centred parameterisation (7.3) and (7.4) and

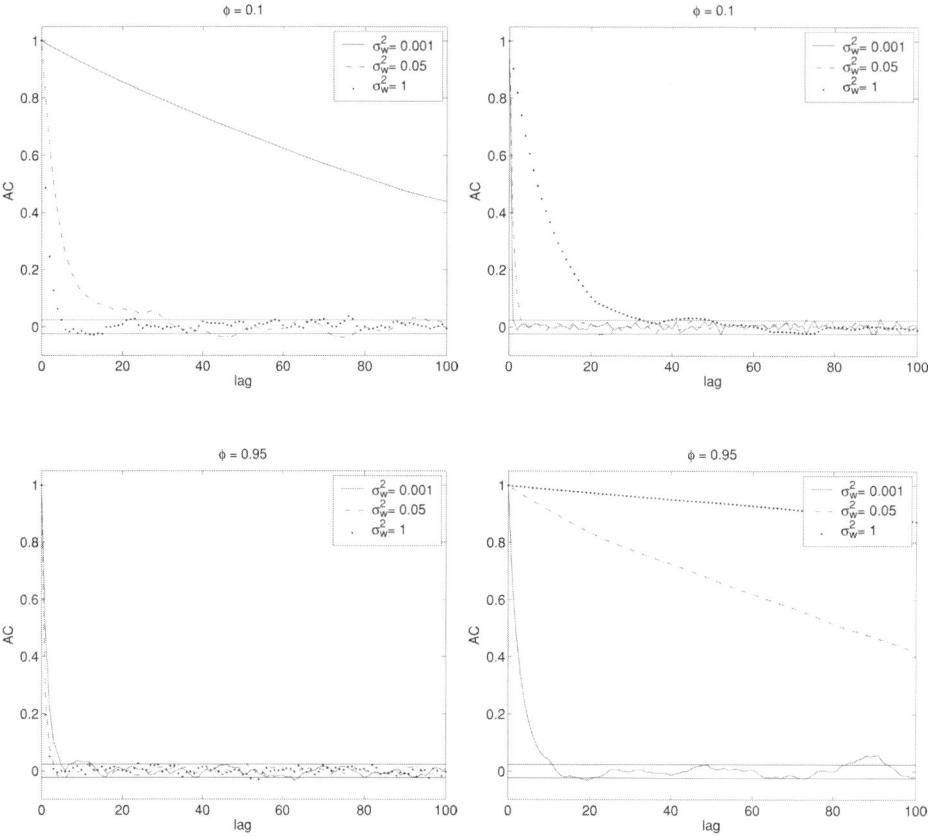

Fig. 7.2. Data simulated from model (7.1) and (7.2) with $\phi = 0.1$ (top) and $\phi = 0.95$ (bottom) for various values of σ_w^2; empirical autocorrelation of the MCMC draws of μ (ϕ, σ_w^2 and σ_ε^2 fixed at their true values) for the centred parameterisation (left hand side) and the parameterisation noncentred in the location (right hand side).

the noncentred parameterisation (7.5) and (7.6). The state vector is defined as a weighted mean of β_t and $\tilde{\beta}_t$:

$$\beta_t^w = W_t \tilde{\beta}_t + (1 - W_t)\beta_t, \qquad (7.11)$$

with $W_t = 0$ corresponding to the centred and $W_t = 1$ corresponding to the noncentred parameterisation, respectively. The state vector of the partially centred parameterisation is related to the state vector β_t of the centred parameterisation through

$$\beta_t^w = \beta_t - W_t \mu, \tag{7.12}$$

whereas the relation to the state vector $\tilde{\beta}$ of the noncentred parameterisation is given by

$$\beta_t^w = \tilde{\beta}_t + (1 - W_t)\mu. \tag{7.13}$$

The partially noncentred model takes the following form:

$$\beta_t^w = (1 - W_t)\mu + w_t, \quad w_t \sim N\left(0, \sigma_w^2\right), \tag{7.14}$$
$$y_t = Z_t \beta_t^w + Z_t W_t \mu + \varepsilon_t, \quad \varepsilon_t \sim N\left(0, \sigma_\varepsilon^2\right). \tag{7.15}$$

Thus for a partially noncentred model the location parameter μ appears both in the transition and the observation equation.

For a model with time invariant predictor $Z_t \equiv Z$, Papaspiliopoulos, Roberts and Sköld (2003) showed that iid samples may be produced by choosing

$$W_t \equiv 1 - D = \frac{1}{1 + Z^2 \sigma_w^2/\sigma_\varepsilon^2}, \tag{7.16}$$

where D is defined in (7.7) and $\sigma_w^2/\sigma_\varepsilon^2$ is the signal-to-noise ratio. Again we find that the centred parameterisation results under a high signal-to-noise ratio whereas the noncentred parametrisation results under a small signal-to-noise ratio. For more general models with time varying Z_t, Papaspiliopoulos, Roberts and Sköld (2003) showed that iid samples may be produced by choosing data dependent weights:

$$W_t = \frac{1}{1 + Z_t^2 \sigma_w^2/\sigma_\varepsilon^2}. \tag{7.17}$$

The partially noncentred parameterisation introduced in Papaspiliopoulos, Roberts and Sköld (2003) is limited to the case of a random-effects model and no results were given for time series models. Nevertheless as above, one may define β_t^w as in (7.12) also for a time series model:

$$\beta_t^w = \beta_t - W_t \mu, \tag{7.18}$$

to obtain the following partially centred parameterisation:

$$\beta_t^w = \phi \beta_{t-1}^w + [(1 - \phi) - (W_t - \phi W_{t-1})]\mu + w_t, \quad w_t \sim N\left(0, \sigma_w^2\right), \tag{7.19}$$
$$y_t = Z_t \beta_t^w + Z_t W_t \mu + \varepsilon_t, \quad \varepsilon_t \sim N\left(0, \sigma_\varepsilon^2\right). \tag{7.20}$$

For $W_t = 0$ we obtain the centred parameterisation and for $W_t = 1$ the noncentred one. It is, however, still unclear how to select W_t for a time series model in such a way that the corresponding two block MCMC sampler produces (nearly) iid draws.

7.3 Reparameterisation of the scale

7.3.1 The effect of unknown variances

In the previous section we showed that under high levels of heterogeneity MCMC estimation of $(\tilde{X}, \mu)$ may be safely based on the centred parameterisation (7.1) and (7.2). With σ_w^2 approaching zero, however, the sampler deteriorates and noncentring the location may lead to a dramatic improvement of the convergence properties of the MCMC sampler. These conclusions, however, were drawn under the assumption of *known* variances, and little is known about the effect of also considering the variances σ_w^2 and σ_ε^2 as unknown parameters especially for time series models with $\phi \neq 0$.

For a random-effects model with unknown variances it has been noted by Meng and van Dyk (1998) that for models with low but unknown levels of heterogeneity noncentring the location, although improving convergence in μ, leads to a sampler that is rather inefficient in sampling σ_w^2. For random-effects models with low (but unknown) heterogeneity, Meng and van Dyk (1998) and Van Dyk and Meng (2001) suggested recentring the model both in location and scale and demonstrated considerable improvement over the parameterisation that is noncentred only in the location. This material will be reviewed in Section 7.3.2. The extension to time series models where ϕ is a known nonzero parameter will be considered in Section 7.3.3.

7.3.2 Noncentring both location and scale for a random effects model

7.3.2.1 Recentring of location and scale

Meng and van Dyk (1998) showed for a random-effects model which is noncentred in the location that sampling both the location and the variances leads to a poor sampler, when the coefficient of determination D defined in (7.7) is small. Note that although the parameterisation is noncentred in the location, which is preferable to the centred one for D small, the parameterisation is obviously not optimal for the scale parameters. As a remedy, Meng and van Dyk (1998) suggested using the following reparameterisation which is based on rescaling the state vector to a variable with unit variance:

$$\beta_t^\star = \frac{\tilde{\beta}_t}{\sigma_w}, \qquad (7.21)$$

leading to the following random-effects model:

$$\beta_t^\star = w_t, \qquad w_t \sim N(0,1), \qquad (7.22)$$
$$y_t = Z_t\mu + Z_t\sigma_w\beta_t^\star + \varepsilon_t, \qquad \varepsilon_t \sim N(0,\sigma_\varepsilon^2). \qquad (7.23)$$

One effect of recentring the scale is that the unknown variance σ_w^2, or rather the standard deviation σ_w, is moved from the transition equation (7.22) to the observation equation (7.23). The model is now parameterised in such a way that the square root of σ_w^2 appears in the observation equation as an unknown regression coefficient with latent, unobserved regressor $\beta_t^\star$. Therefore an MCMC sampler under this parameterisation is easily derived. We will present further details in Section 7.5.

As in (7.22) and (7.23) the scale of the state variable $\beta_t^\star$, measured by $\beta_t^{\star 2}$, is centred around 1, $E(\beta_t^{\star 2}) = 1$, we call this parameterisation noncentred in the scale. In contrast to that parameterisation (7.1) and (7.2) as well as (7.5) and (7.6) are centred in the scale: $E(\tilde{\beta}_t^2) = \sigma_w^2$. Note that parameterisation (7.22) and (7.23) is noncentred both in the location and in the scale.

7.3.2.2 To centre or not to centre?

As for the location parameter, there exists a whole continuum of reparameterisations of the scale. Meng and van Dyk (1997, 1999) were the first to discuss partially recentring of the scale of a *t*-distribution. These results were extended to the univariate random-effects model in Meng and van Dyk (1998, p.574). Starting from a model that is noncentred in the location, they defined

$$\beta_t^a = \frac{\tilde{\beta}_t}{(\sigma_w)^A}, \qquad (7.24)$$

leading to the following state space model:

$$\beta_t^a = (\sigma_w)^{1-A}w_t, \qquad w_t \sim N(0,1), \qquad (7.25)$$
$$y_t = Z_t\mu + Z_t(\sigma_w)^A\beta_t^a + \varepsilon_t, \qquad \varepsilon_t \sim N(0,\sigma_\varepsilon^2). \qquad (7.26)$$

Note that $(\beta_t^a)^2$ is a weighted geometric mean between the state vector $\tilde{\beta}_t^2$, that is centred in the scale, and $(\beta_t^\star)^2$ which is noncentred in the scale:

$$(\beta_t^a)^2 = (\tilde{\beta}_t^2)^{1-A}(\beta_t^{\star,2})^A. \qquad (7.27)$$

The effect of the reparameterisation is that the unknown variance σ_w^2 appears both in the transition equation (7.25) as well as in the observation equation (7.26). For $A = 0$, we obtain a model that is centred in the scale, whereas for $A = 1$ we obtain a model that is noncentred in the scale.

Meng and van Dyk (1998, equation (4.4)), suggested the following choice for A:

$$A = \frac{2(1-D)}{2-D} = \frac{\sigma_\varepsilon^2}{\sigma_\varepsilon^2 + \overline{Z_t^2}\sigma_w^2/2}, \qquad (7.28)$$

where D is defined in (7.7). From (7.28) we obtain that for the small heterogeneity case A will be close to 1, and noncentring in the scale is preferred, whereas under high heterogeneity, A is close to 0 and centring in the variance is preferred.

7.3.3 Noncentring both location and scale for a time series model

7.3.3.1 Recentring of location and scale

Practically nothing is known about the gain of recentring a time series model in the scale parameter. A model that is noncentred both in location and scale is given by

$$\beta_t^\star = \phi\beta_{t-1}^\star + w_t, \qquad w_t \sim N(0,1), \qquad (7.29)$$
$$y_t = Z_t\mu + Z_t\sigma_w\beta_t^\star + \varepsilon_t, \qquad \varepsilon_t \sim N(0,\sigma_\varepsilon^2). \qquad (7.30)$$

As above we will assume that the parameter ϕ is known. As for a random-effects model, the magnitude of the unknown variances plays a crucial role. If the true value of the unknown variance σ_w^2 is not too small, the conclusions drawn in Section 7.2.2.2 about the circumstances under which the centred parameterisation should be preferred remain valid. With σ_w^2 approaching zero, however, these conclusions are no longer valid. Noncentring the location, although improving convergence in μ, leads to a sampler that is rather inefficient in sampling σ_w^2. Noncentring both location and scale improves the sampler considerably, as will be demonstrated below for the simulated data.

7.3.3.2 Illustration simulated for data

We reconsider the simulated data introduced in Section 7.2.2.3. For nonpersistent time series with $\phi = 0.1$, Figure 7.3 compares, for three different levels of heterogeneity, the preferred parameterisation for the location, with

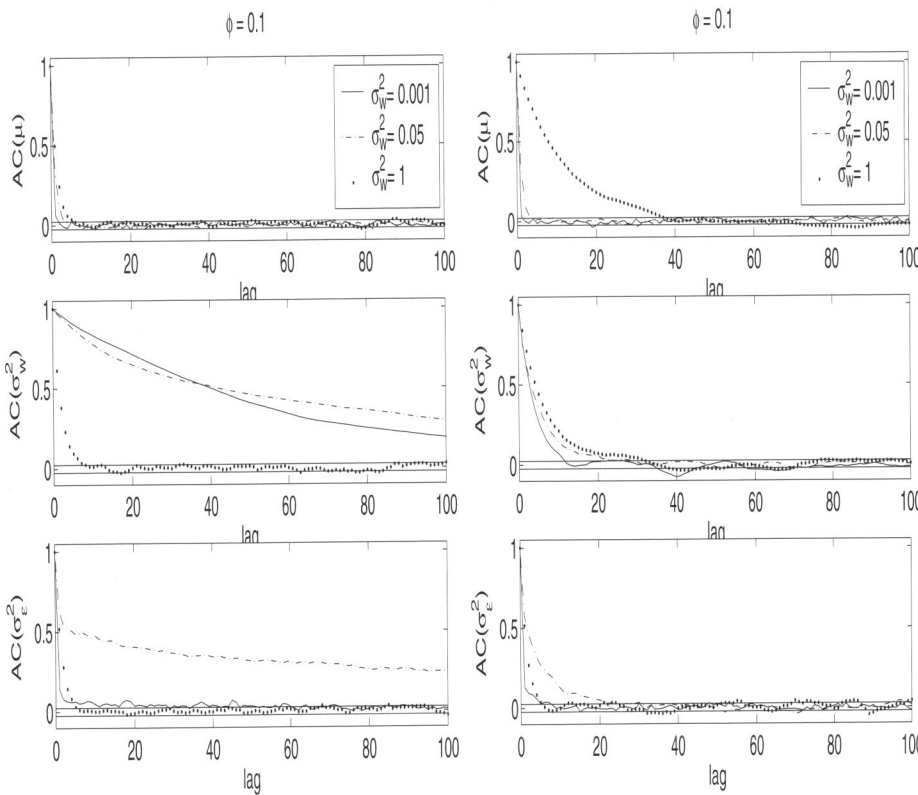

Fig. 7.3. Data simulated from model (7.1) and (7.2) with $\phi = 0.1$ for various values of σ_w^2; empirical autocorrelation of the MCMC draws of μ (top), σ_w^2 (middle) and σ_ε^2 (bottom) (ϕ fixed at the true value) for the preferred parameterisation for the location (left hand side, centred only for $\sigma_w^2 = 1$) and the parameterisation which is noncentred both in location and scale (right hand side)

the variances assumed to be unknown, with the parameterisation that is noncentred both in location and scale. For nonpersistent time series models with unknown small variances noncentring only in the location, although improving convergence in μ, leads to a sampler that is rather inefficient in sampling σ_w^2. Noncentring both location and scale improves the sampler considerably, see especially the example in Figure 7.3 where $\phi = 0.1$ and $\sigma_w^2 = 0.001$.

For persistent time series with $\phi = 0.95$, this comparison is carried out in Figure 7.4. Also for persistent time series model (ϕ close to 1) with unknown small variances the centred parameterisation, although efficient for sampling

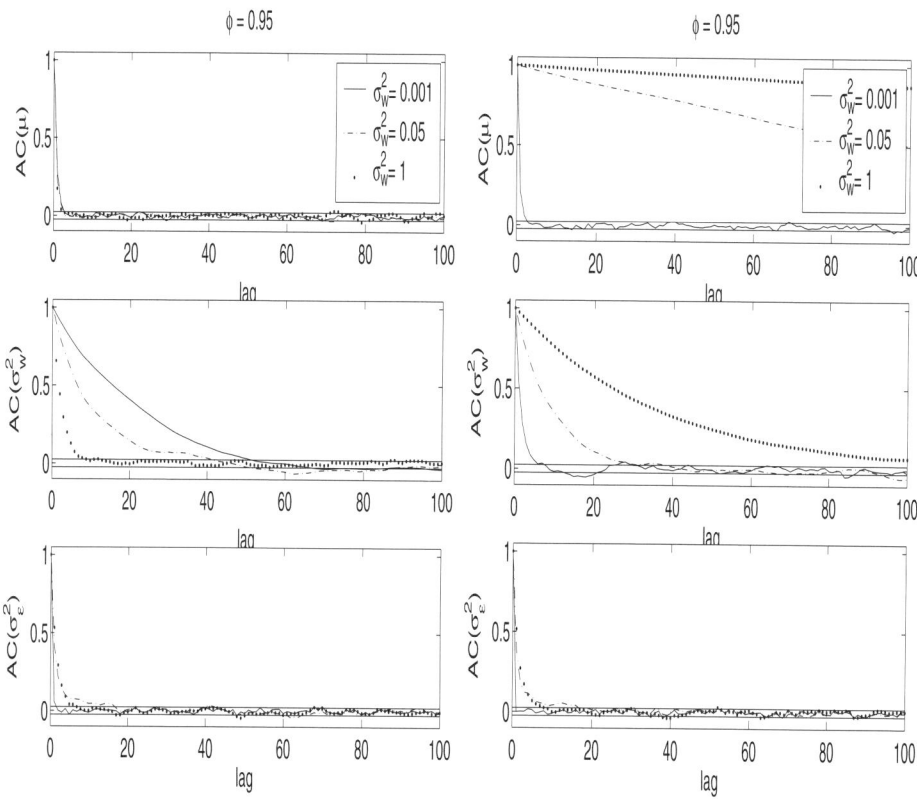

Fig. 7.4. Data simulated from model (7.1) and (7.2) with $\phi = 0.95$ for various values of σ_w^2; empirical autocorrelation of the MCMC draws of μ (top), σ_w^2 (middle) and σ_ε^2 (bottom) (ϕ fixed at the true value) for the centred parameterisation (left hand side) and the parameterisation which is noncentred both in location and scale (right hand side)

μ, leads to a sampler that is rather inefficient for sampling σ_w^2. Again, noncentring both location and scale improves the sampler considerably, see especially the example in Figure 7.4 where $\phi = 0.95$ and $\sigma_w^2 = 0.001$. Compared with nonpersistent time series, however, the range where the noncentred parameterisation is sensible is much smaller for persistent time series; compare Figure 7.3 and Figure 7.4. For highly persistent time series MCMC sampling based on the noncentred parameterisation quickly leads to rather poor results, when σ_w^2 moves away from 0. This result, however, is not surprising in the light of the results of Pitt and Shephard (1999a) discussed in Section 7.2.2.2.

7.4 Reparameterising time series models based on disturbances

7.4.1 The effect of an unknown autocorrelation structure

The results of the previous section hold under the assumption that the parameter ϕ, which determines the autocorrelation structure, is a known quantity. We now discuss what happens if ϕ is an unknown parameter. We will discuss only the case where $\phi > 0$.

The addition of ϕ as an extra unknown parameter does not affect the conclusions drawn so far if the variance σ_w^2 is not too small. However, problems arise when σ_w^2 is too close to 0. In this case the state space model is nearly oversised, making MCMC estimation of the model parameters difficult. Consider for instance, the case where $\sigma_w^2 = 0.001$. For ϕ fixed, the parameterisation (7.29) and (7.30) that is noncentred both in location and scale leads to a very efficient sampler, see again Figures 7.3 and 7.4. When adding ϕ as an unknown parameter, however, the resulting sampler is rather inefficient, see Figures 7.5 and 7.6. Mixing is slow mainly for ϕ, but rather fast for the other parameters. Remember that for this parameterisation ϕ still remains in the transition equation, whereas μ and σ_w^2 are moved to the observation equation. This suggests considering a parameterisation where ϕ is moved from the transition equation to the observation equation as well.

7.4.2 Parameterisation based on noncentred disturbances

A parameterisation in which all unknown parameters are moved into the observation equation and where we use the standardised disturbances w_t in (7.29) rather than $\beta_t^\star$ as missing data is.

$$w_t \sim N(0, 1), \tag{7.31}$$
$$y_t = Z_t \mu + Z_t \sigma_w \beta_t^\star + \varepsilon_t, \qquad \varepsilon_t \sim N(0, \sigma_\varepsilon^2), \tag{7.32}$$

where $\beta_t^\star = \phi \beta_{t-1}^\star + w_t$. The missing data are defined as

$$\tilde{X} = (\beta_0^\star, w_1, \ldots, w_N).$$

The transition equation (7.31) and consequently the prior of $w_1, \ldots, w_N$ is independent of any model parameter. It is no longer possible to sample ϕ through a Gibbs sampler, instead we use a random walk Metropolis Hastings algorithm of the Gibbs step, see Section 7.5 for more details. Note that due

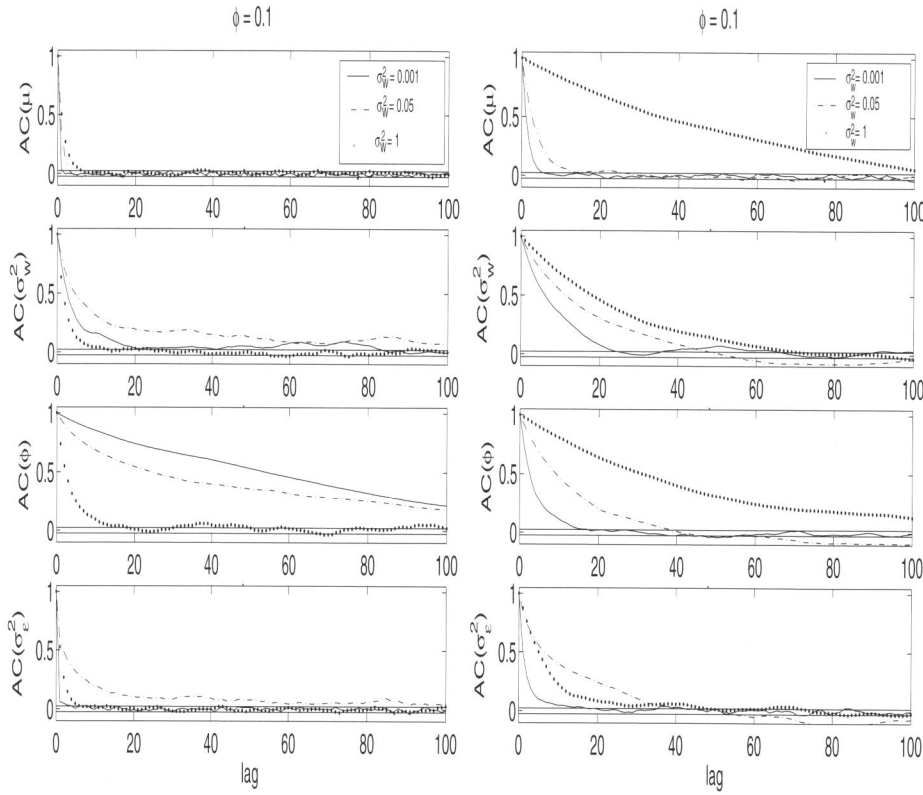

Fig. 7.5. Data simulated from model (7.1) and (7.2) with $\phi = 0.1$ for various values of σ_w^2; empirical autocorrelation of the MCMC draws of μ (top), σ_w^2, ϕ and σ_ε^2 (bottom) for the preferred parameterisation with ϕ known (left hand side, centred in the location and scale for $\sigma_w^2 = 1$, noncentred otherwise) and the parameterisation based on the noncentred disturbances (right hand side)

to the specific choice of $\tilde{X}$, the state process $\beta_1^\star, \ldots, \beta_N^\star$ changes whenever we change ϕ.

The examples in Figures 7.5 and 7.6 show that for σ_w^2 small the parameterisation based on the noncentred disturbances is considerable better than the parameterisation based on the noncentred, but correlated state process $\beta_t^\star$.

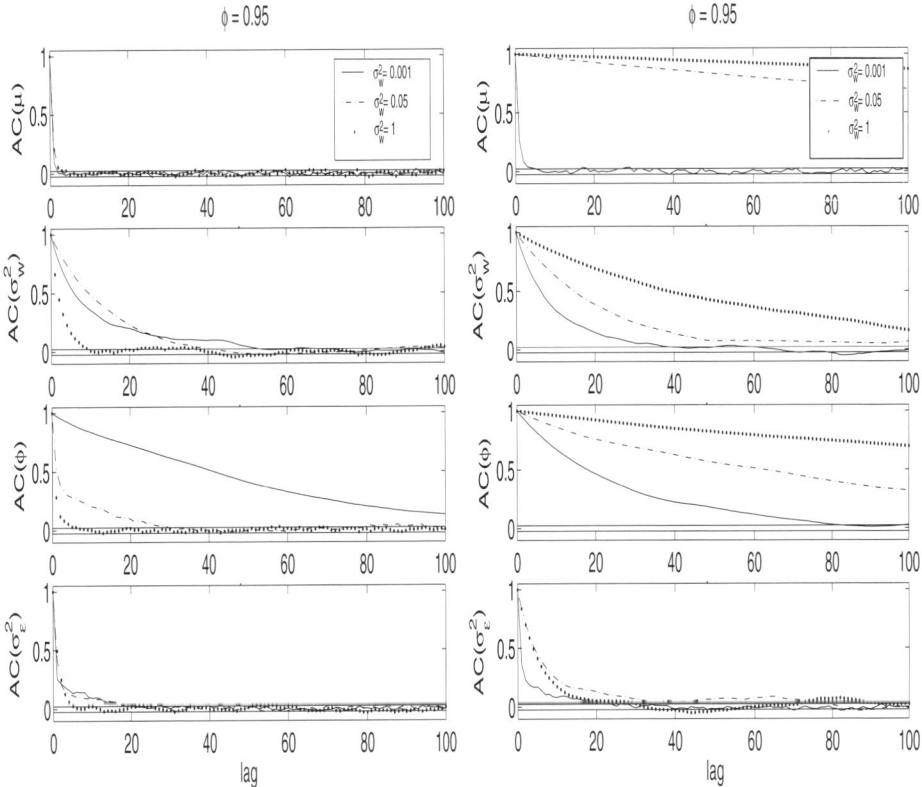

Fig. 7.6. Data simulated from model (7.1) and (7.2) with $\phi = 0.95$ for various values of σ_w^2; empirical autocorrelation of the MCMC draws of μ (top), σ_w^2, ϕ and σ_ε^2 (bottom) for the preferred parameterisation with ϕ known (left hand side, noncentred in the location and scale for $\sigma_w^2 = 0.001$, centered otherwise) and the parameterisation based on the noncentred disturbances (right hand side)

7.4.3 Illustration for simulated data

We reconsider the simulated data introduced in Section 7.2.2.3. First, we consider that parameterisation for the location and scale, which was the preferred one in Section 7.3.3.2 for ϕ fixed, however, this time with ϕ assumed to be unknown. Results are reported in Figure 7.5 for the non-persistent time series with $\phi = 0.1$ and in Figure 7.6 for the persistent time series with $\phi = 0.95$. In both cases, adding ϕ as an unknown parameter leads to a sampler that is inefficient in sampling ϕ for the small heterogeneity case $\sigma_w^2 = 0.001$. For larger values of σ_w^2, however, adding ϕ does not impose any convergence problems. For highly persistent time series the centred parameterisation is valid even for ϕ unknown, as long as σ_w^2 is not too close to 0.

Second, we consider the parameterisation that is based on the noncentred disturbances. A comparison of these two parameterisations is carried out in Figures 7.5 and 7.6 for $\phi = 0.1$ and $\phi = 0.95$, respectively. For a low level of heterogeneity with $\sigma_w^2 = 0.001$, choosing this parameterisation leads to a dramatic improvement of the convergence properties of the sampler compared with the parameterisation discussed above.

This improvement, however, is valid only for nearly oversized models. Figures 7.5 and 7.6 clearly show that the parameterisation based on the noncentred disturbances may be much worse than the centred parameterisation, when the true value of σ_w^2 moves away from 0. For the highly persistent time series in Figures 7.6, the parameterisation based on the noncentred disturbances is worse for all values but $\sigma_w^2 = 0.001$. For the nonpersistent time series in Figure 7.5 the range for which the parameterisation based on the noncentred disturbances is better than or comparable to the other parameterisations is much larger. Nevertheless, none of the parameterisations dominates the other one.

7.4.4 An alternative parameterisation based on disturbances

Various alternative parameterisations based on the disturbances are possible, an example being a parameterisation which is based on choosing the centred shocks u_t in

$$\beta_t = \phi \beta_{t-1} + u_t, \qquad u_t \sim N\left(\gamma, \sigma_w^2\right),$$

where $\gamma = (1 - \phi)\mu$, as missing data. This parameterisation reads

$$u_t \sim N\left(\gamma, \sigma_w^2\right), \qquad (7.33)$$
$$y_t = Z_t \beta_t + \varepsilon_t, \qquad \varepsilon_t \sim N\left(0, \sigma_\varepsilon^2\right), \qquad (7.34)$$

where $\beta_t = \phi \beta_{t-1} + u_t$. The missing data are defined as

$$\tilde{X} = (\beta_0, u_1, \ldots, u_N).$$

The transition equation (7.33) and consequently the prior of $u_1, \ldots, u_N$ is independent of ϕ, whereas the model parameters μ and σ_w^2 remain in the transition equation. A comparable parameterisation has been applied in Roberts, Papaspiliopoulos and Dellaportas (2001) for the stochastic volatility model of Barndorff-Nielsen and Shephard (2001).

For the time-varying parameter model we could not identify any parameter combinations where a parameterisation based on the centred shocks outperformed either the centred parameterisation or the parameterisation based on the noncentred disturbances. For nearly oversised models with $\sigma_w^2 = 0.001$ the resulting sampler is poor, as μ and σ_w^2 still remain in the transition equation. For models that are best estimated with the centered parameterisation (e.g. for models with $\sigma_w^2 = 1$), the effect of moving ϕ from the transition to the observation equation is a most undesirable one. Moving ϕ introduces conditional independence between ϕ and γ, whereas marginally ϕ and γ are highly correlated, see also Section 7.5.4.

7.4.5 Random selection of the parameterisation

From the investigations of the previous sections we identified only two sensible parameterisations, one centred in both location and scale and one based on the noncentred disturbances, with the noncentred parameterisation being the preferred one for nearly oversized models.

As the choice between the two parameterisation depends on the unknown parameters, it is in general not possible to decide *a priori* which parameterisation to choose. A rather pragmatic solution is to try both of them and to select the one with the lower autocorrelations in the MCMC draws. Alternatively, we could randomly select one of these two parameterisations. First experiences with this hybrid sampler for the simulated time series of this paper are rather promising, see also Figure 7.7. For this hybrid sampler we randomly select the parameterisation before each draw. In particular this gives a 0.5 probability that the parameterisation remains unchained and a 0.5 probability that the model is recentred.

7.5 Bayesian estimation and MCMC implementation

7.5.1 Prior distributions

For Bayesian estimation we assume the following prior distributions. First, we assume a flat prior for μ, an inverted Gamma for σ_ε^2 and a Beta distribution for ϕ. The initial state β_0 is assumed to arise from the marginal distribution of the AR(1) process: $\beta_0|\phi, \mu \sim N(\mu, \sigma_w^2/(1-\phi^2))$, see Schotman (1994) for a motivation.

The prior on σ_w^2 depends on the parameterisation. For a parameterisation that is centred in the scale or is based on the centred shocks, we assume

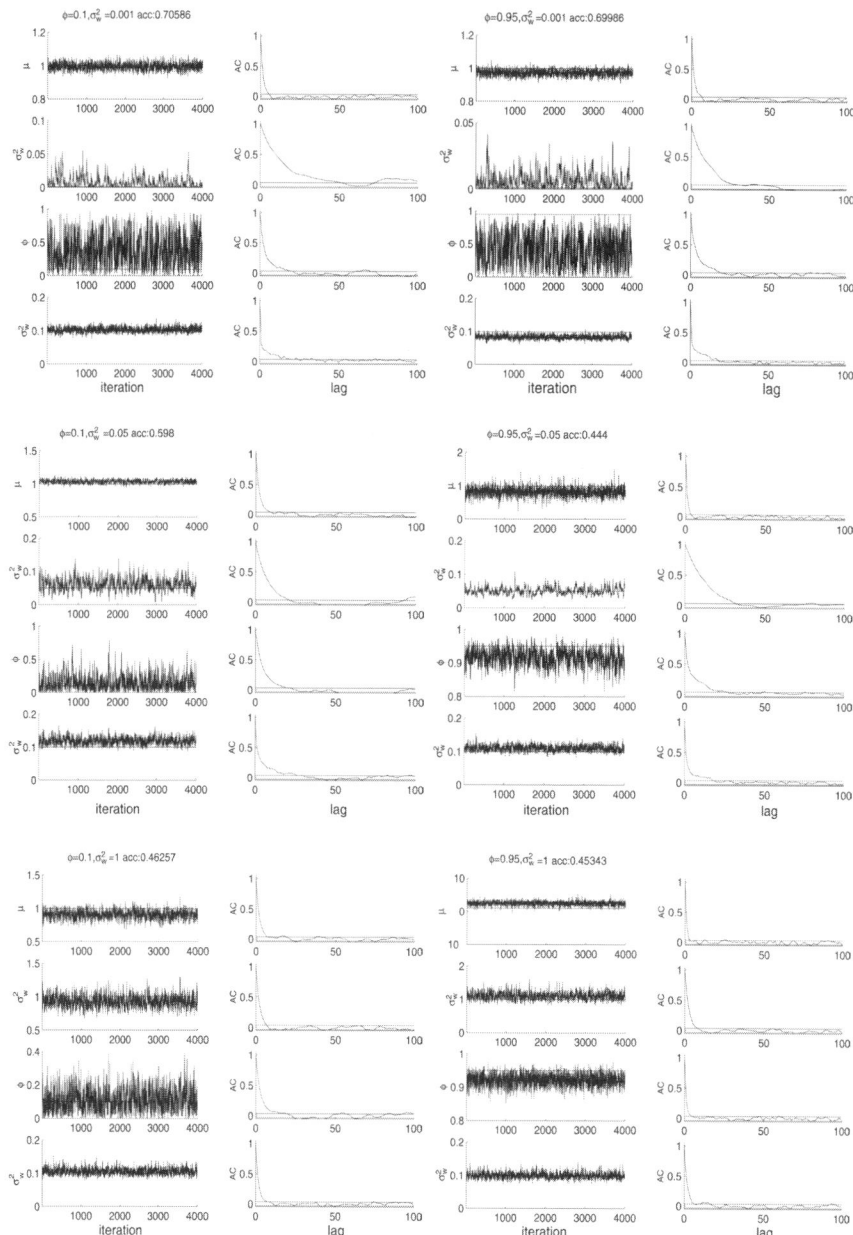

Fig. 7.7. Data simulated from model (7.1) and (7.2) with $\phi = 0.1$ (left hand side) and $\phi = 0.95$ (right hand side); $\sigma_w^2 = 0.001$ (top), $\sigma_w^2 = 0.05$ (middle) and $\sigma_w^2 = 1$ (bottom); MCMC draws with empirical autocorrelation obtained for μ, σ_w^2, ϕ and σ_ε^2 by choosing randomly the centred parameterisation or the parameterisation based on noncentred disturbances and keeping every second draw

that σ_w^2 is inverted Gamma, whereas for a parameterisation that is noncentred in the scale or based on noncentred disturbances we assume that σ_w is normal with prior mean equals 0 and prior variance proportional to σ_ε^2.

7.5.2 MCMC implementation

7.5.2.1 MCMC estimation for parameterisation centred in location and scale

MCMC estimation for a parameterisation that is centred in both location and scale is pretty straightforward and has been already outlined in the introduction. We discuss here more details of step (b). Conditional on $\tilde{X}$, (μ, ϕ, σ_w^2) is independent from σ_ε^2. However, rather than sampling (μ, ϕ, σ_w^2) directly, we sample the transformed parameter $(\gamma, \phi, \sigma_w^2)$ from the 'regression model' model (7.1), where $\gamma = (1 - \phi)\mu$. From (7.1) we obtain a marginal inverted Gamma proposal for σ_w^2, and a conditional normal proposal for $\gamma, \phi | \sigma_w^2$. To include the prior on β_0, which depends on $(\gamma, \phi, \sigma_w^2)$ in a nonconjugate way, we use a Metropolis–Hastings algorithm as in Chib and Greenberg (1994). Finally, from regression model (7.2), σ_ε^2 follows an inverted Gamma posterior.

7.5.2.2 MCMC estimation for parameterisations noncentred in location and centred in scale

For a model that is noncentred in the location, the following modification of step (b) is necessary. Conditional on $\tilde{X}$, the parameters ϕ and σ_w^2 appearing in the state equation (7.8) are independent of the parameters μ and σ_ε^2 appearing in the observation equation (7.9). We sample σ_w^2 and ϕ jointly from the autoregressive model (7.8) using an obvious modification of the Metropolis–Hastings algorithm discussed in Section 7.5.2.1. μ and σ_ε^2 are sampled jointly from the regression model (7.9).

7.5.2.3 MCMC estimation for parameterisations noncentred in location and scale

MCMC estimation for parameterisations that are noncentred in the scale is less straightforward. Van Dyk and Meng (2001) discussed MCMC estimation for random-effects model that are noncentred in the scale. Here we will present further details and discuss the extension to time series models. A straightforward MCMC sampler for parameterisation (7.29) and (7.30) which is noncentred in location and scale, but with ϕ remaining in the transition equation, is the following:

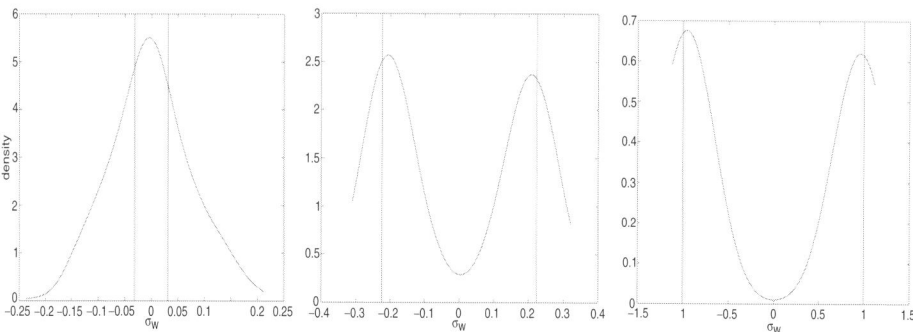

Fig. 7.8. Data simulated from model (7.1) and (7.2) with $\phi = 0.1$ and $\sigma_w^2 = 0.001$ (left), $\sigma_w^2 = 0.05$ (middle) and $\sigma_w^2 = 1$ (right); density of the marginal posterior distribution of σ_w (all model parameters considered unknown) under the parameterisation based noncentred disturbances; each MCMC draw has been concluded by a random sign switch; vertical lines indicate the true values

(a) Sample a path of $\tilde{X} = (\beta_0^\star, \ldots, \beta_N^\star)$ conditional on known parameters using any multimove sampler.

(b) Conditional on $\tilde{X}$, the parameters $(\mu, \sigma_w, \sigma_\varepsilon^2)$ are independent of ϕ. From model (7.29) obtain a normal proposal for ϕ that can be used within a Metropolis–Hastings algorithm to sample ϕ from the conditional posterior $\pi(\phi|\tilde{X}, y)$ which included the prior on $\beta_0^\star$. Finally, sample the parameters $(\mu, \sigma_w, \sigma_\varepsilon^2)$ jointly conditional on $\tilde{X}$ from 'regression model' (7.30).

A (small) variance σ_w now appears in (7.30) as a regression coefficient close to 0. For Bayesian estimation we do not constrain σ_w to be positive, making the whole model unidentified:

$$\psi = (\beta_1^\star, \ldots, \beta_N^\star, \mu, \sigma_w, \phi, \sigma_\varepsilon^2)$$

will result in the same likelihood function $L(y|\psi)$ as $\tilde{\psi}$, defined by:

$$\tilde{\psi} = (-\beta_1^\star, \ldots, -\beta_N^\star, \mu, -\sigma_w, \phi, \sigma_\varepsilon^2).$$

The advantage of this Bayesian unidentifiability is that by allowing σ_w to move freely in the parameter space, we avoid the boundary space problem for small variances. This seems to be the main reason why mixing improves a lot for models with small variances.

The symmetry of the likelihood around $\sigma_w = 0$ causes bimodality of the marginal posterior density $\pi(\sigma_w|y)$ for models where the true value of σ_w is actually different from 0. If the true value of σ_w is close to 0, a unimodal posterior which concentrates around 0, results. Thus the posterior distribution of σ_w may be used to explore how far, for the data at hand, the model is bounded away from a model where $\sigma_w^2 = 0$. Figure 7.8 shows the posterior distributions of σ_w for those data in Section 7.2.2.3 that were simulated from a model with $\phi = 0.1$ under the three different settings of heterogeneity. For MCMC estimation all parameters including ϕ are considered unknown. For the time series simulated under $\sigma_w^2 = 0.001$, the model is not bounded away from 0 and the mode of the posterior lies practically at 0. For the two other time series, however, the bimodality of the posterior is well pronounced and the posterior is bounded away from 0.

As expected theoretically, the posterior densities in Figure 7.8 are symmetric around 0. If the posterior density is estimated from MCMC sampling as described above, it need not be symmetric. The sampler may stay at one mode or may show occasional sign switching to the other mode. This makes it difficult to compare MCMC draws from different runs, to assess convergence and to interpret the posterior density. In order to explore the entire posterior distribution, we performed a random sign switching after each MCMC draw in order to produce Figure 7.8.

7.5.2.4 MCMC estimation for parameterisations based on the noncentred disturbances

Finally, we discuss MCMC estimation for the parameterisation that is based on the noncentred disturbances.

(a) Sample a path of $\tilde{X} = (\beta_0^\star, w_1, \ldots, w_N)$ conditional on known parameters using the disturbance simulation smoother of de Jong and Shephard (1995) and Durbin and Koopman (2002).
(b) Sample the parameters $(\mu, \sigma_w, \phi, \sigma_\varepsilon^2)$ jointly conditional on $\tilde{X}$.

Note that under this parameterisation ϕ is no longer independent of $(\mu, \sigma_w, \sigma_\varepsilon^2)$ conditional on $\tilde{X}$ as in the previous subsection. Hence this setup differs from that in the previous subsection. For joint sampling of $(\mu, \sigma_w, \phi, \sigma_\varepsilon^2)$ we use a Metropolis–Hastings algorithm. First, we use a proposal for ϕ that is either independent of all parameters or depends only on the old value of ϕ. For nearly oversized models ($\sigma_w^2 = 0.001$ in our examples) we prefer to sample from the prior on ϕ, whereas for all other models we use a random walk proposal. Conditional on ϕ, we propose $(\mu, \sigma_w, \sigma_\varepsilon^2)$ from 'regression model' (7.30) as in the previous subsection.

Again, σ_w acts as a regression coefficient in (7.30). As in the previous subsection, we do not constrain σ_w to be positive, leaving the whole model unidentified. To produce posterior densities of σ_w, we again conclude each MCMC draw with a random sign switch.

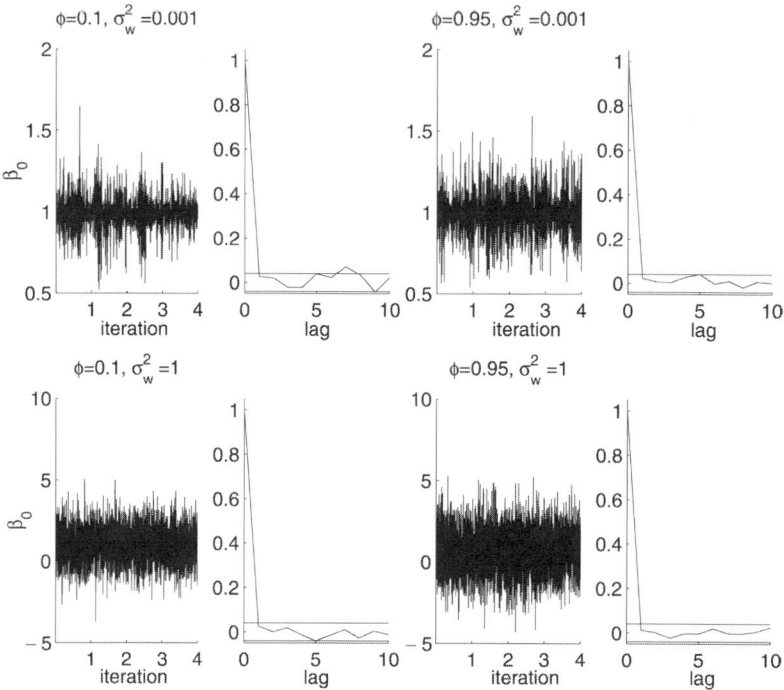

Fig. 7.9. Data simulated from model (7.1) and (7.2) with $\phi = 0.1$ (left hand side) and $\phi = 0.95$ (right hand side); $\sigma_w^2 = 0.001$ (top) and $\sigma_w^2 = 1$ (bottom); 4000 MCMC draws of β_0 with empirical autocorrelation obtained for the preferred parameterisation.

7.5.3 Sampling the unobserved initial state

All MCMC schemes discussed so far consider the unobserved initial parameter β_0 (or appropriate transformations of β_0) as part of the missing data $\tilde{X}$. Alternatively, we may view β_0 as part of the model parameter θ, and sample $\tilde{X}$ conditional on β_0. We found, however, that sampling β_0 (or appropriate transformations of β_0) as part of the missing data $\tilde{X}$ is the most efficient strategy over the whole range of parameters considered in this paper, see also Figure 7.9.

7.5.4 Blocking or transforming model parameters

Step (b) of the MCMC sampler discussed in Sections 7.5.2.1–7.5.2.4 is based on sampling the parameters $(\mu, \sigma_w^2, \phi, \sigma_\varepsilon^2)$ (or transformations of these parameters) jointly within one block rather than separately in two or more blocks. The reason for this is that we might have to deal with considerable or even extreme posterior correlation among the parameters, see for instance Figure 7.10 and Figure 7.11. In this case blocking the parameters is able to prevent slow mixing due to high correlation between parameters in different

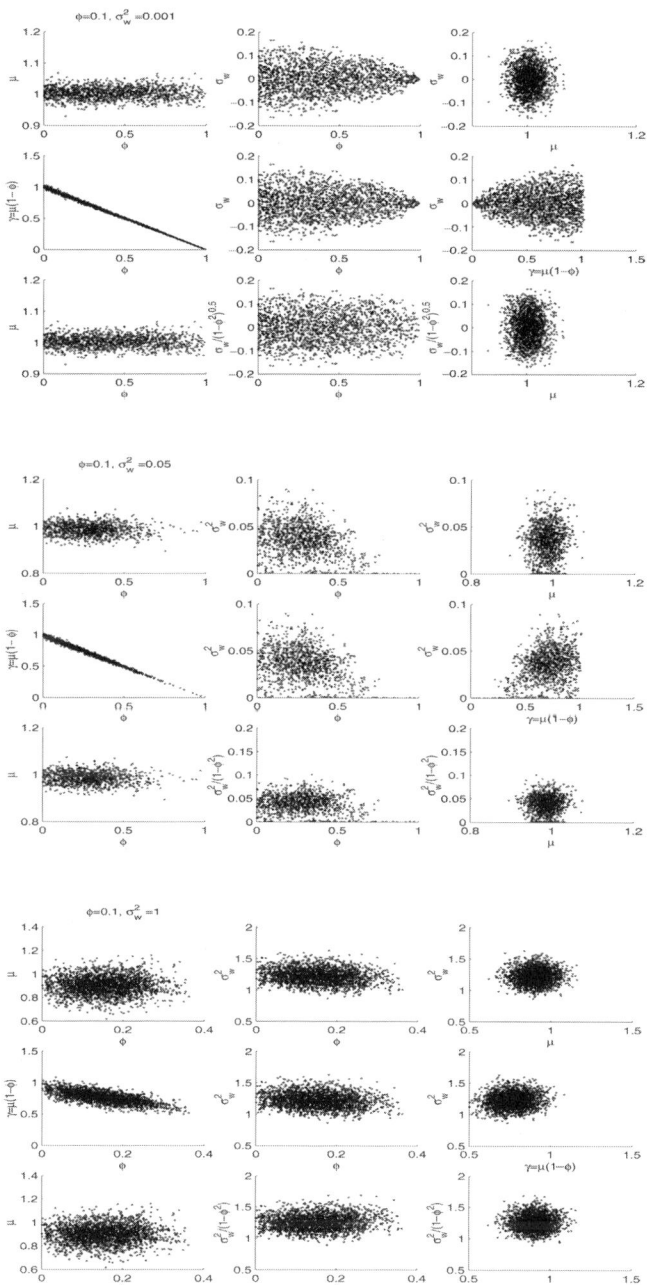

Fig. 7.10. Data simulated from model (7.1) and (7.2) with $\phi = 0.1$ and $\sigma_w^2 = 0.001$ (top), $\sigma_w^2 = 0.05$ (middle) and $\sigma_w^2 = 1$ (bottom); scatter plots of the MCMC draws under different definitions of model parameters

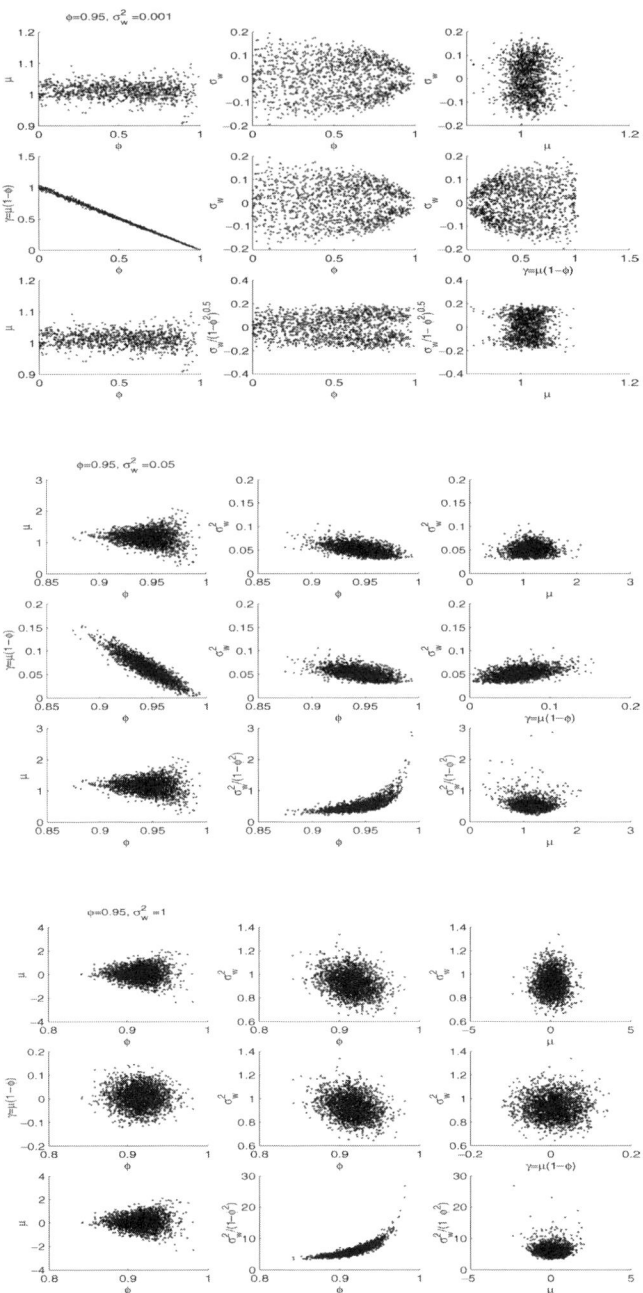

Fig. 7.11. Data simulated from model (7.1) and (7.2) with $\phi = 0.95$ and $\sigma_w^2 = 0.001$ (top), $\sigma_w^2 = 0.05$ (middle) and $\sigma_w^2 = 1$ (bottom); scatter plots of the MCMC draws under different definitions of model parameters

blocks. Under normal disturbances as in (7.1) and (7.2), tailor-made proposals are available that capture potentially strong posterior correlation as long as the model parameters are sampled jointly. Under nonnormal disturbances, however, it is much harder to find a suitable proposal density. It is therefore worthwhile to consider transformations of the parameters of the hidden model that prevent strong posterior correlations.

Starting from the 'natural' model parameter $\theta = (\mu, \sigma_w^2, \phi)$, we considered a transformed model parameter, called θ^C, that is based on the conditional moments of the shocks, $\theta^C = (\gamma, \sigma_w^2, \phi)$, where $\gamma = \mu(1 - \phi)$, as well as another transformed model parameter, called θ^U, that is based on the unconditional moments of the state process, $\theta^U = (\mu, \sigma_\beta^2, \phi)$, where $\sigma_\beta^2 = \sigma_w^2/(1 - \phi^2)$. Figures 7.10 and 7.11 show that transforming the model parameters may have a considerable influence on the shape of the posterior distribution.

For oversized models where the noncentred parameterisation is valid near independence is achieved by selecting the unconditional model parameter θ^U. Under the natural model parameter θ, the parameters ϕ and σ_w^2 suffer from posterior correlation, whereas under the conditional model parameter θ^C, the parameters ϕ and γ are highly correlated. This suggests combining the noncentred parameterisation with the unconditional model parameter θ^U.

For highly persistent time series, where the centred parameterisation is valid, posterior densities with elliptical contours are achieved by selecting the conditional model parameter θ^C. Single-move sampling should be avoided for ϕ and γ; however, the larger σ_w^2 the smaller the posterior correlation. Under the natural parameter θ, the posterior of μ strongly depends on how close ϕ is to the unit root, whereas under the unconditional model parameter θ^U the joint marginal posterior of ϕ and σ_β^2 is banana-shaped. This suggests combining the centred parameterisation with the conditional model parameter θ^C.

7.6 Concluding remarks

Although the discussion here has been confined entirely to the univariate time varying parameter model (7.1) and (7.2), the results seem to be of interest also for more general state space models.

Many other state space models, such as the dynamic factor model or stochastic volatility models, are based on hidden stationary processes and are formulated in such a way that the model parameters governing the hidden process appear only in the transition equation. It appears natural to call such a parameterisation centred also for more general state space model. MCMC estimation under the centred parameterisation considers the hidden process as missing data and uses a sampler that draws the missing data conditional on the model parameters and the model parameters conditional on the missing data.

Sampling a parameter that appears only in the transition equation of the state space model will not cause any problems as long as the randomness of the hidden process is actually of relevance for explaining the variance of the observed process. If the assumed randomness of the hidden process is of no relevance for explaining the variance of the observed process, for instance if the variance of the process is very close to 0, then selecting this process as missing data leads to a sampler that is poorly mixing for the parameters in the transition equation.

A very useful technique is then noncentring of the model in those parameters for which MCMC sampling based on the centred parameterisation is inefficient. Noncentring a model in a certain parameter means choosing the missing data $\tilde{X}$ in such a way that this model parameter no longer appears in the transition equation or equivalently in the prior distribution of the missing data, but is moved to the likelihood of the data given $\tilde{X}$. We have demonstrated in this paper that this may lead to considerable improvement of the resulting MCMC sampler in all cases where the centred parameterisation is performing in a poor way. Although noncentred parameterisations are rather useful, especially for nearly oversized models, the following aspects should be kept in mind.

First, model parameters that are not correlated in cases where the centred parameterisation is valid, may be highly correlated under the noncentred parameterisation especially if the model is oversized and one or more parameters are nearly unidentified. It is therefore often insufficient to substitute the Gibbs sampler that causes problems for a certain parameter under the centred parameterisation simply by a single-move Metropolis–Hastings algorithm under the noncentred parameterisation.

Second, if the randomness of the true underlying hidden process is of considerable relevance for explaining the variance of the observed process, the noncentred parameterisation will be rather poor and does not converge geometrically for the limiting case where all variance of the observed process stems from the hidden process. Furthermore, the range of variance parameters over which the noncentred parameterisation is better than the centred one depends very much on the autocorrelation of the hidden process. For a highly persistent hidden process which is close to a unit root, this range is restricted to processes with extremely small variance. The more one moves away from the unit root process, the bigger is the range of variance parameters for which the noncentred parameterisation is valid. Thus, in a setting with multivariate hidden factors, it might be necessary to use a hybrid parameterisation where a centred parameterisation for a highly persistent and/or highly random factor is combined with a noncentred parameterisation for a nearly constant factor with small variance.

Third, it is still unclear how to extend the results of this paper to state space models with nonnormal disturbances. Papaspiliopoulos, Roberts and Sköld (2003) showed for a random-effects model with all model parameters but μ being known that the centred parameterisation does not converge geometrically if the normal disturbances in the observation equation (7.4) are substituted by disturbances from a Cauchy distribution. On the other hand, the parameterisation which is noncentred in the location does not converge geometrically if this time the normal disturbances in the transition equation (7.4) are substituted by disturbances from a Cauchy distribution. This result would favor the centred parameterisation whenever the tails of the transition equation are fat compared with the tails of the observation distribution.

Finally, for nonstationary state space models such as the local level model or the basic structural model, the conditions (if any) under which the centred parameterisation fails and the way in which a noncentred parameterisation may be obtained still need to be explored.

8

Empirical Bayesian inference in a nonparametric regression model

Gary Koop

Department of Economics, University of Leicester

Dale Poirier

Department of Economics, University of California at Irvine

Abstract

The Normal linear regression model with natural conjugate prior offers an attractive framework for carrying out Bayesian analysis of non- or semi-parametric regression models. The points on the unknown nonparametric regression line can be treated as unobserved components. Prior information on the degree of smoothness of the nonparametric regression line can be combined with the data to yield a proper posterior, despite the fact that the number of parameters in the model is greater than the number of data points. In this paper, we investigate how much prior information is required in order to allow for empirical Bayesian inference about the nonparametric regression line. To be precise, if η is the parameter controlling the degree of smoothness in the nonparametric regression line, we investigate what the minimal amount of nondata information is such that η can be estimated in an empirical Bayes fashion. We show how this problem relates to the issue of estimation of the error variance in the state equation for a simple state space model. Our theoretical results are illustrated empirically using artificial data.

8.1 Introduction

The Normal linear regression model with natural conjugate prior provides an attractive framework for conducting Bayesian inference in non- or semi-parametric regression models. Not only does this well-understood model provide simple analytical results, it can easily be used as a component of a more complicated model (e.g. a nonparametric probit or tobit model or nonparametric regression with non-Normal errors). Since Bayesian analysis of more complicated models often requires posterior simulation (e.g. Markov chain Monte Carlo, MCMC, algorithms), having analytical results for the basic nonparametric regression component of the MCMC algorithm offers great computational simplifications. In this paper, we show how empirical Bayesian inference in a nonparametric regression model can be carried out using empirical Bayesian methods.

Bayesian inference requires a prior. In the nonparametric regression model, the points on the regression line are treated as unknown parameters. Since the number of parameters in the model exceeds the number of observations, the prior can be important and having a prior which is noninformative in all dimensions is not possible. In previous work (see Koop and Poirier (2003)), we stressed that the key prior hyperparameters related to the degree of smoothness in the nonparametric regression line. Hence, it was often possible to make reasonable choices for prior hyperparameters. If this was not possible, we suggested cross-validation, prior sensitivity analysis or extreme bounds analysis might be appropriate tools. Furthermore, we stressed that these prior choices were analogous to the choices (e.g. of kernel and bandwidth) made with nonparametric kernel methods. Thus, we argued that a subjective Bayesian approach was sensible and that, if it was not, cross-validation could be used to select prior hyperparameters.

In this paper, we take a different viewpoint. Empirical Bayesian methods, where prior hyperparameters are estimated from the data, are popular among researchers who wish to avoid subjective prior elicitation. Here we consider a prior hyperparameter which we call η that controls the degree of smoothness in the nonparametric regression line. We are interested in an empirical Bayesian approach where a value for η is selected which maximises the marginal likelihood. We show first that if noninformative priors for all other parameters in the model are used, then an empirical Bayesian approach is impossible. Formally, the marginal likelihood is maximised by letting $\eta \to \infty$. We then consider the question of what is the minimal amount of prior information necessary to implement an empirical Bayesian approach. We show how the problem is very similar to issues addressed in the state space and unobserved components literatures. In particular, for the Bayesian, the problem is closely related to the fact that improper priors can lead to improper posteriors in the local level model. For the non-Bayesian, the problem is closely related to the initialisation of the Kalman filter. In the state space literature, a simple variant of our model implies that η would be the signal-to-noise ratio. Interpreted as such, estimating it from the data is a logical thing to do.

There are a myriad of related Bayesian nonparametric approaches which can be used to address the issues considered in this paper. However, by staying within the framework of the familiar Normal linear regression model with natural conjugate, we feel our approach is computationally and theoretically simple. Furthermore, the use of empirical Bayesian methods allows us to avoid substantive use of subjective prior information. The reader interested in related Bayesian approaches is referred to (among many others) Green and Silverman (1994) for a discussion of the penalised likelihood approach, Silverman (1985) or Wahba (1983) for a discussion of splines, Smith and Kohn (1996) for a discussion of splines implemented using Bayesian model averaging. The recent Bayesian nonparametric literature has been dominated by various approaches which adopt mixtures of distributions. Escobar and West (1995) and Robert (1996) are good citations from this literature. Dey, Müller and Sinha (1998) is an edited volume which has papers on these and other Bayesian nonparametric approaches.

The paper is organised as follows. The second section describes the basic theoretical results for the case where observations of the explanatory variable are equally spaced (that is, for the case most directly comparable to a state space model in discrete time). The third section illustrates these results using artificial data. The fourth section extends the theoretical results of the second section to the case where observations on the explanatory variables are not equally spaced. The fifth section concludes.

8.2 The nonparametric regression model

To draw out the basic insights and results as simply as possible, we work with the nonparametric regression model involving a single explanatory variable:

$$y_i = f(x_i) + \varepsilon_i, \tag{8.1}$$

where y_i is the dependent variable, x_i is a scalar explanatory variable which is treated nonparametrically, $f(.)$ is an unknown function and ε_i is iid. $N(0, \sigma^2)$ for $i = 1, \ldots, N$. The explanatory variables are either fixed or exogenous. Observations are ordered so that $x_1 \leq x_2 \leq \cdots \leq x_N$. For now we will assume that the observations on the explanatory variables are equally spaced (that is, $x_i - x_{i-1} = c$ for $i = 2, \ldots, N$). This assumption is freed up in Section 8.4.

We note here that many extensions can be handled trivially. For instance, treatment of the partial linear model

$$y_i = z_i \beta + f(x_i) + \varepsilon_i,$$

where z_i is a vector of explanatory variables treated parametrically, adds little complexity to the analysis. Furthermore, if x_i is a vector then, provided a meaningful ordering of the explanatory variables can be found (e.g. based on some distance metric), the analysis proceeds in exactly the same manner as for the scalar case. Koop and Poirier (2003) provided additional details on both these cases (see also Yatchew (1998)).

Define $y = (y_1, \ldots, y_N)'$ and $\varepsilon = (\varepsilon_1, \ldots, \varepsilon_N)'$. Letting $\gamma_i = f(x_i)$ and $\gamma = (\gamma_1, \ldots, \gamma_N)'$, we can write (8.1) as

$$y = \gamma + \varepsilon. \tag{8.2}$$

In words, we are treating the points on the nonparametric regression line as unknown parameters. In Koop and Poirier (2003), we developed a Bayesian approach to inference in this model based on a partially informative natural conjugate prior. In particular, we used the standard noninformative prior for σ^{-2}:

$$p(\sigma^{-2}) \propto \sigma^2. \tag{8.3}$$

For the coefficients in the nonparametric part of the model, we used a partially informative natural conjugate prior (see Poirier (1995, p. 535)) on the mth differences of γ (conditional on σ^2):

$$D\gamma \sim N(0_{N-m}, \sigma^2 V(\eta)), \tag{8.4}$$

where $V(\eta)$ is a positive definite matrix which depends on a hyperparameter η, 0_{N-m} is the $N - m$ vector of zeros and D is the $(N - m) \times N$ mth-differencing matrix given by

$$D = \begin{bmatrix} d_0 & d_1 & \ldots & d_m & 0 & \ldots & 0 \\ 0 & d_0 & d_1 & \ldots & d_m & \ldots & 0 \\ \ldots & \ldots & \ldots & \ldots & \ldots & \ldots & \ldots \\ 0 & \ldots & \ldots & 0 & d_0 & \ldots & d_m \end{bmatrix}, \tag{8.5}$$

where $d_0, \ldots, d_m$ are the appropriate differencing weights. For instance, with first differencing $d_0 = -1$, $d_1 = 1$, with second differencing $d_0 = 1$, $d_1 = -2$, $d_2 = 1$, etc. (see also Akaike (1980)). The prior was motivated as being useful in the case where the researcher had prior information about the degree of smoothness in the nonparametric regression function, but wished to be noninformative about the remaining parameters in the model. The degree of smoothness is defined in relation to mth differences. We showed that, despite the fact that the number of parameters exceeded N and the prior distribution was only $N-m$ dimensional, the resulting posterior was proper. Hence, Bayesian nonparametric regression could be carried out within the framework of the familiar Normal linear regression model with natural conjugate prior and only prior information about the mth difference of $f(x_i)$ is required. Thus, loosely speaking, only prior information on the degree of smoothness in the nonparametric regression line is required.

In our previous work, we discussed various ways of choosing $V(\eta)$ such as cross-validation. We argued that choosing a value for η was analogous to the bandwidth choice in traditional kernel smoothing algorithms. However, in the absence of prior information about η, it is not possible to treat η as an unknown parameter and estimate it since, for reasonable choices of $V(\eta)$, the posterior of η, $p(\eta|y)$, becomes infinite as $\eta \to \infty$. Consider, for instance, the case where $V(\eta) = \eta I_{N-m}$. In this case, an empirical Bayes algorithm which chooses η to maximise the marginal likelihood would always result in a choice of $\eta = \infty$, a value which implies no smoothing at all in the nonparametric regression line. Precise details, including citations which provide a proof of these statements, are provided below.

An alternative method for dealing with the problems of empirical Bayes algorithms is to draw on results from the state space literature. The structure of this nonparametric regression model, including likelihood and prior, is identical to a Normal linear state space model. To focus discussion we will consider the case where prior information is available on first differences (that is, $m = 1$) and $V(\eta) = \eta I_{N-1}$, although the basic results derived in this paper hold for any positive definite $V(\eta)$. In this case, it can be seen that the nonparametric regression model given in (8.1)–(8.5) can be written in the form

$$y_i = \gamma_i + \varepsilon_i \qquad (8.6)$$

and

$$\gamma_{i+1} = \gamma_i + u_i, \qquad (8.7)$$

where, for $i = 1, \ldots, N$, ε_i is iid. $N(0, \sigma^2)$, u_i is iid. $N(0, \eta\sigma^2)$ and ε_i and u_j are uncorrelated contemporaneously and at all leads and lags. This is the familiar Normal local level model commonly used in the state space literature (see, e.g., Durbin and Koopman (2001, Chapter 2) or Harvey (1989, p. 37)). The non-Bayesian econometrician would refer to (8.7) as a transition equation which, coupled with (8.6), constitutes a model for which a likelihood can be written. In contrast, most Bayesians would refer to (8.7) as a hierarchical prior. But, provided proper priors are used, this is merely a semantic distinction, and an identical model will be obtained regardless of which terminology is used (see Bayarri, DeGroot and Kadane (1988)).

State space models are typically used with time series data, but the relationship between state space models and nonparametric kernel regression has been noted before (see, e.g., Ansley and Kohn (1985) and Harvey and Koopman (2000)). In the state space literature, η can be estimated from the data using maximum likelihood methods (see, e.g., Durbin and Koopman (2001, Chapter 7)). The question of why state space modellers using maximum likelihood can estimate η, whereas Koop and Poirier (2003) could not use comparable empirical Bayesian methods is directly related to the role of prior information and forms a convenient starting point for our discussion of the role of prior information in Bayesian nonparametric regression. As we shall see, the answer relates to treatment of γ_1 or, in the language of the state space modeller, the initialisation of the Kalman filter.

There is a large literature related to Bayesian analysis of state space models (see, among many others, Akaike (1980), Carter and Kohn (1994), de Jong and Shephard (1995), Frühwirth-Schnatter (1994b,c), Hickman and Miller (1981), Koop and van Dijk (2000), Kim, Shephard and Chib (1998), Shively and Kohn (1997) and West and Harrison 1997)). With the partial exception of Koop and van Dijk (2000), prior elicitation has not been a central focus of this literature. However, prior elicitation can be important since the use of improper priors can lead to the local level model having an improper posterior. Although Fernández, Osiewalski and Steel (1997) is not a paper which discusses state space or nonparametric regression models, the class of models it does discuss includes them as a special case. The interested reader is referred to their paper for several results. Suffice it to note here that some sorts of improper priors lead to improper posteriors. Posterior impropriety is, of course, implied by an infinite marginal likelihood. Thus, the impropriety of the posterior in the local level model is simply another way of looking at the failure of empirical Bayesian methods in the nonparametric regression model given in (8.2), (8.3) and (8.4).

In this paper we are interested in investigating how much prior information is necessary to allow for valid posterior inference on η or, equivalently, to use empirical Bayesian methods. We find that the amount of necessary prior information is minimal. Of course, in many applications, the researcher may not wish to adopt the minimally informative approach described in this paper. If prior information is available about any or all of the model parameters, then it can be used and the issues addressed in this paper are irrelevant. Alternatively, one of the many approaches in the state space literature relating to initialising the Kalman filter can be used. Even more simply, we could work with differenced data throughout and the issues we discuss would not be relevant (although such a procedure would mean the intercept in the nonparametric regression line would not be identifiable).

In order to draw out the role of prior information in the selection of η, it is useful to focus on the case $m = 1$ (cases where $m > 1$ can be handled as a trivial extension) and parameterise (8.2) as

$$y = W\theta + \varepsilon, \qquad (8.8)$$

where $\theta = (\gamma_1, \gamma_2 - \gamma_1, \ldots, \gamma_N - \gamma_{N-1})'$ and

$$W = \begin{pmatrix} 1 & 0'_{N-1} \\ \iota_{N-1} & D^{-1} \end{pmatrix},$$

where ι_{N-1} is an $(N - 1)$ vector of 1s. For the case under consideration, with $m = 1$, D^{-1} is an $(N - 1) \times (N - 1)$ lower triangular matrix with all nonzero elements equalling one. Note that γ_1 plays the role of an intercept in the nonparametric regression model. An alternative interpretation, relevant if this model were interpreted in a time series context, is that γ_1 plays the role of an initial condition.

Our strategy is to begin with a fully informative prior and then gradually make it noninformative in certain dimensions in order to see how far we can become noninformative, without losing propriety of the posterior. To this end, we assume for the moment a proper natural conjugate, Normal-Gamma prior for θ and σ^{-2} conditional on a known hyperparameter, η. Using the standard notation for the Normal-Gamma (e.g. Poirier (1995, p. 526)), we write this prior as NG($\underline{\theta}, \underline{V}, \underline{s}^{-2}, \underline{\nu}$), where $\underline{\theta}$ is a given $N \times 1$ vector, $\underline{V} = \underline{V}(\eta)$ is a given positive definite matrix which depends on a hyperparameter, η, and the two scalars $\underline{s}^{-2}$ and $\underline{\nu}$ satisfy $\underline{\nu} > 0$ and $\underline{s}^{-2} > 0$. These assumptions, which will be freed up shortly, make the prior proper and informative with $\theta | \sigma^{-2} \sim N(\underline{\theta}, \sigma^2 \underline{V})$ and σ^{-2} having a Gamma distribution with mean $\underline{s}^{-2}$ and variance $2/\underline{\nu} \underline{s}^4$. Using standard results (e.g. Poirier (1995, p. 527)), it follows that the posterior for θ and σ^{-2} given η, denoted by $p(\theta, \sigma^{-2} | y, \eta)$

is $\text{NG}(\underline{\theta}, \underline{V}, \underline{s}^{-2}, \underline{\nu})$, where

$$\overline{\theta} = \overline{V}\left(\underline{V}^{-1}\underline{\theta} + W'y\right), \tag{8.9}$$

$$\overline{V} = \left(\underline{V}^{-1} + W'W\right)^{-1}, \tag{8.10}$$

$$\overline{\nu} = \underline{\nu} + N \tag{8.11}$$

and

$$\overline{\nu s}^2 = \underline{\nu s}^2 + \left(y - W\overline{\theta}\right)'\left(y - W\overline{\theta}\right) + \left(\overline{\theta} - \underline{\theta}\right)'\underline{V}^{-1}\left(\overline{\theta} - \underline{\theta}\right). \tag{8.12}$$

The properties of the Normal-Gamma distribution imply that it is trivial to transform back from the parameterisation in (8.8) to the original parameterisation given in (8.2). That is, if the posterior for $(\theta, \sigma^{-2})\,|\eta$ is $\text{NG}(\overline{\theta}, \overline{V}, \overline{s}^{-2}, \overline{\nu})$ then the posterior for $(\gamma, \sigma^{-2})\,|\eta$ is $\text{NG}(\overline{\gamma}, \overline{V}_\gamma, \overline{s}^{-2}, \overline{\nu})$, where

$$\overline{\gamma} = W\overline{\theta}$$

and

$$\overline{V}_\gamma = W\overline{V}W'.$$

We refer to the posterior mean of γ (that is, $\overline{\gamma}$) as the fitted nonparametric regression line.

It is also worth mentioning that, in this paper, we directly invert the $N \times N$ matrix in (8.10) using a standard matrix inversion subroutine. However, in the state space literature smoothing algorithms exist which implicitly do this inversion in a much more computationally efficient manner.

The marginal likelihood associated with this model takes the form (e.g. Poirier (1995, p. 543)) of a multivariate t-distribution, written

$$t\left(y|\overline{\gamma}, \overline{s}^2\left[I_N + W\overline{V}_\gamma W'\right], \overline{\nu}\right),$$

where

$$p(y|\eta) = c \left(\frac{|\overline{V}|}{|\underline{V}|}\right)^{\frac{1}{2}} \left(\overline{\nu s}^2\right)^{-\frac{\overline{\nu}}{2}}, \tag{8.13}$$

with

$$c \equiv c\left(\underline{s}^{-2}, \underline{\nu}\right) = \frac{\Gamma\left(\overline{\nu}/2\right)\left(\underline{\nu s}^2\right)^{\frac{\underline{\nu}}{2}}}{\Gamma\left(\underline{\nu}/2\right)\pi^{\frac{N}{2}}}. \tag{8.14}$$

It can be directly verified that, despite the fact that the number of parameters in this model is $N+1$ and the number of observations N, the posterior is a proper density and $p(y|\eta)$ is finite. Hence, Bayesian inference in the nonparametric regression model using a proper natural conjugate prior can be carried out in a straightforward manner (e.g. $\overline{\theta}$ can be used to provide an estimate of the fitted nonparametric regression model and $p(y|\eta)$ can be used to compare the nonparametric regression model against a parametric alternative). Empirical Bayesian methods could be used to estimate η in cases where it is unknown. The value of η which maximised $p(y|\eta)$ in (8.13) could then be used as a basis for further Bayesian analysis of the nonparametric regression model.

However, in some cases, the researcher may feel uncomfortable eliciting $\underline{\theta}, \underline{s}^{-2}, \underline{V}$ and $\underline{\nu}$ and the question arises as to what extent we can choose noninformative values for some or all of these prior hyperparameters and still use empirical Bayesian methods for dealing with the unknown η. We have seen above that becoming fully noninformative is unacceptable, but we shall show below that being noninformative in certain directions is acceptable.

We begin by considering a structure for $\underline{V}$ which uses the intuition that neighbouring points on the nonparametric regression line should not be too different from one another. Thus we set

$$\underline{V}(\eta) = \begin{pmatrix} \underline{V}_{11} & 0'_{N-1} \\ 0_{N-1} & V(\eta) \end{pmatrix}, \qquad (8.15)$$

where $V(\eta) = \eta I_{N-1}$ and $\underline{\theta} = (\underline{\theta}_1, 0, \ldots, 0)'$. In other words, this prior implies $\gamma_{i+1} - \gamma_i$ is $N(0, \eta\sigma^2)$ and the hyperparameter η can be used to control the desired degree of smoothness in the nonparametric regression line.

Koop and Poirier (2003) sought to be noninformative over the remaining parameters (that is, the intercept and the error variance, γ_1 and σ^2) and set $\underline{\nu}$ and $\underline{V}_{11}^{-1}$ at their limiting values $\underline{\nu} = \underline{V}_{11}^{-1} = 0$. With these choices, the values of $\underline{s}^2$ and $\underline{\theta}_1$ are irrelevant. For these noninformative choices, Koop and Poirier (2003) showed that $p(\theta, \sigma^{-2}|y, \eta)$ is a well defined posterior. However, with regards to the marginal likelihood, two problems arise. First, the integrating constant in (8.14) is indeterminate. Insofar as interest centres on η, or the marginal likelihood is used for comparing the present model with another with the same noninformative prior for the error variance, this first problem is not a serious one. The constant c either does not enter or cancels out of any derivation and can be ignored. Second, the term $\overline{\nu s}^2$ goes to zero as $\eta \to \infty$. To see this, note that with all the hyperparameters set to noninformative values $\overline{\theta} = (W'W)^{-1}W'y$ and $y - W\overline{\theta} = 0_N$. Iwata (1996, Lemma 3, p. 246) provided a proof that this degeneracy is enough to overcome the countervailing tendency for $|V(\eta)^{-1}|$ to go to zero and implies that the marginal likelihood in (8.13) becomes infinite as $\eta \to \infty$.

One way to carry out empirical Bayesian estimation of η is to use a hierarchical prior for η. If η is treated as an unknown parameter, then Bayes theorem implies $p(\eta|y) \propto p(y|\eta)p(\eta)$, where $p(\eta)$ is the prior and we can write

$$p(\eta|y) \propto c \left(\frac{|\overline{V}|}{|V|}\right)^{\frac{1}{2}} \left(\overline{\nu s^2}\right)^{-\frac{\overline{\nu}}{2}} p(\eta), \tag{8.16}$$

and we are assuming that, conditional on θ and σ^{-2}, η and y are independent (that is, $\eta \perp\!\!\!\perp y|\theta, \sigma^{-2}$). If a Uniform prior, unbounded above, for η were used, then $p(\eta|y)$ would become infinite as $\eta \to \infty$ if $\underline{\nu} = \underline{V}_{11}^{-1} = 0$. However, informative priors for η which either rule out the $\eta \to \infty$ case or go to zero very quickly as $\eta \to \infty$ will rule out the pathology caused by the noninformative prior and allow for posterior inference on η. For instance, a Gamma prior for η will imply that (8.16) is the kernel of a valid posterior density even if $\underline{\nu} = \underline{V}_{11}^{-1} = 0$. Thus, even a small amount of prior information about η will allow us to treat η as an unknown parameter and obtain a valid posterior for it. If $p(\eta|y)$ is well defined, we can either use this marginal posterior to obtain an empirical Bayes estimate of η (e.g. $\hat{\eta} = \arg\max[p(\eta|y)]$) or do a fully Bayesian analysis of $p(\theta, \sigma^{-2}, \eta|y)$. The latter can be done in a straightforward fashion using simple Monte Carlo integration. That is,

$$p(\theta, \sigma^{-2}, \eta|y) = p(\theta, \sigma^{-2}|y, \eta) p(\eta|y)$$

and $p(\theta, \sigma^{-2}|y, \eta)$ is Normal-Gamma (see (8.9)–(8.12)) and $p(\eta|y)$ is one dimensional. Thus, drawing from $p(\eta|y) \propto p(y|\eta)p(\eta)$ and, conditional upon this draw, drawing from $p(\theta, \sigma^{-2}|y, \eta)$ yields a draw from the joint posterior.

However, prior information on η is not the only path which allows for valid posterior inference on this parameter. Prior information on γ_1 (which is θ_1 in the parameterisation of (8.8)) or σ^2 also can be used to yield a proper $p(\eta|y)$, even if $p(\eta)$ is a Uniform improper prior (or, equivalently, the marginal likelihood in (8.16) will be finite). In particular, if $\underline{V}_{11}^{-1} > 0$ or $\underline{\nu} > 0$ then $\overline{\nu s^2} > 0$ even if $\eta \to \infty$. The proof of these statements follows from direct examination of all quantities in (8.9)–(8.12).

In summary, in the natural conjugate framework for nonparametric regression, prior information on the degree of smoothness in the nonparametric regression line is enough to allow for valid posterior inference. However, we require more prior information if we wish to estimate the degree of smoothness parameter, η, rather than simply select a value subjectively or use cross-validation. In order to ensure a proper posterior for η we need prior information on η or prior information on the initial condition, γ_1, or prior information on σ^2. *We stress that only one of these three possible sources of prior information is required.*

How does this relate to what is done by state space modellers when they estimate η in the local level model? They make assumptions which are of the same nature as our prior choices relating to γ_1. However, they do not always explicitly refer to this as prior elicitation, but rather *initialisation of the Kalman filter*. For instance, Durbin and Koopman (2001, Chapter 7) assumed the initial condition, γ_1, is $N(\underline{\theta}_1, \underline{V}_{11})$ (using our notation) and noted that if $\underline{\theta}_1$ and $\underline{V}_{11}$ are known then the likelihood function is well behaved (that is, prior information on the initial condition is available which allows for valid likelihood-based inference on θ, σ^{-2} and η). However, they pointed out that the likelihood function will not converge as $\underline{V}_{11} \to \infty$. To counter this problem they suggested working with what they called the *diffuse loglikelihood* which adds to the loglikelihood an extra penalty term involving $\underline{V}_{11}$. Such a penalty is mathematically identical to a prior for γ_1. That is, instead of having a Normal prior for γ_1 and letting $\underline{V}_{11} \to \infty$, the diffuse loglikelihood is equivalent to using a prior for γ_1 which is Normal times the penalty term with $\underline{V}_{11} \to \infty$ (see also Ansley and Kohn (1985)). An alternative approach (e.g. Harvey (1989, pp. 121–2)) is simply to choose a large but finite value for $\underline{V}_{11}$. All such approaches relate to prior elicitation for γ_1, but as we have noted above prior information on η or σ^2 can also be used to obtain a proper pdf for η.

8.3 Empirical illustration

The theoretical results in the previous section show that a fully noninformative empirical Bayesian analysis of the nonparametric regression model is impossible. However, prior information about one or more of the smoothing parameter (η), the initial condition (γ_1) or the error variance (σ^2) can be used to prevent the marginal likelihood from becoming infinite. The latter theoretical result might be of little practical importance if a large amount of prior information about these parameters is required, or if the fitted nonparametric regression line is very sensitive to prior information or if the fitted nonparametric regression line has some other undesirable property. These issues can best be addressed empirically in the context of a particular data set. Accordingly, in order to illustrate the role of various types of prior information in empirical Bayesian nonparametric regression methods, we simulate a single data set with $N = 100$ observations from

$$y_i = x_i \cos(4\pi x_i) + \varepsilon_i, \qquad (8.17)$$

where ε_i is iid $N(0, 0.09)$ and $x_i = i/N$ (that is, observations on the explanatory variable are an equally spaced grid between 0 and 1) for $i = 1, \ldots, N$. This data generating process is the almost same as that used in Koop and Poirier (2003) and Yatchew (1998, Figure 3).

Throughout this section, the prior and posterior densities will be

$$\mathrm{NG}(\underline{\theta}, \underline{V}, \underline{s}^{-2}, \underline{\nu})$$

and

$$\mathrm{NG}(\overline{\theta}, \overline{V}, \overline{s}^{-2}, \overline{\nu}),$$

respectively, where the posterior arguments are defined in (8.9)–(8.12). In all cases, we will assume $\underline{V}$ has the form given in (8.15). We use a Gamma prior for η with mean $\underline{\mu}_\eta$ and degrees of freedom $\underline{\nu}_\eta$. That is

$$p(\eta) = f_G\left(\eta | \underline{\mu}_\eta, \underline{\nu}_\eta\right). \qquad (8.18)$$

Empirical Bayesian estimation involves treating η as an unknown parameter and estimating it by maximising (8.16). Here this maximisation is done using a grid search. In improper, noninformative cases, we follow the standard practice of simply ignoring the corresponding term in the integrating constant given in (8.14) as it is irrelevant in the context of empirical Bayesian estimation of η (see the discussion in the paragraph after (8.15)). Noninformative prior choices for the parameters σ^{-2}, γ_1 and η are $\underline{\nu} = 0$, $\underline{V}_{11}^{-1} = 0$ and $\underline{\nu}_\eta = 0$, respectively. Throughout this section, we set $\underline{\theta}_1 = 0$, $\underline{\mu}_\eta = 1$ and $\underline{s}^2 = 0.09$, values which become irrelevant for the noninformative cases.

We begin by carrying out an empirical Bayesian analysis of the fully noninformative case where $\underline{\nu} = 0$, $\underline{V}_{11}^{-1} = 0$ and $\underline{\nu}_\eta = 0$. Of course, theoretical considerations imply that the marginal likelihood should become infinite as $\eta \to \infty$ and, thus, the empirical Bayes estimate of η, denoted by $\widehat{\eta}$, is ∞. However, an examination of the shape of $p(\eta|y)$ is quite informative. Figure 8.1, which plots the log of $p(\eta|y)$, reveals an interior mode. Furthermore, $p(\eta|y)$ is increasing with η in such a manner as to eventually become infinite. However, $p(\eta|y)$ is increasing very gradually with η, suggesting that only a small amount of prior information may be necessary to ensure that the interior mode dominates. Note that Figure 8.1 is truncated at $\eta = 1000$.

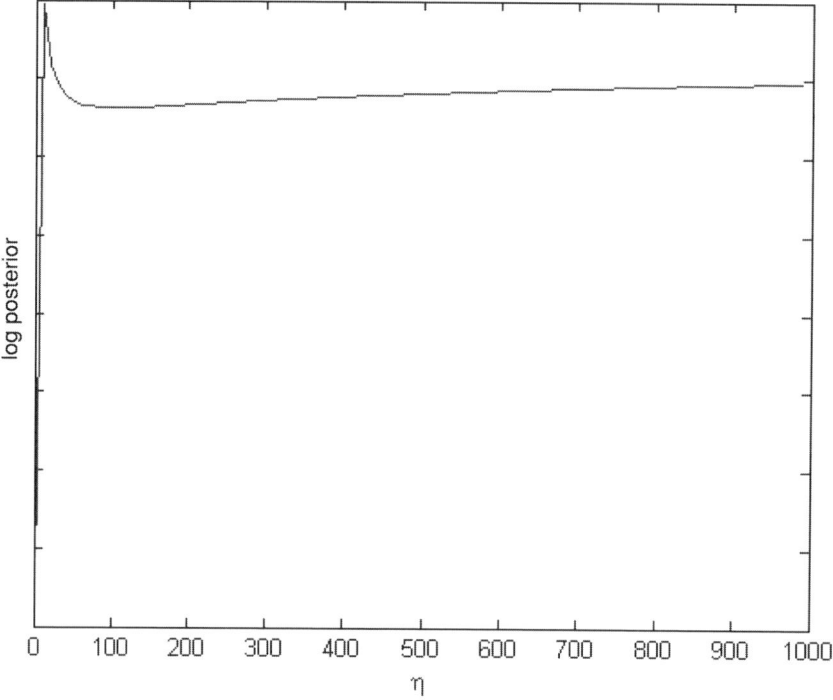

Fig. 8.1. Log of posterior for η with noninformative prior.

Our theoretical discussion implies that prior information about one of σ^{-2}, γ_1 and η is necessary to ensure $p(\eta|y)$ is a proper pdf and, thus, that meaningful empirical Bayesian inference in the nonparametric regression model is possible. The researcher who dislikes prior information would be interested in finding out what is the minimal amount of prior information possible. To this end, we put priors which are proper but relatively noninformative on the three relevant parameters, one at a time. To be precise, the solid line in Figure 8.2 presents the log of $p(\eta|y)$ where $\underline{\nu} = 0$ and $\underline{V}_{11}^{-1} = 0$ but we have a weakly informative prior for η. That is, we set $\underline{\nu}_\eta = 1$, a relatively noninformative choice. With this prior, $\widehat{\eta} = 0.31$. The line with long dashes in Figure 8.2 presents the log of $p(\eta|y)$ for the case where only weak prior information about the initial condition, γ_1, is used. In particular, we set the prior variance of the initial condition to be 10 (that is, $\underline{V}_{11} = 10$). With this prior, $\widehat{\eta} = 0.35$. The line with short dashes in Figure 8.2 presents the log of $p(\eta|y)$ for the case where only weak prior information about the error variance, σ^2, is used. In particular, we set $\underline{\nu} = 1$. Since $\underline{\nu}$ can be interpreted informally as a prior 'sample size', the choice of $\underline{\nu} = 1$ implies the prior contains much less information than the data. With this prior, $\widehat{\eta} = 0.35$, as well.

The results in Figures 8.1 and 8.2 provide us with a clear interpretation.

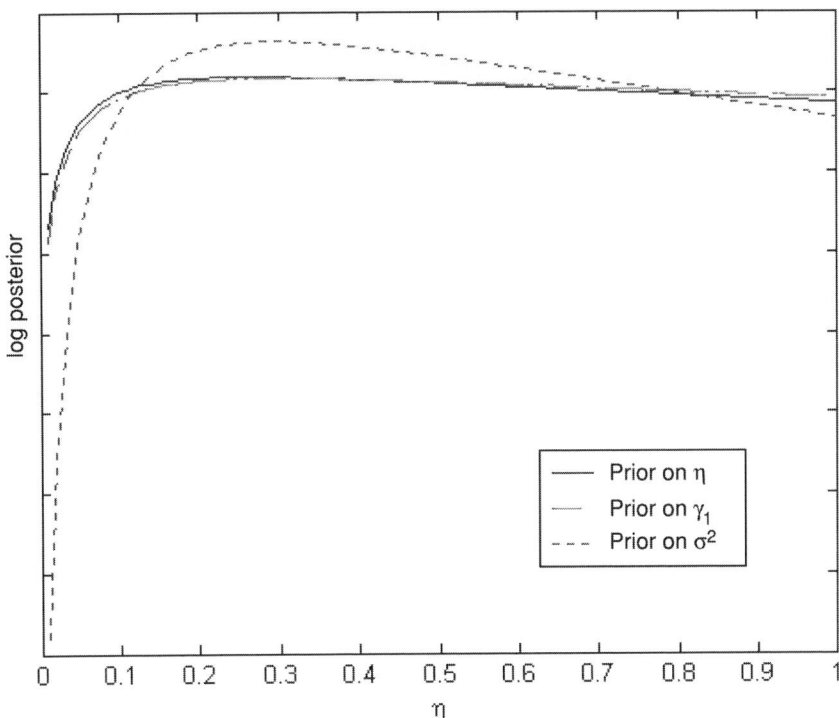

Fig. 8.2. Log of posterior for η with informative prior.

Empirical Bayesian analysis of the nonparametric smoothing parameter, η, is impossible using a prior which is fully noninformative, since the marginal likelihood goes to infinity as η goes to infinity. However, at least for this data set, an interior mode does exist and the increase of the marginal likelihood is very gradual. This suggests that only a very small amount of prior information is necessary to correct this pathology. Figure 8.2 indicates that this is the case. Very weak prior information about η or the initial condition or the error variance is all that is required to make the interior mode noted in Figure 8.1 predominant. Thus, a finite empirical Bayes estimate of η is obtained. Furthermore, $\hat{\eta}$ is roughly the same in all cases. That is, weak prior information about three different parameters leads to empirical Bayes estimates which are basically the same, indicating a high degree of prior robustness.

Figure 8.3 plots the data set, the actual regression line from (8.17) along with the fitted nonparametric regression line (that is, the posterior mean of γ) for the case where weakly informative prior information is used about the initial condition (that is, the case used to produce the line with short dashes in Figure 8.2 which had $\hat{\eta} = 0.29$, the other cases yield essentially the same result). It can be seen that our empirical Bayesian nonparametric regression procedure works well, with the fitted nonparametric regression line tracking the true regression line quite well. Note that this success is achieved even though a moderate degree of error has been added in the data generating process. Remember that we are smoothing the nonparametric regression line by using prior information on first differences, $\gamma_i - \gamma_{i-1}$. This is done so that we can illustrate basic concepts in the simplest way possible. However, smoothness is often defined in terms of second differences. The researcher who finds the slight irregularities in Figure 8.3 disturbing would find that putting a prior on second differences yields a smoother fitted nonparametric regression line.

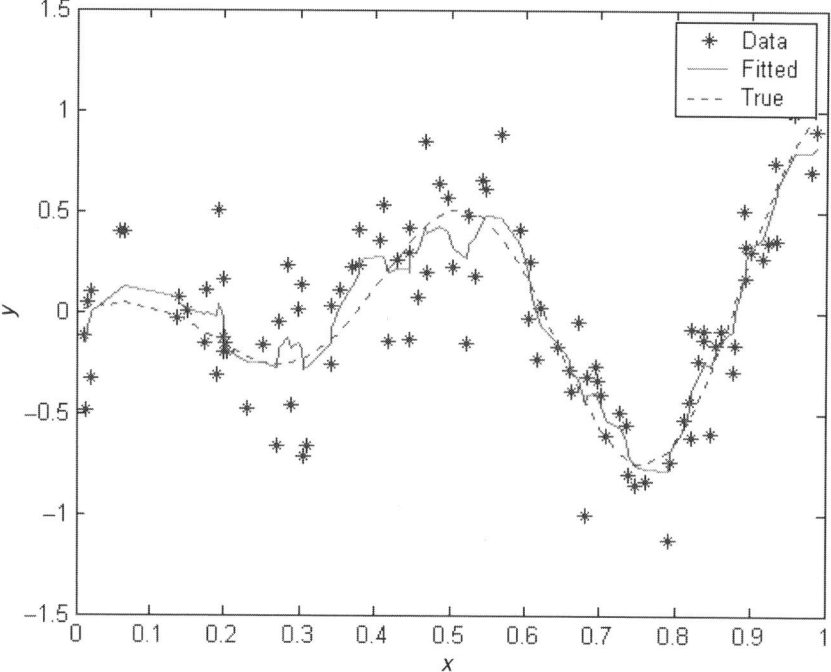

Fig. 8.3. Fitted and true nonparametric regression line for $\eta = 0.29$.

Another question of possible interest is how sensitive empirical Bayes estimates are to prior information. Figure 8.2 suggests a high degree of prior robustness. To investigate this issue more deeply, we calculated empirical Bayes estimates for a wide range of priors on the initial condition. We considered a grid of values of $\underline{V}_{11}$ over the huge interval $[0.001, 1 \times 10^6]$. We do not provide a graph of the empirical Bayes estimates of η for the various prior variances, since it is effectively a horizontal line. In fact, the line begins at $\widehat{\eta} = 0.28$ for very tight priors (that is, $\eta = 0.001$) before rising to $\widehat{\eta} = 0.30$ at $\eta = 0.01$ and staying at this value for the rest of the interval. Thus, empirical Bayesian estimation appear to be very robust to changes in the prior for the initial condition. A similar robustness was found using the weakly informative priors for either η or σ^2, although results are not presented here for the sake of brevity.

These results are similar in spirit to those found in the non-Bayesian state space literature (see the discussion at the end of the previous section). It seems that only a small (even minuscule) degree of nondata information is required to offset the pathology in the likelihood function which occurs as $\eta \to \infty$. However, it is worth stressing that the state space literature focuses on nondata information about the initial condition in the state equation. Here we show that nondata information about η or σ^2 can also be used to remove the pathology.

We have, of course, framed our discussion in terms of the nonparametric regression model. But, since this is equivalent to a state space model, we note that our results are of relevance for the Bayesian state space literature.

8.4 The nonparametric regression model with irregularly spaced explanatory variables

Above we have assumed that the x_is are equally spaced. This assumption would usually hold with time series data, but is unlikely to hold otherwise. Accordingly, it is useful to briefly discuss methods for relaxing this assumption. The bottom line is that several different methods in which one might treat irregularly spaced data can be shown to be very similar to one another. In terms of the notation of Section 8.2, these methods change the model only through specifying somewhat different forms for $V(\eta)$. The proofs of the results in Section 8.2 relied only on $V(\eta)$ being positive definite. Provided this is the case (as it is for the suggestions below), the extension to allow for irregular spacing of the explanatory variables does not change any of the issues discussed above with regards to the role of prior information in nonparametric regression. Accordingly, in this section we offer only a brief, largely verbal discussion, of this issue.

The simplest manner to deal with irregularly spaced data points is to incorporate the distance between the points into the prior. This amounts to saying the degree of smoothing should depend on the distance between points. If x_i and x_{i+1} are very close to one another, then we would expect $f(x_i)$ and $f(x_{i+1})$ to be very close to one another. If x_i and x_{i+1} are far apart, then $f(x_i)$ and $f(x_{i+1})$ might be much farther apart. This information can be easily incorporated in our setup by defining some measure of distance (e.g. $(x_{i+1} - x_i)^2$) and then having this affect the degree of smoothing. A simple way of doing this is defining everything as in Section

8.2 except for $V(\eta)$. The latter is modified so that

$$v_i = \eta\,(x_i - x_{i-1}),$$

where v_i is the (i,i)th element of $V(\eta)$. Since the data are ordered so that $x_1 \leq x_2 \leq \cdots \leq x_N$, v_i is always nonnegative. If the explanatory variable is continuously distributed, then $x_i \neq x_j$ for $i \neq j$ and $V(\eta)$ is guaranteed to be positive definite in theory. When working with an actual data set, it is possible that $x_i = x_j$ and some correction would have to be added to all the approaches described in this section. Simple *ad hoc* corrections such as perturbing x_j by a very small amount should have a minimal effect on the empirical results.

An alternative way of approaching this problem is to draw on the spline literature (e.g. Wahba (1983), Silverman (1985) and Green and Silverman (1994)) where the knots are given by $x_1, \ldots, x_N$. Green and Silverman (1994, Chapter 3) showed that, in this case, a natural cubic spline is equivalent to a Bayesian approach involving a Normal likelihood identical to ours and a prior given by

$$p(\gamma) \propto \exp\left(-\frac{\eta}{\sigma^2}\gamma' Q \gamma\right)$$

for a certain matrix Q which is of rank $N - 2$. This approach interprets smoothness of the nonparametric regression line in terms of second derivatives and is very similar to our approach with $m = 2$. In fact, if we had used $m = 2$ and the fully noninformative prior given in (8.3) and (8.4), then our prior covariance matrix for γ would also be of rank $N - 2$. Our prior covariance matrix has similar properties (although it is not identical to that used in the natural cubic spline literature) and the same considerations apply. To be precise, the Bayesian formulation of the natural cubic spline literature as outlined in Green and Silverman (1994) does not allow for an empirical Bayesian estimation of the smoothing parameter η. In fact, Green and Silverman (1994) recommend plugging in an estimate of σ^2 and choosing η through cross-validation. However, the results of Section 8.2 imply σ^2 can be integrated out analytically and that empirical Bayesian methods can be used to estimate η if an informative prior is used for η, the error variance or the initial conditions. In summary, spline methods can be used to treat irregularly spaced data. Interpreted in a Bayesian fashion, they motivate a particular prior which falls in the category of those considered in Section 8.2 and, thus, all the issues raised in Section 8.2 are relevant.

An alternative way of dealing with irregularly spaced data is to draw on ideas from the state space literature. In this context, data which are recorded at irregular time periods are treated by working in continuous time.

To be precise, a continuous time model is derived with values observed at discrete (irregularly spaced) time periods. However, it has been shown in the literature that the resulting model is equivalent to a spline model (see, e.g., Durbin and Koopman (2001, pp. 57–63)). Accordingly, the issues relating to a continuous time formulation are equivalent to those discussed in the preceding paragraph.

Yet another approach is due to O'Hagan (1978). This uses a prior covariance matrix which implies prior correlations between points on the nonparametric regression line that depend on the distance they are from one another (that is, prior correlations depend on $|x_i - x_j|$ for $i \neq j$). We do not provide further details here, but note only that all of these approaches implicitly or explicitly handle irregularly spaced data by allowing the distance between observations to enter $V(\eta)$ (or $\underline{V}$). Thus, they all fall into the framework of Section 8.2 and issues relating to empirical Bayesian analysis are the same as discussed previously.

8.5 Conclusions and discussion

In this paper, we have presented a general framework for carrying out Bayesian nonparametric regression. The advantage of this framework is its theoretical (as compared to much of the frequentist nonparametric regression literature) and computational (relative to alternative Bayesian approaches which require posterior simulation) simplicity. It is based solely on the Normal linear regression model with natural conjugate prior. The focus of the present paper is on the role of prior information. Of critical importance is a prior hyperparameter we call η which controls the degree of smoothness of the nonparametric regression line. If we select a particular value for η, then valid posterior inference can be carried out, even if we use improper, noninformative priors over all other parameters in the model. However, with such a fully noninformative prior, we cannot do an empirical Bayesian analysis. That is, we must choose a value for η; we cannot estimate it from the data. We then show what kind of prior information is required in order to estimate η. We show how information on the error variance or the initial condition or η itself is enough to allow for an empirical Bayesian analysis to be carried out. An empirical illustration using artificial data indicates that the amount of prior information required is minimal.

It is worth stressing that, in many possible applications of nonparametric methods, the nonparametric regression model discussed in this paper would form a single component of a larger models. Examples include nonparametric censored data (e.g. tobit) or nonparametric qualitative choice (e.g. probit) models. Bayesian inference in such models typically uses computationally-intensive MCMC methods. Having analytical results relating to the nonparametric component of the MCMC algorithm would yield substantial computational savings. Furthermore, the focus of the present paper is on empirical Bayesian methods for nonparametric regression which seek to find $\widehat{\eta}$ which maximises $p(\eta|y)$, where the latter depends on the marginal likelihood of the model (see (8.16)). For many of the more complicated models one might be interested in, evaluation of the marginal likelihood is quite difficult. Hence, one might expect it would be difficult to carry out empirical Bayesian analysis in these more complicated models. However, many of them (e.g. nonparametric probit or tobit) have a structure which depends upon latent data which we will denote by l (e.g. the latent utilities in the probit model or the unknown values of observations which were censored in the tobit model). For such models, $p(\eta|y,l)$ will take on the simple form described in Section 8.2 (suitably augmented with latent data). Thus, $p(\eta|y,l)$ can be handled in the context of the MCMC algorithm and empirical Bayesian inference carried out in a straightforward manner. So, for instance, our framework implies that empirical Bayesian analysis of a nonparametric probit model is only slight more difficult than a traditional analysis of a parametric probit model.

Acknowledgements

We would like to thank participants at the conference in honour of Professor J. Durbin on state space models and unobserved components and, in particular, Bill Bell, Aart de Vos, Andrew Harvey and Peter Young for their helpful comments.

9
Resampling in state space models

David S. Stoffer
Department of Statistics, University of Pittsburgh

Kent D. Wall
Defense Resources Management Institute, Naval Postgraduate School

Abstract

Resampling the innovations sequence of state space models has proved to be a useful tool in many respects. For example, while under general conditions, the Gaussian MLEs of the parameters of a state space model are asymptotically normal, several researchers have found that samples must be fairly large before asymptotic results are applicable. Moreover, problems occur if any of the parameters are near the boundary of the parameter space. In such situations, the bootstrap applied to the innovation sequence can provide an accurate assessment of the sampling distributions of the parameter estimates. We have also found that a resampling procedure can provide insight into the validity of the model. In addition, the bootstrap can be used to evaluate conditional forecast errors of state space models. The key to this method is the derivation of a reverse-time innovations form of the state space model for generating conditional data sets. We will provide some theoretical insight into our procedures that shows why resampling works in these situations, and we provide simulations and data examples that demonstrate our claims.

State Space and Unobserved Component Models: Theory and Applications, eds. Andrew C. Harvey, Siem Jan Koopman and Neil Shephard. Published by Cambridge University Press. © Cambridge University Press 2004

9.1 Introduction

A very general model that seems to subsume a whole class of special cases of interest is the state space model or the dynamic linear model, which was introduced in Kalman (1960) and Kalman and Bucy (1961). Although the model was originally developed as a method primarily for use in aerospace-related research, it has been applied to modelling data from such diverse fields as economics (e.g. Harrison and Stevens (1976), Harvey and Pierse (1984), Harvey and Todd (1983), Kitagawa and Gersch (1984), Shumway and Stoffer (1982)), medicine (e.g. Jones (1984)) and molecular biology (e.g. Stultz, White and Smith (1993)). An excellent modern treatment of time series analysis based on the state space model is the text by Durbin and Koopman (2001). We note, in particular, that autoregressive moving average models with exogenous variables (ARMAX) can be written in state space form (see e.g. Shumway and Stoffer (2000, Section 4.6)), so anything we say and do here regarding state space models applies equally to ARMAX models.

Here, we write the state space model as

$$\mathbf{x}_{t+1} = \Phi \mathbf{x}_t + \Upsilon \mathbf{u}_t + \mathbf{w}_t, \qquad t = 0, 1, \ldots, n, \tag{9.1}$$

$$\mathbf{y}_t = A_t \mathbf{x}_t + \Gamma \mathbf{u}_t + \mathbf{v}_t, \qquad t = 1, \ldots, n, \tag{9.2}$$

where $\mathbf{x}_t$ represents the p-dimensional state vector, and $\mathbf{y}_t$ represents the q-dimensional observation vector. In the state equation (9.1), the initial state $\mathbf{x}_0$ has mean μ_0 and variance-covariance matrix Σ_0; Φ is $p \times p$, Υ is $p \times r$, and $\mathbf{u}_t$ is an $r \times 1$ vector of fixed inputs. In the observation equation (9.2), A_t is $q \times p$ and Γ is $q \times r$. Here, $\mathbf{w}_t$ and $\mathbf{v}_t$ are white noise series (both independent of $\mathbf{x}_0$), with $\text{var}(\mathbf{w}_t) = Q$, $\text{var}(\mathbf{v}_t) = R$, but we also allow the state noise and observation noise to be correlated at time t; that is, $\text{cov}(\mathbf{w}_t, \mathbf{v}_t) = S$, and zero otherwise. Note, S is a $p \times q$ matrix. Throughout, we assume the model coefficients and the correlation structure of the model are uniquely parameterised by a $k \times 1$ parameter vector Θ; thus, $\Phi = \Phi(\Theta)$, $\Upsilon = \Upsilon(\Theta)$, $Q = Q(\Theta)$, $A_t = A_t(\Theta)$, $\Gamma = \Gamma(\Theta)$, $R = R(\Theta)$ and $S = S(\Theta)$.

We denote the best linear predictor of $\mathbf{x}_{t+1}$ given the data $\{\mathbf{y}_1, \ldots, \mathbf{y}_t\}$ as $\mathbf{x}_{t+1}^t$, and denote the covariance matrix of the prediction error, $(\mathbf{x}_{t+1} - \mathbf{x}_{t+1}^t)$, as P_{t+1}^t. The Kalman filter (e.g. Anderson and Moore (1979)) can be used to obtain the predictors and their covariance matrices successively as new observations become available. The innovation sequence, $\{\epsilon_t; t = 1, \ldots, n\}$, is defined to be the sequence of errors in the best linear prediction of $\mathbf{y}_t$ given the data $\{\mathbf{y}_1, \ldots, \mathbf{y}_{t-1}\}$. The innovations are

$$\epsilon_t = \mathbf{y}_t - A_t \mathbf{x}_t^{t-1} - \Gamma \mathbf{u}_t, \qquad t = 1, \ldots, n, \tag{9.3}$$

where the innovation variance-covariance matrix is given by

$$\Sigma_t = A_t P_t^{t-1} A_t' + R, \qquad t = 1, \ldots, n. \tag{9.4}$$

The innovations form of the Kalman filter, for $t = 1, \ldots, n$, is given by the following equations with initial conditions $\mathbf{x}_1^0 = \Phi \mu_0 + \Upsilon \mathbf{u}_0$ and $P_1^0 = \Phi \Sigma_0 \Phi' + Q$:

$$\mathbf{x}_{t+1}^t = \Phi \mathbf{x}_t^{t-1} + \Upsilon \mathbf{u}_t + K_t \epsilon_t, \tag{9.5}$$

$$P_{t+1}^t = \Phi P_t^{t-1} \Phi' + Q - K_t \Sigma_t K_t', \tag{9.6}$$

$$K_t = (\Phi P_t^{t-1} A_t' + S) \Sigma_t^{-1}. \tag{9.7}$$

In this article, we will work with the standardised innovations

$$\mathbf{e}_t = \Sigma_t^{-1/2} \epsilon_t, \tag{9.8}$$

so we are guaranteed these innovations have, at least, the same first two moments. In (9.8), $\Sigma_t^{1/2}$ denotes the unique square root matrix of Σ_t defined by $\Sigma_t^{1/2} \Sigma_t^{1/2} = \Sigma_t$. We now define the $(p+q) \times 1$ vector

$$\xi_t = \begin{bmatrix} \mathbf{x}_{t+1}^t \\ \mathbf{y}_t \end{bmatrix}.$$

Combining (9.3) and (9.5) results in a vector first-order equation for ξ_t given by

$$\xi_\mathbf{t} = F_t \xi_{t-1} + G \mathbf{u}_t + H_t \mathbf{e}_t, \tag{9.9}$$

where

$$F_t = \begin{bmatrix} \Phi & 0 \\ A_t & 0 \end{bmatrix}, \quad G = \begin{bmatrix} \Upsilon \\ \Gamma \end{bmatrix}, \quad H_t = \begin{bmatrix} K_t \Sigma_t^{1/2} \\ \Sigma_t^{1/2} \end{bmatrix}.$$

Estimation of the model parameters Θ is accomplished by Gaussian quasi-maximum likelihood. The innovations form of the Gaussian likelihood (ignoring a constant) is

$$\begin{aligned} -\ln L_Y(\Theta) &= \frac{1}{2} \sum_{t=1}^n \left[\ln |\Sigma_t(\Theta)| + \epsilon_t(\Theta)' \Sigma_t(\Theta)^{-1} \epsilon_t(\Theta) \right] \\ &= \frac{1}{2} \sum_{t=1}^n \left[\ln |\Sigma_t(\Theta)| + \mathbf{e}_t(\Theta)' \mathbf{e}_t(\Theta) \right], \end{aligned} \tag{9.10}$$

where $L_Y(\Theta)$ denotes the likelihood of Θ given the data $\mathbf{y}_1, \ldots, \mathbf{y}_n$ assuming normality; note that we have emphasised the dependence of the innovations on the parameters Θ. We stress the fact that it is not necessary for the data to be Gaussian to consider (9.10) as the criterion function to be used for parameter estimation. Furthermore, under certain rare conditions, the Gaussian quasi-MLE of Θ when the process is non-Gaussian is asymptotically optimal; details can be found in Caines (1988, Chapter 8).

9.2 Assessing the finite sample distribution of parameter estimates

Although, under general conditions (which we assume to hold in this section), the MLEs of the parameters of the model, Θ, are consistent and asymptotically normal, time series data are often of short or moderate length. Several researchers have found evidence that samples must be fairly large before asymptotic results are applicable (e.g. Dent and Min (1978) and Ansley and Newbold (1980)). Moreover, it is well known that problems occur if the parameters are near the boundary of the parameter space. In this section, we discuss an algorithm for bootstrapping state space models to assess the finite sample distribution of the model parameters. This algorithm and its justification, including the non-Gaussian case, along with examples and simulations, can be found in Stoffer and Wall (1991).

Let $\widehat{\Theta}$ denote the Gaussian quasi-MLE of Θ, that is, $\widehat{\Theta} = \operatorname{argmax}_\Theta L_Y(\Theta)$, where $L_Y(\Theta)$ is given in (9.10); of course, if the process is Gaussian, $\widehat{\Theta}$ is the MLE. Let $\epsilon_t(\widehat{\Theta})$ and $\Sigma_t(\widehat{\Theta})$ be the innovation values obtained by running the filter under $\widehat{\Theta}$. Once this has been done, the bootstrap procedure is accomplished by the following steps:

(i) Construct the standardised innovations

$$\mathbf{e}_t(\widehat{\Theta}) = \Sigma_t^{-1/2}(\widehat{\Theta})\epsilon_t(\widehat{\Theta}).$$

(ii) Sample, with replacement, n times from the set $\{\mathbf{e}_1(\widehat{\Theta}), \ldots, \mathbf{e}_n(\widehat{\Theta})\}$ to obtain $\{\mathbf{e}_1^*, \ldots, \mathbf{e}_n^*\}$, a bootstrap sample of standardised innovations.

(iii) To construct a bootstrap data set $\{\mathbf{y}_1^*, \ldots, \mathbf{y}_n^*\}$, solve (9.9) using $\mathbf{e}_t^*$ in place of $\mathbf{e}_t$; that is, solve

$$\xi_t^* = F_t(\widehat{\Theta})\xi_{t-1}^* + G(\widehat{\Theta})\mathbf{u}_t + H_t(\widehat{\Theta})\mathbf{e}_t^*, \qquad (9.11)$$

for $t = 1, \ldots, n$. The exogenous variables $\mathbf{u}_t$ and the initial conditions of the Kalman filter remain fixed at their given values, and the parameter vector is held fixed at $\widehat{\Theta}$. Note that a bootstrapped observation $\mathbf{y}_t^*$ is obtained from the final q rows of the $(p+q) \times 1$ vector ξ_t^*. Because of startup irregularities, it is sometimes a good idea to set $\mathbf{y}_t^* \equiv \mathbf{y}_t$ for the first few values of t, say $t = 1, 2, \ldots, t_0$, where t_0 is small, and to sample from $\{\mathbf{e}_{t_0+1}(\widehat{\Theta}), \ldots, \mathbf{e}_n(\widehat{\Theta})\}$. That is, do not bootstrap the first few data points; typically setting t_0 to 4 or 5 will suffice.

(iv) Using the bootstrap data set $\{\mathbf{y}_t^*;\ t = 1,\ldots,n\}$, construct a likelihood, $L_{Y^*}(\Theta)$, and obtain the MLE of Θ, say, $\widehat{\Theta}^*$.

(v) Repeat steps (ii)–(iv), a large number, B, of times, obtaining a bootstrapped set of parameter estimates $\{\widehat{\Theta}_b^*;\ b = 1,\ldots,B\}$. The finite sample distribution of $(\widehat{\Theta} - \Theta)$ may be approximated by the distribution of $(\widehat{\Theta}_b^* - \widehat{\Theta})$, for $b = 1,\ldots,B$.

9.2.1 Stochastic regression

An interesting application of the state space model was given in Newbold and Bos (1985, pp. 61–73). Of the several alternative models they investigated, we focus on the one specified by their equations (4.7a) and (4.7b). Their model had one output variable, the nominal interest rate recorded for three-month treasury bills, y_t. The output equation is specified by

$$y_t = \alpha + \beta_t z_t + v_t,$$

where z_t is the quarterly inflation rate in the Consumer Price Index, α is a fixed constant, β_t is a stochastic regression coefficient and v_t is white noise with variance σ_v^2. The stochastic regression term, which comprises the state variable, is specified by a first-order autoregression,

$$(\beta_{t+1} - b) = \phi(\beta_t - b) + w_t,$$

where b is a constant, and w_t is white noise with variance σ_w^2. The noise processes, v_t and w_t, are assumed to be uncorrelated. Using the notation of the state space model (9.1) and (9.2), we have in the state equation, $\mathbf{x}_t = \beta_t$, $\Phi = \phi$, $\mathbf{u}_t \equiv 1$, $\Upsilon = (1-\phi)b$, $Q = \sigma_w^2$, and in the observation equation, $A_t = z_t$, $\Gamma = \alpha$, $R = \sigma_v^2$, and $S = 0$. The parameter vector is $\Theta = (\phi, \alpha, b, \sigma_w, \sigma_v)'$.

We consider the first estimation exercise reported in Table 4.3 of Newbold and Bos. This exercise covers the period from the first quarter of 1953 through the second quarter of 1965, $n = 50$ observations. We repeat their analysis so our results can be compared to their results. In addition, we focus on this analysis because it demonstrates that the bootstrap applied to the innovation sequence can provide an accurate assessment of the sampling distributions of the parameter estimates when analysing short time series. Moreover, this analysis demonstrates that a resampling procedure can provide insight into the validity of the model.

The results of the Newton–Raphson estimation procedure are listed in

Table 9.1. The MLEs obtained in Newbold and Bos (1985) are in agreement with our values, and differ only in the fourth decimal place; the differences are attributed to the fact that we use a different numerical optimisation routine. Included in Table 9.1 are the asymptotic standard errors reported in Newbold and Bos (1985). Also shown in the Table 1 are the corresponding standard errors obtained from $B = 500$ runs of the bootstrap. These standard errors are simply the square root of $\sum_{b=1}^{B}(\widehat{\Theta}_{ib}^{*} - \widehat{\Theta}_{i})^{2}/(B-1)$, where Θ_i, represents the ith parameter, $i = 1, \ldots, 5$, and $\widehat{\Theta}_i$ is the MLE of Θ_i.

Table 9.1. *Comparison of asymptotic standard errors (SE) and bootstrapped standard errors (B = 500)*

parameter	MLE	asymptotic SE	Newbold & Bos SE	bootstrap SE
ϕ	0.841	0.200	0.212	0.304
α	−0.771	0.645	0.603	0.645
b	0.858	0.278	0.259	0.277
σ_w	0.127	0.092	NA	0.182
σ_v	1.131	0.142	NA	0.217

The asymptotic standard errors listed in Table 9.1 are typically smaller than those obtained from the bootstrap. This result is the most pronounced in the estimates of ϕ, σ_w and σ_v, where the bootstrapped standard errors are about 50% larger than the corresponding asymptotic value. Also, asymptotic theory prescribes the use of normal theory when dealing with the parameter estimates. The bootstrap, however, allows us to investigate the small sample distribution of the estimators and, hence, provides more insight into the data analysis.

For example, Figure 9.1 shows the bootstrap distribution of the estimator of ϕ. This distribution is highly skewed with values concentrated around 0.8, but with a long tail to the left. Some quantiles of the bootstrapped distribution of ϕ are −0.09 (2.5%), 0.03 (5%), 0.16 (10%), 0.87 (90%), 0.92 (95%), 0.94 (97.5%), and they can be used to obtain confidence intervals. For example, a 90% confidence interval for ϕ would be approximated by (0.03, 0.92). This interval is rather wide, and we will interpret this after we discuss the results of the estimation of σ_w.

Figure 9.2 shows the bootstrap distribution of the estimator of σ_w. The distribution is concentrated at two locations, one at approximately $\widehat{\sigma}_w^* = 0.15$ and the other at $\widehat{\sigma}_w^* = 0$. The cases in which $\widehat{\sigma}_w^* \approx 0$ correspond to deterministic state dynamics. When $\sigma_w = 0$ and $|\phi| < 1$, then $\beta_t \approx b$ for large t, so the approximately 25% of the cases in which $\widehat{\sigma}_w^* \approx 0$ suggest a fixed

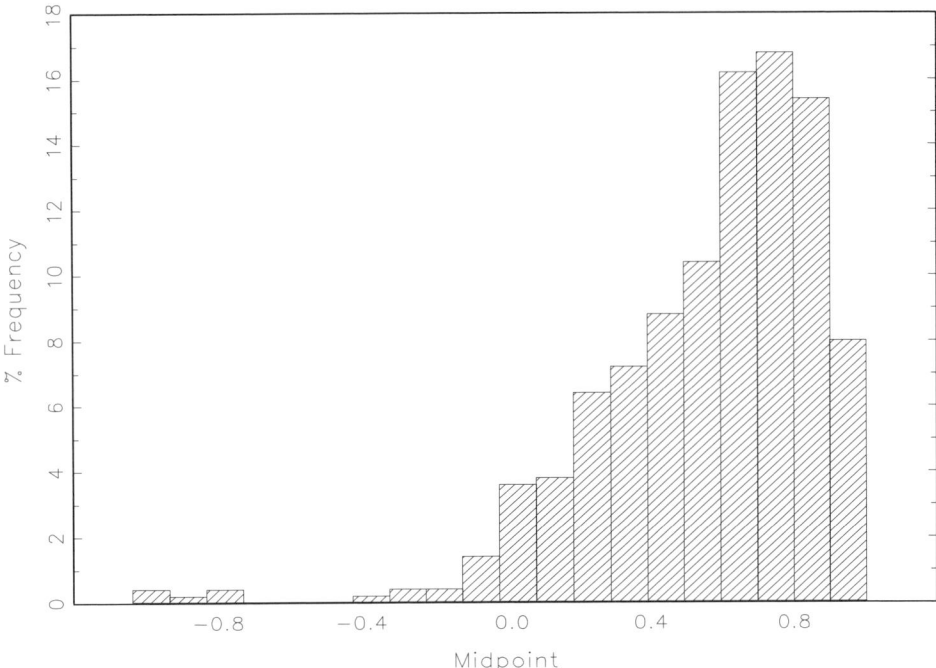

Fig. 9.1. Bootstrap distribution, $B = 500$, of the estimator of ϕ.

state, or constant coefficient model. The cases in which $\widehat{\sigma}_w^*$ is away from zero would suggest a truly stochastic regression parameter. To investigate this matter further, Figure 9.3 shows the joint bootstrapped estimates, $(\widehat{\phi}^*, \widehat{\sigma}_w^*)$, for nonnegative values of $\widehat{\phi}^*$. The joint distribution suggests $\widehat{\sigma}_w^* > 0$ corresponds to $\widehat{\phi}^* \approx 0$. When $\phi = 0$, the state dynamics are given by $\beta_t = b + w_t$. If, in addition, σ_w is small relative to b (as it appears to be in this case), the system is nearly deterministic; that is, $\beta_t \approx b$. Considering these results, the bootstrap analysis leads us to conclude the dynamics of the data are best described in terms of a *fixed*, rather than stochastic, regression effect.

If, however, we use the same model for the entire data set presented in Newbold and Bos (1985) (that is, 110 quarters of three-month treasury bills and inflation rate, covering 1953:I to 1980:II), stochastic regression appears to be appropriate. In this case the estimates using Newton–Raphson with estimated standard errors ('asymptotic' | 'bootstrap') are:

$$\widehat{\phi} = 0.896 \ (0.067 \mid 0.274), \quad \widehat{\alpha} = -0.970 \ (0.475 \mid 0.538),$$

$$\widehat{b} = 1.090 \ (0.158 \mid 0.221),$$

$$\widehat{\sigma}_w = 0.117 \ (0.037 \mid 0.122), \quad \widehat{\sigma}_v = 1.191 \ (0.108 \mid 0.171).$$

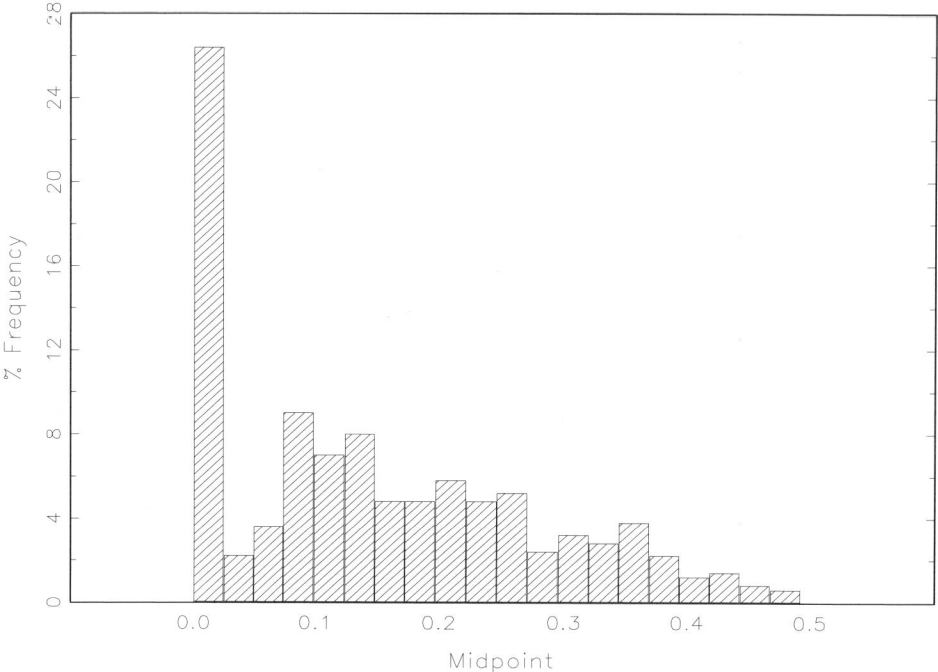

Fig. 9.2. Bootstrap distribution, $B = 500$, of the estimator of σ_w.

We note that the asymptotic standard error estimates are still too small, and the bootstrapped distribution of $\widehat{\phi}$ is still markedly skewed. In particular, a 90% bootstrap confidence interval for ϕ is (0.46, 0.92).

9.2.2 Stochastic volatility

This problem is somewhat different than the previous section in that it is not a straightforward application of the algorithm. In this example, we consider the stochastic volatility model discussed in Harvey, Ruiz and Shephard (1994). Let r_t denote the return or growth rate of a process of interest. For example, if s_t is the value of a stock at time t, the return or relative gain of the stock is $r_t = \ln(s_t/s_{t-1})$. Typically, it is the time changing variance of the returns that is of interest. In the stochastic volatility model, we model $h_t = \ln \sigma_t^2$ as an AR(1), that is,

$$h_{t+1} = \phi_0 + \phi_1 h_t + w_t, \qquad (9.12)$$

where w_t is white Gaussian noise with variance σ_w^2; this comprises the state equation. The observations are taken to be $y_t = \ln r_t^2$, and y_t is related to

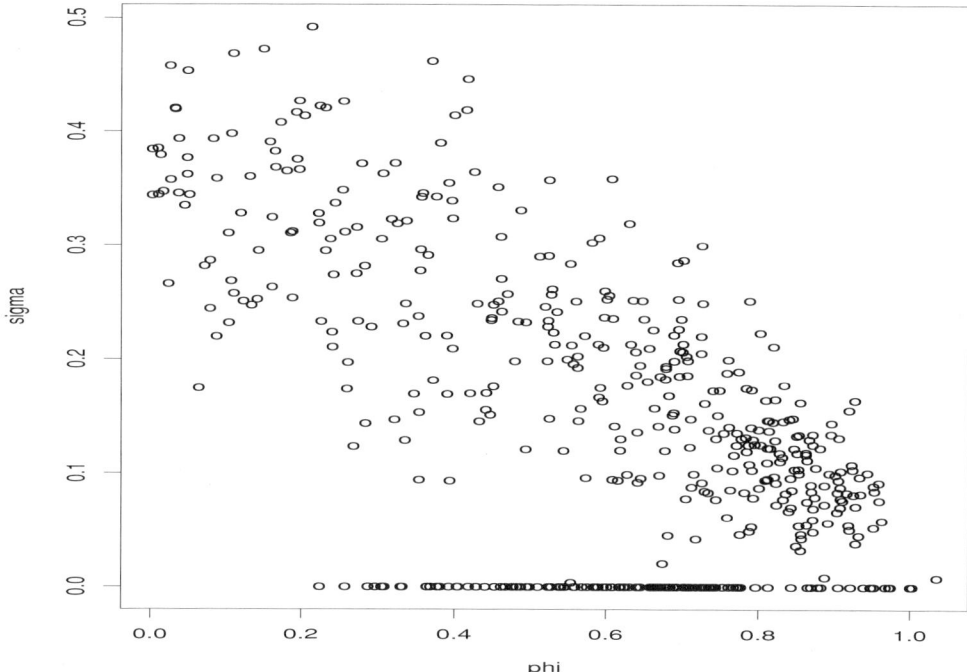

Fig. 9.3. Joint bootstrap distribution, $B = 500$, of the estimators of ϕ and σ_w. Only the values corresponding to $\widehat{\phi}^* \geq 0$ are shown.

the state via

$$y_t = \alpha + h_t + v_t. \qquad (9.13)$$

Together, (9.12) and (9.13) make up the stochastic volatility model, where h_t represents the unobserved volatility of the process y_t. If v_t were Gaussian white noise, (9.12)–(9.13) would form a Gaussian state space model, and we could then use standard results to fit the model to data. Unfortunately, $y_t = \ln r_t^2$ is rarely normal, so one typically assumes that $v_t = \ln z_t^2$, where z_t is standard Gaussian white noise. In this case, $\ln z_t^2$ is distributed as the log of a chi-squared random variable with one degree of freedom. Kim, Shephard and Chib (1998) proposed modelling the log of a chi-squared random variable by a mixture of normals.

Various approaches to the fitting of stochastic volatility models have been examined; these methods include a wide range of assumptions on the observational noise process. A good summary of the proposed techniques, both Bayesian (via MCMC) and non-Bayesian approaches (such as quasi-maximum likelihood estimation and the EM algorithm), can be found in Jacquier, Polson and Rossi (1994), and Shephard (1996). Simulation methods for classical inference applied to stochastic volatility models are discussed in Danielsson (1994) and Sandmann and Koopman (1998).

In an effort to keep matters simple, our method (see Shumway and Stoffer (2000, Section 4.10)) of fitting stochastic volatility models is to retain the Gaussian state equation, (9.12), but in the observation equation, (9.13), we consider v_t to be white noise, and distributed as a mixture of two normals, one centred at zero. In particular, we write

$$v_t = (1 - \eta_t) z_{t0} + \eta_t z_{t1}, \tag{9.14}$$

where η_t is an iid Bernoulli process, $\Pr\{\eta_t = 0\} = \pi_0$, $\Pr\{\eta_t = 1\} = \pi_1$, with $\pi_0 + \pi_1 = 1$, and where $z_{t0} \sim$ iid $N(0, \sigma_0^2)$, and $z_{t1} \sim$ iid $N(\mu_1, \sigma_1^2)$.

The advantage of this model is that it is fairly easy to fit because it uses normality. The model specified by (9.12)–(9.14), and the corresponding filter are similar to those presented in Peña and Guttman (1988), who used the idea to obtain a robust Kalman filter, and, as previously mentioned, Kim, Shephard and Chib (1998). In addition, this technique is similar to the technique discussed in Shumway and Stoffer (2000, Section 4.8). In particular, the filtering equations for this model are:

$$h_{t+1}^t = \phi_0 + \phi_1 h_t^{t-1} + \sum_{j=0}^{1} \pi_{tj} K_{tj} \epsilon_{tj}, \tag{9.15}$$

$$P_{t+1}^t = \phi_1^2 P_t^{t-1} + \sigma_w^2 - \sum_{j=0}^{1} \pi_{tj} K_{tj}^2 \Sigma_{tj}, \tag{9.16}$$

$$\epsilon_{t0} = y_t - \alpha - h_t^{t-1}, \tag{9.17}$$

$$\epsilon_{t1} = y_t - \alpha - h_t^{t-1} - \mu_1, \tag{9.18}$$

$$\Sigma_{t0} = P_t^{t-1} + \sigma_0^2, \tag{9.19}$$

$$\Sigma_{t1} = P_t^{t-1} + \sigma_1^2, \tag{9.20}$$

$$K_{t0} = \phi_1 P_t^{t-1} / \Sigma_{t0}, \tag{9.21}$$

$$K_{t1} = \phi_1 P_t^{t-1} / \Sigma_{t1}. \tag{9.22}$$

To complete the filtering, we must be able to assess the probabilities $\pi_{t1} = \Pr(\eta_t = 1 | y_1, \ldots, y_t)$ for $t = 1, \ldots, n$; of course, $\pi_{t0} = 1 - \pi_{t1}$. Let $f_j(t|t-1)$ denote the conditional density of y_t given the past $y_1, \ldots, y_{t-1}$, and $\eta_t = j$ ($j = 0, 1$). Then,

$$\pi_{t1} = \frac{\pi_1 f_1(t \mid t-1)}{\pi_0 f_0(t \mid t-1) + \pi_1 f_1(t \mid t-1)}, \tag{9.23}$$

where we assume the distribution π_j, for $j = 0, 1$ has been specified *a priori*.

If the investigator has no reason to prefer one state over another the choice of uniform priors, $\pi_1 = 1/2$, will suffice. Unfortunately, it is computationally difficult to obtain the exact values of $f_j(t|t-1)$; although we can give an explicit expression of $f_j(t|t-1)$, the actual computation of the conditional density is prohibitive. A viable approximation, however, is to choose $f_j(t|t-1)$ to be the normal density, $N(h_t^{t-1} + \mu_j, \Sigma_{tj})$, for $j = 0, 1$ and $\mu_0 = 0$; see Shumway and Stoffer (2000, Section 4.8) for details.

The innovations filter given in (9.15)–(9.23) can be derived from the Kalman filter by a simple conditioning argument. For example, to derive (9.15), we write

$$E\left(h_{t+1}|y_1,\ldots,y_t\right) = \sum_{j=0}^{1} E\left(h_{t+1}|y_1,\ldots,y_t, \eta_t = j\right) \Pr(\eta_t = j|y_1,\ldots,y_t)$$

$$= \sum_{j=0}^{1} \left(\phi_0 + \phi_1 h_t^{t-1} + K_{tj}\epsilon_{tj}\right) \pi_{tj}$$

$$= \phi_0 + \phi_1 h_t^{t-1} + \sum_{j=0}^{1} \pi_{tj} K_{tj} \epsilon_{tj}.$$

Estimation of the parameters $\Theta = (\phi_0, \phi_1, \sigma_0^2, \mu_1, \sigma_1^2, \sigma_w^2)'$ is accomplished via MLE based on the likelihood given by

$$\ln L_Y(\Theta) = \sum_{t=1}^{n} \ln \left(\sum_{j=0}^{1} \pi_j\, f_j(t|t-1) \right), \qquad (9.24)$$

where the densities for $f_j(t|t-1)$ are approximated by the normal densities previously mentioned.

To perform the bootstrap, we develop a vector first-order equation, as was done in (9.9). First, using (9.17)–(9.18), and noting that $y_t = \pi_{t0} y_t + \pi_{t1} y_t$, we may write

$$y_t = \alpha + h_t^{t-1} + \pi_{t0}\epsilon_{t0} + \pi_{t1}(\epsilon_{t1} + \mu_1). \qquad (9.25)$$

Consider the standardised innovations

$$e_{tj} = \Sigma_{tj}^{-1/2} \epsilon_{tj}, \qquad j = 0, 1, \qquad (9.26)$$

and define the 2×1 vector

$$\mathbf{e}_t = \begin{bmatrix} e_{t0} \\ e_{t1} \end{bmatrix}.$$

Also, define the 2×1 vector

$$\xi_t = \begin{bmatrix} h_{t+1}^t \\ y_t \end{bmatrix}.$$

Combining (9.15) and (9.25) results in a vector first-order equation for ξ_t given by

$$\xi_t = F\xi_{t-1} + G_t + H_t \mathbf{e}_t, \qquad (9.27)$$

where

$$F = \begin{bmatrix} \phi_1 & 0 \\ 1 & 0 \end{bmatrix}, \quad G_t = \begin{bmatrix} \phi_0 \\ \alpha + \pi_{t1}\mu_1 \end{bmatrix}, \quad H_t = \begin{bmatrix} \pi_{t0} K_{t0} \Sigma_{t0}^{1/2} & \pi_{t1} K_{t1} \Sigma_{t1}^{1/2} \\ \pi_{t0} \Sigma_{t0}^{1/2} & \pi_{t1} \Sigma_{t1}^{1/2} \end{bmatrix}.$$

Hence, the steps in bootstrapping for this case are the same as steps (i)–(v) previously described, but with (9.11) replaced by the following first-order equation:

$$\xi_t^* = F(\widehat{\Theta})\xi_{t-1}^* + G_t(\widehat{\Theta}; \widehat{\pi}_{t1}) + H_t(\widehat{\Theta}; \widehat{\pi}_{t1})\mathbf{e}_t^*, \qquad (9.28)$$

where $\widehat{\Theta} = (\widehat{\phi}_0, \widehat{\phi}_1, \widehat{\sigma}_0^2, \widehat{\alpha}, \widehat{\mu}_1, \widehat{\sigma}_1^2, \widehat{\sigma}_w^2)'$ is the MLE of Θ, and $\widehat{\pi}_{t1}$ is estimated via (9.23), replacing $f_1(t|t-1)$ and $f_0(t|t-1)$ by their respective estimated normal densities ($\widehat{\pi}_{t0} = 1 - \widehat{\pi}_{t1}$).

To examine the efficacy of the bootstrap for the stochastic volatility model, we generated $n = 200$ observations from the following stochastic volatility model:

$$h_t = 0.95 h_{t-1} + w_t, \qquad (9.29)$$

where w_t is white Gaussian noise with variance $\sigma_w^2 = 1$. The observations were then generated as

$$y_t = h_t + v_t, \qquad (9.30)$$

where the observational white noise process, v_t, is distributed as the log of a chi-squared random variable with one degree of freedom. The density of v_t is given by

$$f_v(x) = \frac{1}{\sqrt{2\pi}} \exp\left\{-\frac{1}{2}(e^x - x)\right\} \quad -\infty < x < \infty, \qquad (9.31)$$

and its mean and variance are -1.27 and $\pi^2/2$, respectively; the density (9.31) is highly skewed with a long tail on the left. The data are shown in Figure 9.4. Then, we assumed the true error distribution was unknown to us, and we fitted the model (9.12)–(9.14) using the Gauss BFGS variable metric algorithm to maximise the likelihood. The results for the state parameters are given in Table 9.2 in the columns marked *MLE* and *asymptotic SE*. Next,

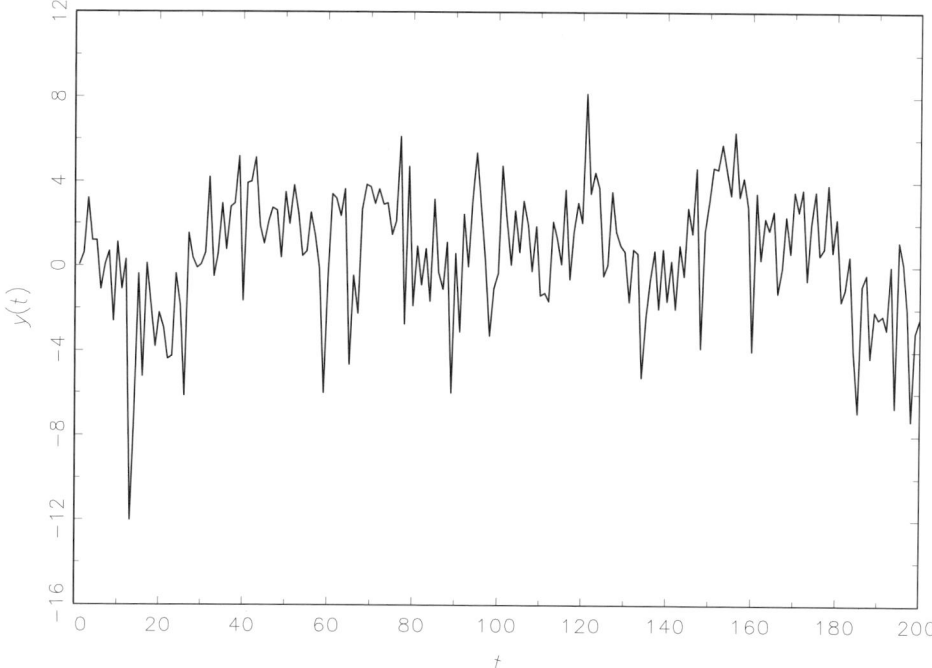

Fig. 9.4. Simulated data, $n = 200$, from the stochastic volatility model (9.29)–(9.30).

we bootstrapped the data, $B = 500$ times, using the incorrect model (9.12)–(9.14) to assess the finite sample standard errors (SE). The results are listed in Table 2 in the column marked *Bootstrap SE*. Finally, using the correct model, (9.29)–(9.30), we simulated 500 processes, estimated the parameters based on the model (9.12)–(9.14) also via a BFGS variable metric algorithm, and assessed the SEs of the estimates of the actual state parameters. These values are listed in Table 9.2 in the column labeled *'True' SE*.

Table 9.2. *Stochastic volatility simulation results*

state parameter	actual value	MLE	asymptotic SE	bootstrap SE[a]	'true' SE[b]
ϕ	0.95	0.963	0.032	0.032	0.036
σ_w	1	1.042	0.279	0.215	0.252

[a] Based on 500 bootstrapped samples. [b] Based on 500 replications.

In Table 9.2 we notice that the bootstap SE and the asymptotic SE of ϕ are about the same; also, both estimates are slightly smaller than the 'true' value. The interest here, however, is not so much in the SEs, but in the actual sampling distribution of the estimates. To explore the finite sample distribution of the estimate of ϕ, Figure 9.5 shows the centred bootstrap histogram: $(\widehat{\phi}_b^* - \widehat{\phi})$, for $b = 1, \ldots, 500$ bootstrapped replications (the bars are filled with lines of positive slope), the centered 'true' histogram: $(\widehat{\phi}_j - \phi)$, where $\widehat{\phi}_j$ is the MLE obtained on the jth iteration, for $j = 1, \ldots, 500$ Monte Carlo replications (the bars are filled with flat lines), and the centred asymptotic normal distribution of $(\widehat{\phi} - \phi)$ (appropriately scaled for comparison with the histograms), superimposed on eachother. Clearly, the bootstrap distribution is closer to the 'true' distribution than the estimated asymptotic normal distribution; the bootstrap distribution captures the positive kurtosis (peakedness) and asymmetry of the 'true' distribution.

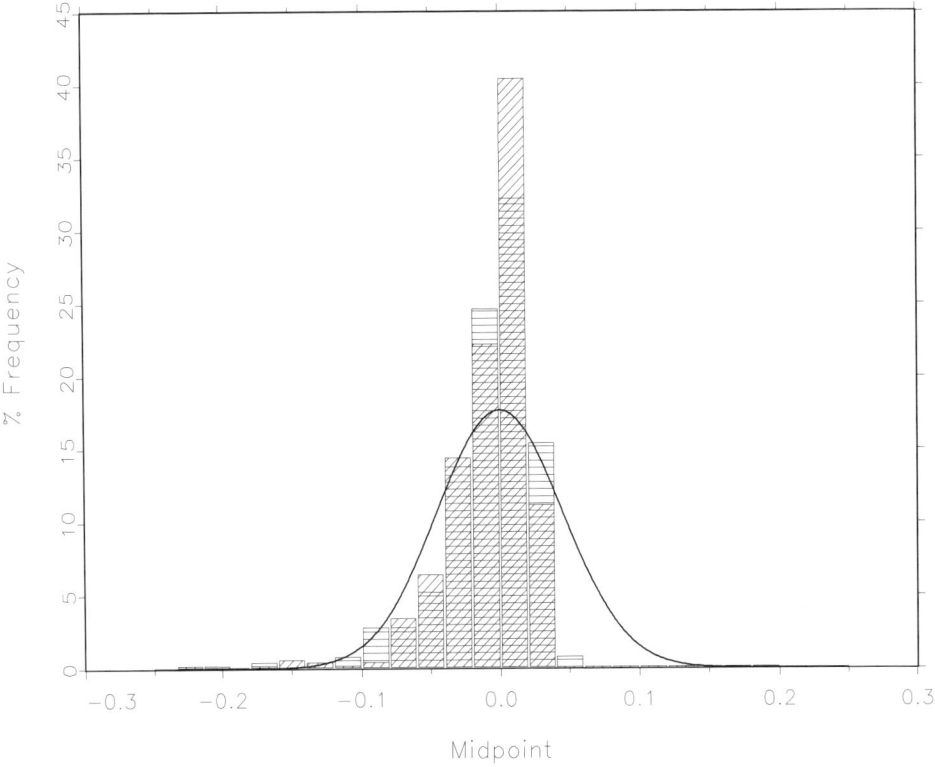

Fig. 9.5. Sampling distributions of the estimate of ϕ; simulated data example: The centred bootstrap histogram (lines with positive slope), the centred 'true' histogram (flat lines), and the centred asymptotic normal distribution.

In an example using actual data, we consider the analysis of quarterly US GNP from 1947(1) to 2002(3), $n = 223$. The data are seasonally

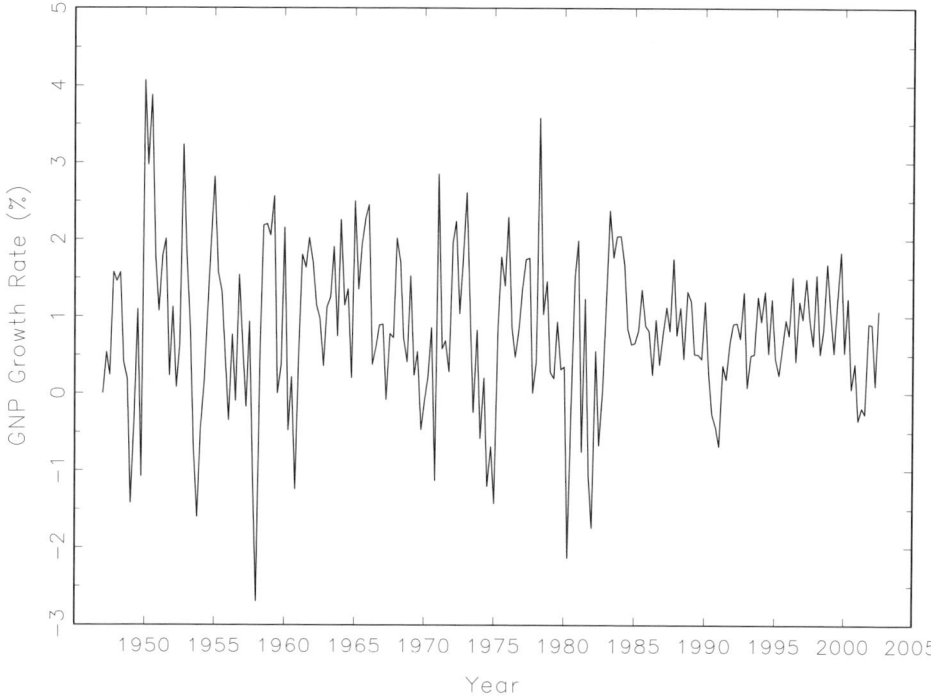

Fig. 9.6. US GNP quarterly growth rate.

adjusted and were obtained from the Federal Reserve Bank of St Louis (http://research.stlouisfed.org/fred/data/gdp/gnpc96). The growth rate is plotted in Figure 9.6 and appears to be a stable process. Analysis of the data indicates the growth rate is an MA(2) (for more details of this part of the analysis, see Shumway and Stoffer (2000, Section 2.8)), however, the residuals of that fit, which appear to be white, suggest that there is volatility clustering.

Figure 9.7 shows the log of the squared residuals, say y_t, from the MA(2) fit on the US GNP series. The stochastic volatility model (9.12)–(9.14) was then fitted to y_t. Table 9.3 shows the MLEs of the model parameters along with their asymptotic SEs assuming the model is correct. Also displayed in Table 9.3 are the means and SEs of $B = 500$ bootstrapped samples. As in the simulation, there is some agreement between the asymptotic values and the bootstrapped values. Based on the previous simulation, we would be more prone to focus on the actual sampling distributions, rather than assume normality. For example, Figure 9.8 compares the bootstrap histogram and asymptotic normal distribution of $\widehat{\phi}_1$. In this case, as in the simulation, the bootstrap distribution exhibits positive kurtosis and skew-

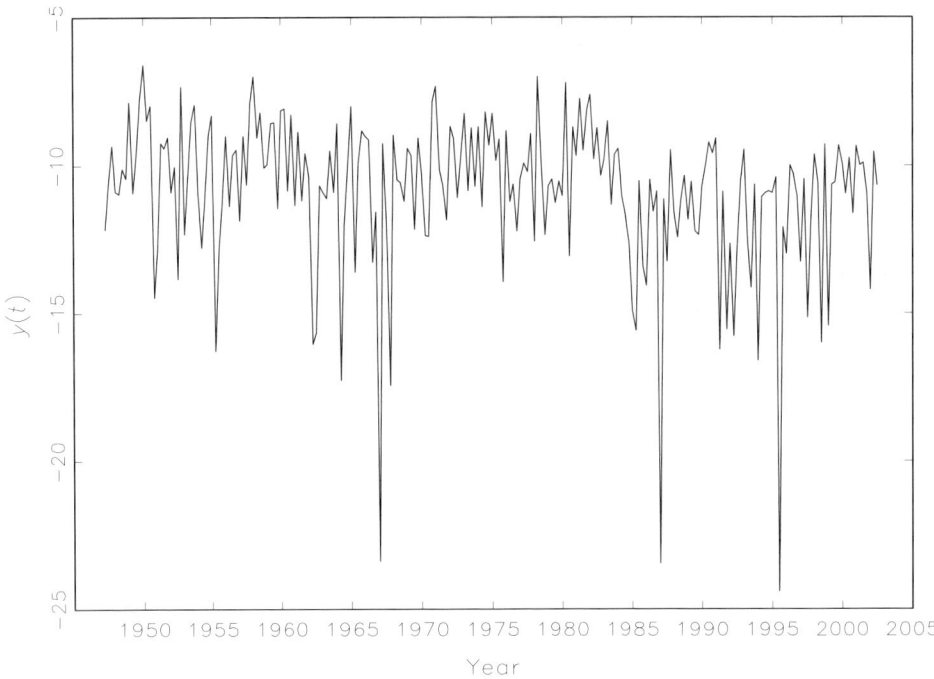

Fig. 9.7. Log of the squared residuals from an MA(2) fit on GNP growth rate.

ness which is missed by the assumption of asymptotic normality. Based on the simulation, we would be prone to believe the results of the bootstrap are fairly accurate.

Table 9.3. *Estimates and their asymptotic and bootstrap standard errors for US GNP example*

parameter	MLE	asymptotic SE	bootstrap mean[a]	bootstrap SE[a]
ϕ_0	0.068	0.274	−0.010	0.353
ϕ_1	0.900	0.099	0.864	0.102
σ_w	0.378	0.208	0.696	0.375
α	−10.524	2.321	−10.792	0.748
μ_1	−2.164	0.567	−1.941	0.416
σ_1	3.007	0.377	2.891	0.422
σ_0	0.935	0.198	0.692	0.362

[a] Based on 500 bootstrapped samples.

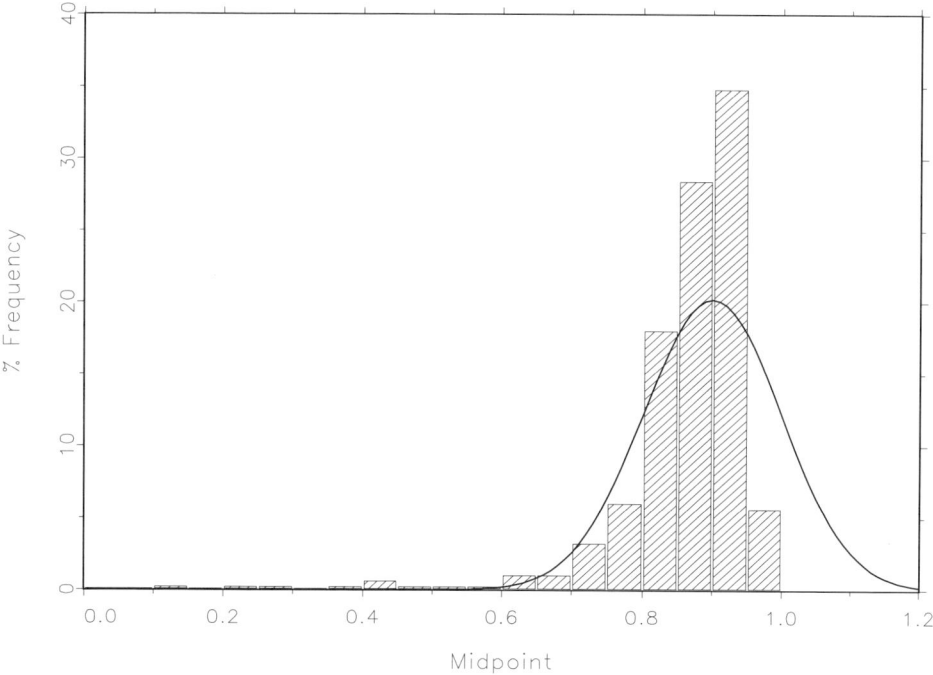

Fig. 9.8. Bootstrap histogram and asymptotic distribution of $\widehat{\phi}_1$ for the US GNP example.

9.3 Assessing the finite sample distribution of conditional forecasts

In this section we focus on assessing the conditional forecast accuracy of time series models using a state space approach and resampling methods. Our work is motivated by the following considerations. First, the state space model provides a convenient unifying representation for various models, including ARMA(p,q) models. Second, the actual practice of forecasting involves the prediction of a future point based on an observed sample path, thus conditional forecast error assessment is of most interest. Third, real-life applications involving time series data are often characterised by short data sets and lack of distributional information. Asymptotic theory provides little help here and often there are no compelling reasons to assume Gaussian distributions apply. Finally, the utility and applicability already demonstrated by the bootstrap for prediction of AR processes suggests that it has much to offer in the prediction of other processes.

Early applications of the bootstrap to assess conditional forecast errors can be found in Findley (1986), Stine (1987), Thombs and Schuchany (1990),

Kabaila (1993) and McCullough (1994, 1996). Interest in the evaluation of confidence intervals for conditional forecast errors has led to methodological problems because a backward, or reverse-time, set of residuals must be generated. Findley (1986) first discussed this problem and Breidt, Davis and Dunsmuir (1992, 1995) offered a solution that is implemented in the work of McCullough (1994, 1996). To date there is a well-grounded methodology for AR models and this work has established the utility of the bootstrap.

A similar state of affairs appears not to exist for other time series models. We suspect this is due to the difficulty with which one can identify mechanisms required to generate bootstrap data sets, whether forwards or backwards in time. For AR models this is easily accomplished because the required initial, or terminal (in the case of conditional forecasts), conditions are given in terms of the observed series. With other time series models this may not be the case because the models require solutions of difference equations involving unobserved disturbances.

The state space model and its related innovations filter offer a way around this difficulty. It is worthwhile, therefore, to investigate how well this can be done in practice. In Section 9.2, such a combination was of use in assessing parameter estimation error, and this naturally leads to the same question being asked in relation to conditional prediction errors. We find that the bootstrap is as useful in evaluating conditional forecast errors as it has proven to be in assessing parameter estimation errors, particularly in a non-Gaussian environment. Our presentation is based on the work of Wall and Stoffer (2002).

9.3.1 Generating reverse time datasets

As seen in Section 9.2, the generation of bootstrap data sets in forward time is easy. Given an initial condition or prior, (9.11) is solved recursively for $t = 1, \ldots, n$ to produce realisations passing through the given initial condition. Such computations are all that is required in obtaining bootstrap estimates of parameter estimation error statistics or unconditional forecast error statistics. The generation of bootstrap data sets for assessing *conditional* forecast errors is not so straightforward because they must be generated backward and this requires a *backward-time state space model*.

An early discussion of the problems related to backward-time models in assessing conditional forecast errors is found in Findley (1986). Further consideration of the problem is found in Breidt, Davis and Dunsmuir (1992, 1995). This literature stresses the need to properly construct a set of 'backward' residuals and Breidt, Davis and Dunsmuir (1992, 1995) provide an algorithm for this that solves the problem for AR(p) models. A similar result is needed for state space models, but development of backward-time representations has not received much attention in the literature. Notable exceptions are the elegant presentation found in Caines (1988, Chapter 4) and a derivation in Aoki (1989, Chapter 5). Our work requires an extension of their results to the time-varying case.

The key system in generating bootstrap data sets is the innovations filter form, (9.9); recall

$$\xi_t = F_t \xi_{t-1} + G\mathbf{u}_t + H_t \mathbf{e}_t, \qquad (9.9)$$

where

$$\xi_t = \begin{bmatrix} \mathbf{x}_{t+1}^t \\ \mathbf{y}_t \end{bmatrix}, \quad F_t = \begin{bmatrix} \Phi & 0 \\ A_t & 0 \end{bmatrix}, \quad G = \begin{bmatrix} \Upsilon \\ \Gamma \end{bmatrix}, \quad H_t = \begin{bmatrix} K_t \Sigma_t^{1/2} \\ \Sigma_t^{1/2} \end{bmatrix}.$$

We require a backward-time representation of this system. All the problems highlighted by Findley (1986) and Breidt, Davis and Dunsmuir (1992, 1995) appear here. For example, the first p rows of (9.9) cannot be solved backwards in time by simply expressing $\mathbf{x}_t^{t-1}$ in terms of $\mathbf{x}_{t+1}^t$. First, Φ is not always invertible; e.g., MA(q) models. Second, even when Φ is invertible, Φ^{-1} has characteristic roots outside the unit circle whenever Φ has its characteristic roots inside the unit circle. This situation is intolerable in generating reverse time trajectories because of the explosive nature of the solutions for ξ_t. In addition, we now have a time-varying system.

These difficulties are overcome by building on the method found in Caines (1988, pp. 236–7). Special attention must be given to the way in which the time-varying matrices propagate through the derivations and proper account must be taken of the effects of the known, or observed input sequence $\mathbf{u}_t$. For ease, we will assume here that $\mathbf{u}_t \equiv \mathbf{0}$; the general case is presented in Wall and Stoffer (2002). Application of the symmetry of minimal splitting subspaces yields the following reverse-time state space representation for $t = n-1, n-2, \ldots, 1$:

$$\mathbf{r}_t = \Phi' \mathbf{r}_{t+1} + B_t \mathbf{x}_t^{t-1} - C_t \mathbf{e}_t, \qquad (9.32)$$

$$\mathbf{y}_t = N_t \mathbf{r}_{t+1} - L_t \mathbf{x}_t^{t-1} + M_t \mathbf{e}_t, \qquad (9.33)$$

where

$$B_t = V_t^{-1} - \Phi'V_{t+1}^{-1}\Phi,$$
$$C_t = \Phi'V_{t+1}^{-1}K_t\Sigma_t^{-1/2},$$
$$D_t = I - \Sigma_t^{-1/2}K_t'V_{t+1}^{-1}K_t\Sigma_t^{-1/2},$$
$$L_t = \Sigma_t^{-1/2}C_t' - A_tV_tB_t,$$
$$M_t = \Sigma_t^{-1/2}D_t - A_tV_tC_t,$$
$$N_t = A_tV_t\Phi' + \Sigma_t^{-1}K_t'$$

and

$$V_{t+1} = \Phi V_t\Phi' + K_t\Sigma_t^{-1}K_t'. \tag{9.34}$$

The reverse-time state vector is $\mathbf{r}_t$. The backward recursion is initialised by $\mathbf{r}_n = V_n^{-1}\mathbf{x}_n^{n-1}$. Details of the derivation are given in Wall and Stoffer (2002).

The above recursion specifies a three-step procedure for the generation of backward time data sets (written here for $\mathbf{u}_t \equiv \mathbf{0}$):

(i) Generate $V_t, B_t, C_t, D_t, L_t, M_t$ and N_t forwards in time, $t = 1, \ldots, n$, with initial condition

$$V_1 = P_1^0. \tag{9.35}$$

(ii) For given $\{\mathbf{e}_t^*; 1 \leq t \leq n-1\}$, set $\mathbf{x}_1^* = \mathbf{0}$ and generate $\{\mathbf{x}_t^*; 1 \leq t \leq n\}$ forwards in time, $t = 1, \ldots, n$, via

$$\mathbf{x}_{t+1}^* = \Phi\mathbf{x}_t^* + K_t\Sigma_t^{1/2}\mathbf{e}_t^*. \tag{9.36}$$

(iii) Set $\mathbf{r}_n^* = \mathbf{r}_n = V_n^{-1}\mathbf{x}_n^{n-1}$ and generate $\{\mathbf{y}_t^*; 1 \leq t \leq n\}$ backwards in time, $t = n-1, n-2, \ldots, 1$, via the reverse-time state space model

$$\mathbf{r}_t^* = \Phi'\mathbf{r}_{t+1}^* + B_t\mathbf{x}_t^* - C_t\mathbf{e}_t^*, \tag{9.37}$$
$$\mathbf{y}_t^* = N_t\mathbf{r}_{t+1} - L_t\mathbf{x}_t^* + M_t\mathbf{e}_t^*. \tag{9.38}$$

This procedure assumes one already has drawn randomly, with replacement, from the model estimated standardised residuals to obtain a set of $n-1$ residuals denoted $\{\mathbf{e}_t^*; 1 \leq t \leq n-1\}$. The last residual is kept set at $\mathbf{e}_n^* = \mathbf{e}_n$ in order to ensure the conditioning requirement is met on ξ_n^*; that is, $\xi_n^* = \xi_n$. This requirement follows from the autoregressive structure of (9.9). The creation of an arbitrary number of bootstrap data sets is accomplished by repeating the above for each set of bootstrap residuals $\{\mathbf{e}_t^*; 1 \leq t \leq n-1; \mathbf{e}_n^* = \mathbf{e}_n\}$.

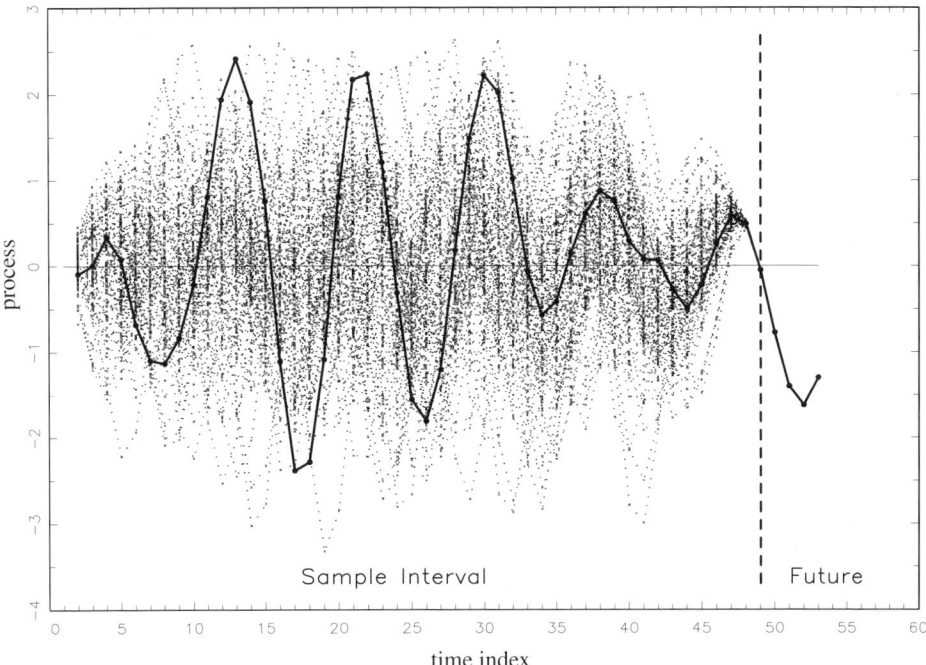

Fig. 9.9. Reverse time realizations of the ARMA(2, 1) process given in (9.41).

As an example, consider the univariate ARMA(p,q) process given by

$$y_t + a_1 y_{t-1} + \cdots + a_p y_{t-p} = v_t + b_1 v_{t-1} + \cdots + b_q v_{t-q}, \tag{9.39}$$

where v_t is an iid process with variance σ_v^2. This process can be represented in state space form, (9.1)–(9.2), in various ways. For example, let $m = \max\{p,q\}$, let $\mathbf{x}_t$ be an m-dimensional state vector, and write the state space coefficient matrices as

$$\Phi = \begin{bmatrix} 0 & \cdots & 0 & 0 & 0 & -a_m \\ 1 & \cdots & 0 & 0 & 0 & -a_{m-1} \\ \vdots & \ddots & \vdots & \vdots & \vdots & \vdots \\ 0 & \cdots & 1 & 0 & 0 & -a_3 \\ 0 & \cdots & 0 & 1 & 0 & -a_2 \\ 0 & \cdots & 0 & 0 & 1 & -a_1 \end{bmatrix},$$

$$A = \begin{bmatrix} 0 & \cdots & 0 & 0 & 0 & 1 \end{bmatrix},$$

$\Upsilon = 0$ and $\Gamma = 0$. The state noise process is defined by the $\mathbf{w}_t = \mathbf{g}\, v_t$, where

$$\mathbf{g} = \begin{bmatrix} b_m - a_m, & b_{m-1} - a_{m-1}, & \ldots, & b_3 - a_3, & b_2 - a_2, & b_1 - a_1 \end{bmatrix}'. \tag{9.40}$$

If $m > p$ then $a_\ell = 0$ for $\ell > p$, and if $m > q$ then $b_\ell = 0$ for $\ell > q$. The variance-covariance matrices are given by

$$Q = \sigma_v^2\, \mathbf{g}\mathbf{g}', \quad R = \sigma_v^2, \quad S = \sigma_v^2\, \mathbf{g}.$$

Figure 9.9 presents a sample of 100 reverse-time trajectories for the Gaussian ARMA(2, 1) model

$$y_t = 1.4 y_{t-1} - 0.85 y_{t-2} + v_t + 0.6 v_{t-1}, \tag{9.41}$$

with $\sigma_v = 0.2$ and $n = 49$. The original, observed sample is plotted with the bold line.

9.3.2 Computing forecast errors via the bootstrap

At this point we assume we have n observations, $\mathbf{y}_1, \ldots, \mathbf{y}_n$, and we wish to forecast m time points into the future. In addition, we have the MLEs of the model parameters Θ, say $\widehat{\Theta}$, based on the data. The associated standardised innovation values are denoted by $\{\mathbf{e}_t(\widehat{\Theta});\ 1 \leq t \leq n\}$; note, to avoid any possible confusion, we emphasise the dependence of the values on the parameters. For $b = 1, 2, \ldots, B$ (where B is the number of bootstrap replications) we execute the following six steps:

(i) Construct a sequence of $n + m$ standardized residuals

$$\{\mathbf{e}_t^b(\widehat{\Theta});\ 1 \leq t \leq n + m\} \tag{9.42}$$

via $n+m-1$ random draws, with replacement, from the standardised residuals $\{\mathbf{e}_t(\widehat{\Theta});\ 1 \leq t \leq n\}$. This sequence is formed as follows: (a) use $n-1$ vectors to form $\{\mathbf{e}_t^b(\widehat{\Theta});\ 1 \leq t \leq n-1\}$; (b) fix $\mathbf{e}_n^b(\widehat{\Theta}) = \mathbf{e}_n(\widehat{\Theta})$; and (c) use the remaining m vectors to form $\{\mathbf{e}_t^b(\widehat{\Theta});\ n+1 \leq t \leq n+m\}$.

(ii) Generate data

$$\{\mathbf{y}_t^b(\widehat{\Theta});\ 1 \leq t \leq n-1\} \tag{9.43}$$

via the backward state space model (9.37) and (9.38) with $\Theta = \widehat{\Theta}$ using the residuals $\{\mathbf{e}_t^b(\widehat{\Theta});\ 1 \leq t \leq n-1\}$. Set $\mathbf{y}_n^b(\widehat{\Theta}) = \mathbf{y}_n$.

(iii) Generate data

$$\{\mathbf{y}_t^b(\widehat{\Theta});\ n+1 \leq t \leq m+n\} \tag{9.44}$$

via the forward state space model (9.9) with $\Theta = \widehat{\Theta}$ and with $\mathbf{x}_t^{t-1;b} = \mathbf{x}_t^{t-1}(\widehat{\Theta})$ and using the residuals $\mathbf{e}_t^b(\widehat{\Theta})$, for $n+1 \leq t \leq n+m$.

(iv) Compute model parameter estimates Θ^b via MLE using the data $\{\mathbf{y}_t^b(\widehat{\Theta}); 1 \leq t \leq n\}$.

(v) Compute the bootstrap conditional forecasts

$$\{\mathbf{y}_t^b(\Theta^b); n+1 \leq t \leq m+n\} \qquad (9.45)$$

via the forward-time state space model (9.9) with $\Theta = \Theta^b$, and with $\mathbf{x}_t^{t-1;b} = \mathbf{x}_t^{t-1}(\Theta^b)$ and $\mathbf{e}_t^b = \mathbf{0}$ for $n+1 \leq t \leq n+m$.

(vi) Compute the bootstrap conditional forecast errors via:

$$\mathbf{d}_\ell^b = \mathbf{y}_{n+\ell}^b(\widehat{\Theta}) - \mathbf{y}_{n+\ell}^b(\Theta^b); \quad 1 \leq \ell \leq m. \qquad (9.46)$$

The extent to which the bootstrap captures the behavior of the actual forecast errors derives from the extent to which these errors mimic the stochastic process $\mathbf{d}_\ell = \mathbf{y}_{n+\ell}(\Theta) - \mathbf{y}_{n+\ell}(\widehat{\Theta}); \ 1 \leq \ell \leq m$.

As an example, consider the univariate $ARMA(1, 1)$ process given by

$$y_t = 0.7 y_{t-1} + v_t + 0.10 v_{t-1}, \qquad (9.47)$$

where $v_t = 0.2 z_t$ and z_t is a mixture of 90% $N(\mu = -1/9, \sigma = 0.15)$ and 10% $N(\mu = 1, \sigma = 0.15)$. To demonstrate the benefits of resampling, we will assume that we do not know the true distribution of v_t and will act as if it were normal. The model is first order with

$$\Phi = [0.70] \quad A = [1] \quad \text{and} \quad \mathbf{g} = [0.80], \qquad (9.48)$$

in the notation of previous example.

In this simulation we use $B = 2000$ and $m = 4$. The approximate 'true' distribution is then given by the relative frequency histogram of the observed conditional forecast errors. The results of the simulation are summarised by two sets of four histograms. One set (Figure 9.10) presents the approximate 'true' relative frequency histograms for each forecast lead time, while the other set (Figure 9.11) presents the relative frequency histograms obtained from application of the bootstrap. Superimposed on each is the Gaussian density that follows from application of the asymptotic Gaussian theory. The simulation uses a short data set with $n = 49$ to emphasise the efficacy of the bootstrap when the use of asymptotics is questionable and where bias is a factor in the forecasts. Prediction intervals follow immediately from the data summarised in the histograms. Although we choose to present only the histograms, the percentile, the bias-corrected (BC), and the accelerated bias-corrected (BC_a) method all are applicable for generating confidence intervals using the generated data (see Efron (1987)).

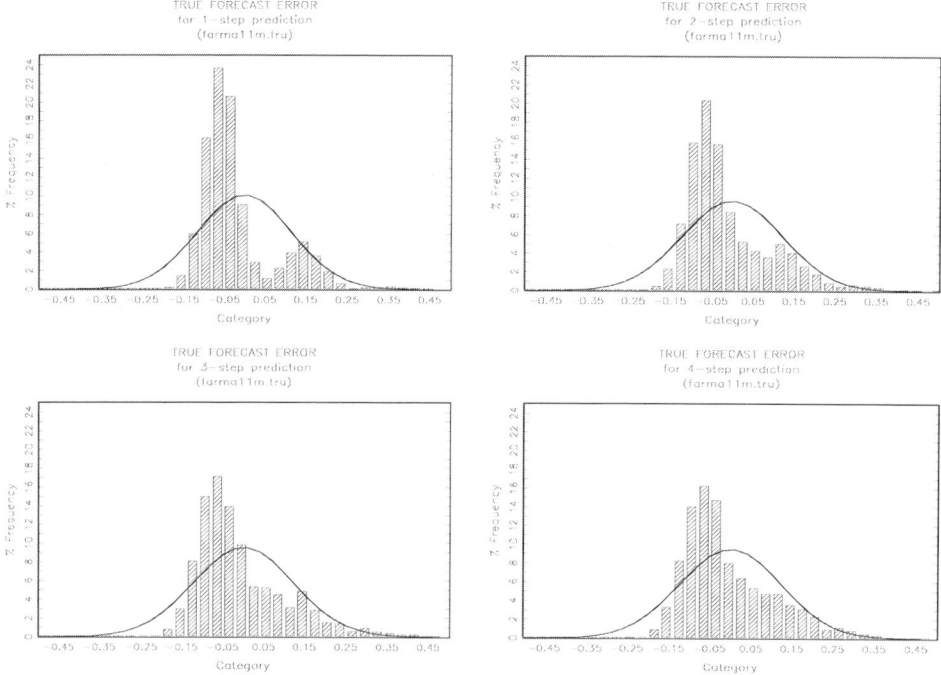

Fig. 9.10. 'True' forecast histograms for the ARMA(1, 1) process given in (9.47).

Figures 9.10 and 9.11 reveal the value of the bootstrap. Indication of the mixture distribution is striking in both the 'true' and the bootstrap distributions; the bimodality and asymmetry are clearly evident.

9.3.3 Stochastic regression

We now illustrate the use of the bootstrap in assessing forecast errors in the data set analysed in Section 2.1. Recall, the treasury bill interest rate is modelled as being linearly related to quarterly inflation as

$$y_t = \alpha + \beta_t z_t + v_t,$$

where α is a fixed constant, β_t is a stochastic regression coefficient and v_t is white noise with variance σ_v^2. The stochastic regression term, which comprises the state variable, is specified by a first-order autoregression,

$$(\beta_t - b) = \phi(\beta_{t-1} - b) + w_t,$$

where b is a constant and w_t is white noise with variance σ_w^2. The noise processes, v_t and w_t, are assumed to be uncorrelated.

The model parameter vector contains five elements, $\Theta = (\phi, \alpha, b, \sigma_w, \sigma_v)'$

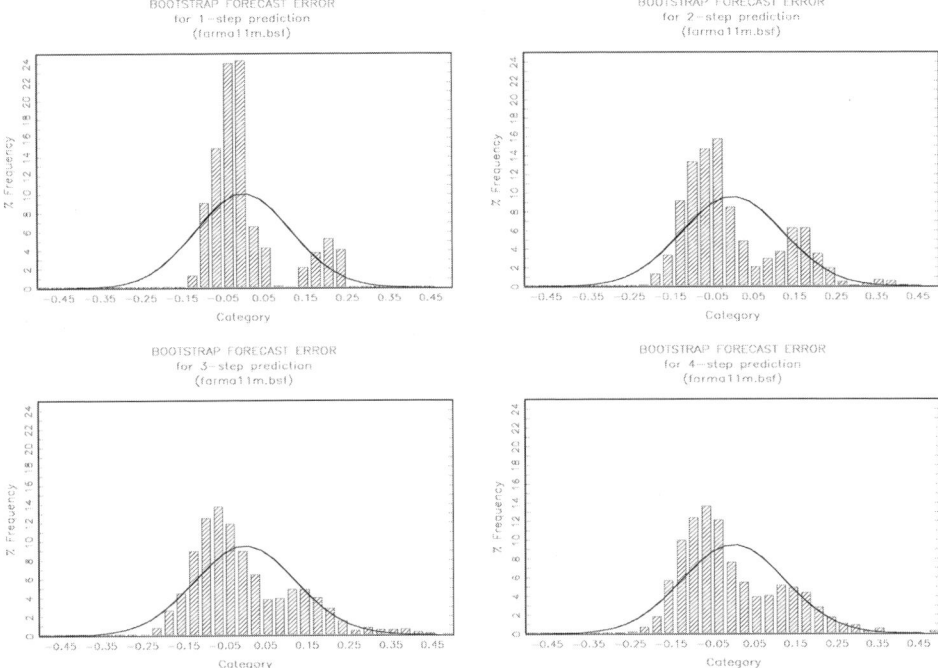

Fig. 9.11. Bootstrap forecast histograms for the ARMA(1,1) process given in (9.47).

and is estimated via Gaussian quasi-maximum likelihood using data from the first quarter of 1967 through the second quarter of 1979 (49 observations). The MLEs and their estimated standard errors (in parentheses) were:

$$\widehat{\phi} = 0.898_{(0.101)}, \quad \widehat{\alpha} = -0.615_{(1.457)}, \quad \widehat{b} = 1.195_{(0.278)},$$

$$\widehat{\sigma}_w = 0.092_{(0.049)}, \quad \widehat{\sigma}_v = 1.287_{(0.197)}.$$

Among the many forecast error assessment questions that can be asked concerning this model are ones concerning the properties of the conditional forecast error distribution *assuming that we know the future values of the inflation rate*. In particular, is a Gaussian assumption warranted when future values of z_t are assumed to be known? Such questions may arise within the context of a 'rational expectations' framework wherein economic agents are assumed so well informed that they 'know' the inflation rate. The bootstrap, coupled with our methodology here, can shed some light on just such a question as this.

Figure 9.12 depicts the bootstrap results with $B = 2000$. The upper left panel presents the dynamic behavior of the quantiles (specifically, 2.5%, 5%,

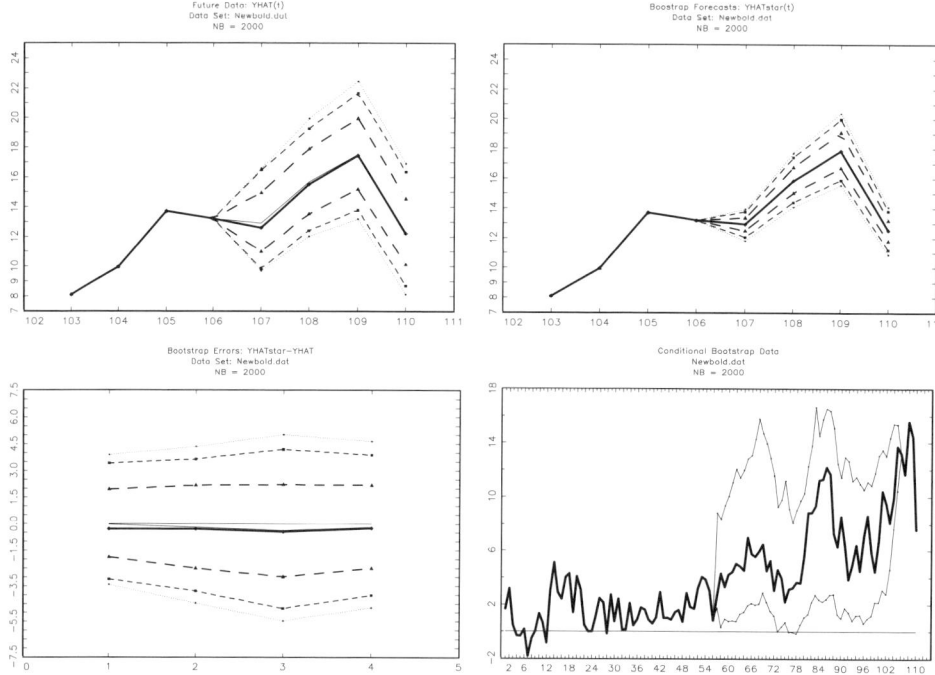

Fig. 9.12. Dymanic behaviour of the quantiles of $y_{n+\ell}^b(\widehat{\Theta})$ (upper left panel), the quantiles of $y_{n+\ell}^b(\Theta^b)$ (upper right panel), and the quantiles of the bootstrap conditional forecast errors $y_{n+\ell}^b(\widehat{\Theta}) - y_{n+\ell}^b(\Theta^b)$, for $\ell = 1,2,3,4$ (lower left panel). The y_t series (lower right panel) is displayed as a bold line and the envelope of the backward data series as fine lines above and below the observed sample in the stochastic regression example.

16%, 50%, 84%, 95%, 97.5%) of $y_{n+\ell}^b(\widehat{\Theta})$ and the upper right panel presents the quantiles of $y_{n+\ell}^b(\Theta^b)$. Given the significant variability in the upper right panel, it is clear that the variability due to the additive disturbances (upper left panel) is not the dominant factor in the forecast uncertainty that it is so often assumed to be. The lower left panel depicts the dynamic behavior of the quantiles of the bootstrap conditional forecast errors $y_{n+\ell}^b(\widehat{\Theta}) - y_{n+\ell}^b(\Theta^b)$, for $\ell = 1,2,3,4$. The lower right panel plots the y_t series as a bold line and the envelope of the backward data series as fine lines above and below the observed sample. We find the backward generated series to be highly representative of the stochastic properties of the observed series.

Figure 9.13 presents histograms of the conditional forecast errors, $y_{n+\ell}^b(\widehat{\Theta}) - y_{n+\ell}^b(\Theta^b)$, for $\ell = 1,2,3,4$, when $B = 2000$ and Figure 9.14 presents the histograms when $B = 10\,000$. Each picture gives an indication of the problems in assuming that the asymptotic theory applies. Negative bias is indicated

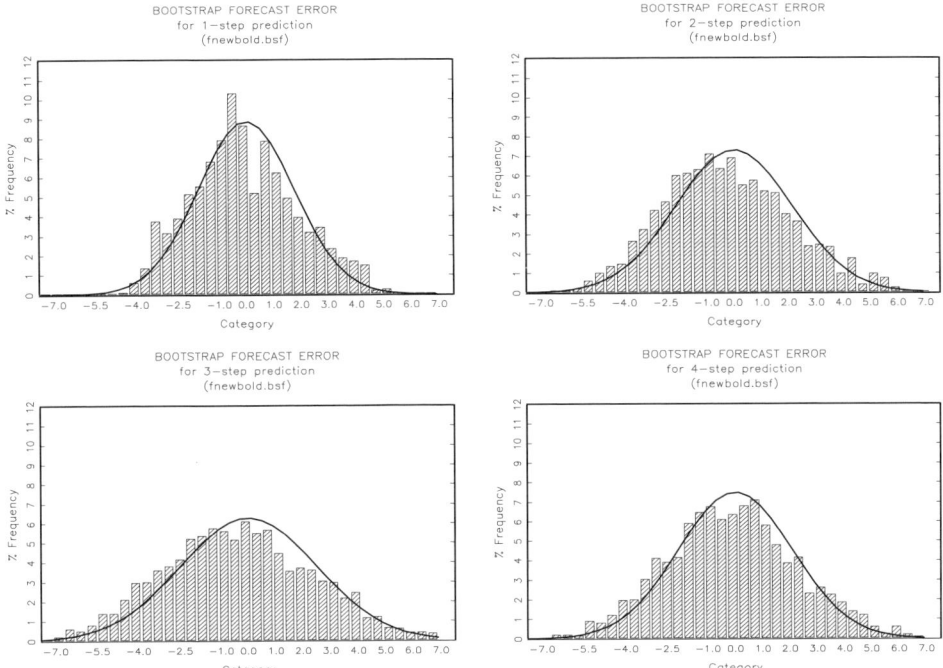

Fig. 9.13. Histograms of four conditional forecast errors, $B = 2000$, in the stochastic regression example.

and t-tests reject zero means for $\ell = 2, 3, 4$, in both bootstrap experiments. A Kolmogorov–Smirnov test rejects the asymptotic Gaussian distribution (which are also displayed in the figures) for all forecast lead times for both values of B. It appears little is gained in extending the bootstrap replications beyond $B = 2000$, other than the more 'smooth' appearance of the histograms.

9.4 Discussion

The state space model provides a convenient unifying representation for various time domain models. This article demonstrates the utility of resampling the innovations of time domain models via state space models and the Kalman (innovations) filter. We have based our presentation primarily on the material in two articles, Stoffer and Wall (1991) and Wall and Stoffer (2002).

In Stoffer and Wall (1991) we developed a resampling scheme to assess the finite sample distribution of parameter estimates for general time domain models. This algorithm uses the elegance of the state space model in

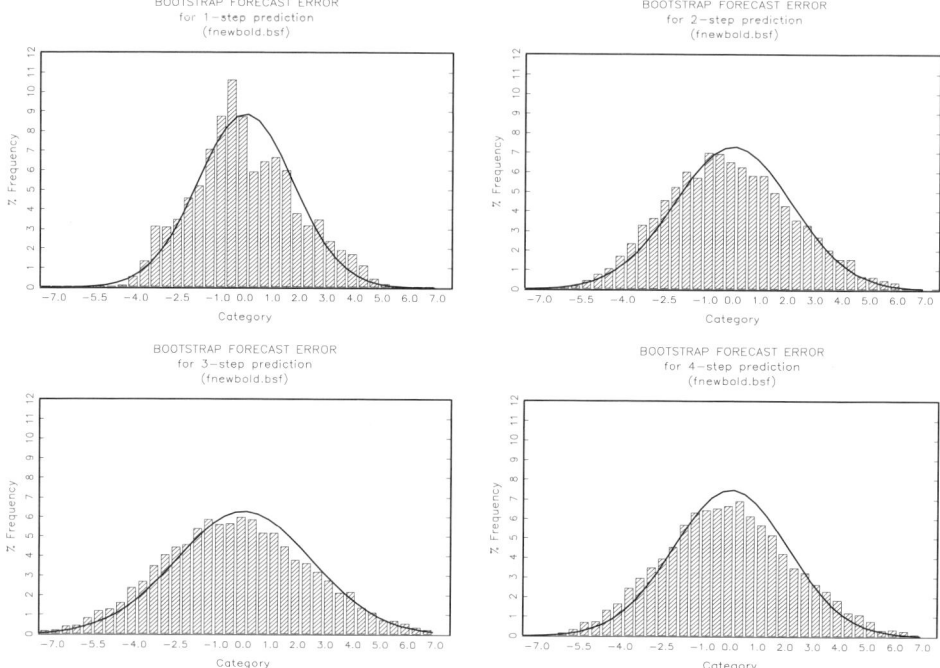

Fig. 9.14. Histograms of four conditional forecast errors, $B = 10\,000$, in the stochastic regression example.

innovations form to construct a simple resampling scheme. The key point is that while under general conditions, the MLEs of the model parameters are consistent and asymptotically normal, time series data are often of short or moderate length so that the use of asymptotics may lead to wrong conclusions. Moreover, it is well known that problems occur if the parameters are near the boundary of the parameter space. We have provided additional examples here that emphasise the usefulness of the algorithm. We have also explained, heuristically, why the resampling scheme is asymptotically correct under appropriate conditions.

We have also discussed conditional forecast accuracy of time domain models using a state space approach and resampling methods that were first presented in Wall and Stoffer (2002). Applications involving time series data are often characterised by short data sets and a lack of distributional information; asymptotic theory provides little help here and frequently there are no compelling reasons to assume Gaussian distributions apply. Interest in the evaluation of confidence intervals for conditional forecast errors in AR models led to methodological problems because a backward, or reverse-time, set of residuals must be generated. This problem was eventually solved and there is now a well-grounded methodology for AR models. Practitioners were confined to AR models because the required initial, or terminal (in the case of conditional forecasts), conditions are given in terms of the observed series. With other time series models this may not be the case because the models require solutions of difference equations involving unobserved disturbances. The state space model and its related innovations filter offered a way around this difficulty. We have exhibited a reverse-time state space in innovations form. We have presented additional examples here that demonstrate resampling as useful in evaluating conditional forecast errors as it has proven to be in assessing parameter estimation errors, particularly in a non-Gaussian environment. In the Appendix, we explain, heuristically, why resampling works in large samples.

Appendix

In Section 9.2, resampling techniques were used to determine the finite sample distributions of the parameter estimates when the use of asymptotics was questionable. In Section 9.3, we used resampling to assess the finite sample distributions of the forecast errors. The extent to which resampling the innovations does what it is supposed to do can be measured in various ways. In the finite sample case, we can perform simulations – where the true distributions are known – and compare the bootstrap results with the known results. If the bootstrap works well in simulations, we may feel confident that the bootstrap will work well in similar situations, but, of course, we have no guarantee that it works in general. In this way, the examples in Sections 9.2 and 9.3 help demonstrate the validity of the resampling procedures discussed in those sections.

Another approach is to ask if the bootstrap will give the correct asymptotic answer. That is, if we have an infinite amount of data and can resample an infinite amount of times, do we get the correct asymptotic distribution (typically, we require asymptotic normality). If the answer is no, we can assume that resampling will not work with small samples. If the answer is yes, we can only hope that resampling will work with small samples, but again, we have no guarantee. For state space models, how well the resampling techniques perform in finite samples hinges on at least three things. First, the techniques are conditional on the data, so the success of the resampling depends on how typical the data set is for the particular model. Second, we assume the model is correct (at least approximately); if the proposed model is far from the truth, the results of the resampling will also be incorrect. Finally, assuming the data set is typical and the model is correct, the success of the resampling depends on how close the empirical distribution of the innovations is to the actual distribution of the innovations. We are guaranteed such closeness in large samples if the innovations are stable and mixing in the sense of Gastwirth and Rubin (1975).

Heuristics for Section 9.2

Stoffer and Wall (1991) established the asymptotic justification of the procedure presented in Section 9.2 under general conditions (including the case where the process is non-Gaussian). To keep matters simple, we assume here that the state space model, (9.1)–(9.2) with $A_t \equiv A$, is Gaussian, observable and controllable, and the eigenvalues of Φ are within the unit circle. We denote the true parameters by Θ_0, and we assume the dimension of Θ_0 is the dimension of the parameter space. Let $\widehat{\Theta}_n$ be the consistent estimator of Θ_0 obtained by maximising the Gaussian innovations likelihood, $L_Y(\Theta)$, given in (9.10). Then, under general conditions ($n \to \infty$),

$$\sqrt{n}\left(\widehat{\Theta}_n - \Theta_0\right) \sim \text{AN}\left[0,\ \mathcal{I}_n(\Theta_0)^{-1}\right],$$

where $\mathcal{I}_n(\Theta)$ is the information matrix given by

$$\mathcal{I}_n(\Theta) = n^{-1} E\left[-\partial^2 \ln L_Y(\Theta)/\partial\Theta\ \partial\Theta'\right].$$

Precise details and the proof of this result are given in Caines (1988, Chapter 7) and in Hannan and Deistler (1988, Chapter 4).

Let $\widehat{\Theta}_n^*$ denote the parameter estimates obtained from the resampling procedure of Section 9.2. Let B_n be the number of bootstrap replications and, for ease, we take $B_n = n$. Then, Stoffer and Wall (1991) established that, under certain regularity conditions ($n \to \infty$),

$$\sqrt{n}\left(\widehat{\Theta}_n^* - \widehat{\Theta}_n\right) \sim \text{AN}\left[0,\ \mathcal{I}_n^*(\widehat{\Theta}_n)^{-1}\right],$$

where $\mathcal{I}_n^*(\Theta)$ is the information matrix given by

$$\mathcal{I}_n^*(\Theta) = n^{-1} E_*\left[-\partial^2 \ln L_Y(\Theta)/\partial\Theta\ \partial\Theta'\right],$$

and E_* denotes expectation with respect to the empirical distribution of the innovations. It was then shown that

$$\mathcal{I}_n(\Theta_0) - \mathcal{I}_n^*(\widehat{\Theta}_n) \to 0 \qquad (9.49)$$

almost surely, as $n \to \infty$; hence, the resampling procedure is asymptotically correct.

It is informative to examine, at least partially, why (9.49) holds. Let $Z_{ta} \equiv Z_{ta}(\Theta) = \partial(\mathbf{e}_t' \mathbf{e}_t)/\partial\theta_a$, where $\mathbf{e}_t \equiv \mathbf{e}_t(\Theta)$ is the standardised innovation, (9.8), and θ_a is the ath component of Θ. Similarly, let $Z_{ta}^* \equiv Z_{ta}^*(\Theta) = \partial(\mathbf{e}_t^{*\prime} \mathbf{e}_t^*)/\partial\theta_a$, where $\mathbf{e}_t^* \equiv \mathbf{e}_t^*(\Theta)$ is the resampled standardised innovation. The (a,b)th element of $\mathcal{I}_n(\Theta_0)$ is

$$n^{-1} \sum_{t=1}^n \{E(Z_{ta} Z_{tb}) - E(Z_{ta})E(Z_{tb})\}\Big|_{\Theta=\Theta_0} \qquad (9.50)$$

whereas the (a,b)th element of $\mathcal{I}_n^*(\widehat{\Theta}_n)$ is

$$n^{-1} \sum_{t=1}^n \{E_*(Z_{ta}^* Z_{tb}^*) - E_*(Z_{ta}^*)E_*(Z_{tb}^*)\}\Big|_{\Theta=\widehat{\Theta}_n}. \qquad (9.51)$$

The terms in (9.51) are

$$E_*(Z_{ta}^*) = n^{-1} \sum_{j=1}^n Z_{ja} \quad \text{and} \quad E_*(Z_{ta}^* Z_{tb}^*) = n^{-1} \sum_{j=1}^n Z_{ja} Z_{jb}. \qquad (9.52)$$

Hence, (9.50) contains population moments, whereas (9.51) contains the corresponding sample moments. It should be clear that under appropriate conditions, (9.50) and (9.51) are asymptotically ($n \to \infty$) equivalent. Details of these results can be found in Stoffer and Wall (1991, Appendix).

Heuristics for Section 9.3

As previously, to keep matters simple, we assume the state space model (9.1)–(9.2), with $A_t \equiv A$, is observable and controllable, and the eigenvalues of Φ are within the unit circle; these assumptions ensure the asymptotic stability of the filter. We assume that we have N observations, $\{\mathbf{y}_{n-N+1}, \ldots, \mathbf{y}_n\}$ available, and that N is large. We let $\widehat{\Theta}_N$ denote the (assumed consistent as $N \to \infty$) Gaussian MLE of Θ, and let $\widehat{\Theta}_N^*$ denote a bootstrap parameter estimate.

For one-step-ahead forecasting the model specifies that the process ξ, which we assume is in steady state at time $n+1$, is given by

$$\xi_{n+1} = F(\Theta)\xi_n + G(\Theta)\mathbf{u}_n + H(\Theta)\epsilon_{n+1}, \tag{9.53}$$

where

$$\xi = \begin{bmatrix} \mathbf{x}_{t+1}^t \\ \mathbf{y}_t \end{bmatrix} \tag{9.54}$$

and

$$F = \begin{bmatrix} \Phi & 0 \\ A & 0 \end{bmatrix}, \quad G = \begin{bmatrix} \Upsilon \\ \Gamma \end{bmatrix}, \quad H = \begin{bmatrix} K\Sigma^{1/2} \\ \Sigma^{1/2} \end{bmatrix};$$

K and Σ represent the steady state gain and innovation variance-covariance matrices, respectively. Recall that $\{\mathbf{u}_t\}$ is a fixed and known input process.

For convenience, we have dropped the parameter from the notation when representing a filtered value that depends on Θ. For example, in (9.53) we wrote $\xi \equiv \xi(\Theta)$ and $\epsilon_t \equiv \epsilon_t(\Theta)$. The process ϵ_t is the standardised, steady state innovation sequence so that $E\{\epsilon_t\} = \mathbf{0}$ and $E\{\epsilon_t \epsilon_t'\} = I_q$.

The one-step-ahead conditional forecast estimate is given by

$$\widetilde{\xi}_{n+1} = F(\widehat{\Theta}_N)\widehat{\xi}_n + G(\widehat{\Theta}_N)\mathbf{u}_n, \tag{9.55}$$

where, in keeping with the notation, we have written $\widehat{\xi}_n \equiv \xi(\widehat{\Theta}_N)$. The conditional forecast estimate is labelled with a tilde. Watanabe (1985) showed that, under the assumed conditions and notation, $\mathbf{x}_{n+1}^n(\widehat{\Theta}_N) = \mathbf{x}_{n+1}^n(\Theta) + \mathbf{o}_p(1)$ ($N \to \infty$), and consequently we write $\widehat{\xi}_n = \xi + \mathbf{o}_p(1)$, noting that the final q elements of $\widehat{\xi}_n$ and ξ are identical. Hence, the conditional prediction error can be written as

$$\begin{aligned} \Delta_N &\equiv \xi_{n+1} - \widetilde{\xi}_{n+1} \\ &= [F(\Theta) - F(\widehat{\Theta}_N)]\widehat{\xi}_n + F(\Theta)\mathbf{o}_p(1) \\ &\quad + [G(\Theta) - G(\widehat{\Theta}_N)]\mathbf{u}_n + H(\Theta)\epsilon_{n+1}. \end{aligned} \tag{9.56}$$

From (9.56) we see the two sources of variation, namely the variation due to estimating the parameter Θ by $\widehat{\Theta}_N$, and the variation due to the predicting the innovation value ϵ_{n+1} by zero.

In the conditional bootstrap procedure, we mimic (9.53) and obtain a pseudo observation

$$\xi^*_{n+1} = F(\widehat{\Theta}_N)\widehat{\xi}_n + G(\widehat{\Theta}_N)\mathbf{u}_n + H(\widehat{\Theta}_N)\epsilon^*_{n+1}, \qquad (9.57)$$

where we hold $\widehat{\xi}_n$ fixed throughout the resampling procedure. Note that because the filter is in steady state, the data, $\{\mathbf{y}_{n-N+1}, \ldots, \mathbf{y}_n\}$, completely determine $\widehat{\Theta}_N$ and consequently $\widehat{\xi}_n$. For finite sample lengths, the data and the initial conditions determine $\widehat{\Theta}_N$. As a practical matter, if precise initial conditions are unknown, one can drop the first few data points from the estimation of Θ so that changing the initial state conditions does not change $\widehat{\Theta}_N$ nor $\widehat{\xi}_n$. We remark that while the data completely determine $\widehat{\xi}_n$, the reverse is not true; that is, fixing $\widehat{\xi}_n$ in no way fixes the entire data sequence $\{\mathbf{y}_{n-N+1}, \ldots, \mathbf{y}_n\}$. For example, in the AR(1) case, fixing $\widehat{\xi}_n$ is equivalent to fixing $\mathbf{y}_n$ only. In addition, ϵ^*_{n+1} is a random draw from the empirical distribution of the standardised, steady state innovations. Under the mixing conditions of Gastwirth and Rubin (1975), the empirical distribution of the standardised, steady state innovations converges weakly ($N \to \infty$) to the standardised, steady state innovation distribution.

To mimic the forecast in (9.55), the bootstrap estimated conditional forecast is given by

$$\widehat{\xi}^*_{n+1} = F(\widehat{\Theta}^*_N)\widehat{\xi}^*_n + G(\widehat{\Theta}^*_N)\mathbf{u}_n, \qquad (9.58)$$

which yields the bootstrapped conditional forecast error

$$\begin{aligned}\Delta^*_N &\equiv \xi^*_{n+1} - \widehat{\xi}^*_{n+1} \\ &= [F(\widehat{\Theta}_N) - F(\widehat{\Theta}^*_N)]\widehat{\xi}_n + [G(\widehat{\Theta}_N) - G(\widehat{\Theta}^*_N)]\mathbf{u}_n + H(\widehat{\Theta}_N)\epsilon^*_{n+1}.\end{aligned} \qquad (9.59)$$

Comparison of (9.56) and (9.59) shows why, in finite samples, the bootstrap works; that is, (9.59) is a sample-based imitation of (9.56). Letting $N \to \infty$ in (9.56), while holding $\widehat{\xi}_n$ fixed, we have that, if $\widehat{\Theta}_N \to_p \Theta$, then $\Delta_N \Rightarrow H(\Theta)\mathbf{e}$, where $\mathbf{e}$ is a random vector that is distributed according to the steady state standardised innovation distribution ($\Rightarrow$ denotes weak convergence). In addition, if conditional on the data, $\widehat{\Theta}^*_N - \widehat{\Theta}_N \to_p \mathbf{0}$, then $\Delta^*_N \Rightarrow H(\Theta)\mathbf{e}$ as $N \to \infty$. Extending these results to m-step-ahead forecasts follows easily by induction. Stoffer and Wall (1991) established conditions under which $\widehat{\Theta}^*_N - \widehat{\Theta}_N \to_p \mathbf{0}$ as $N \to \infty$ when the forward bootstrapped samples are used. It remains to determine the conditions under which this result holds when the backward bootstrap data are used.

Acknowledgments

The work of D. S. Stoffer was supported, in part, by the National Science Foundation under Grant No. DMS-0102511.

Part IV

Applications

10

Measuring and forecasting financial variability using realised variance

Ole E. Barndorff-Nielsen
Department of Mathematical Sciences, University of Aarhus

Bent Nielsen
Nuffield College, University of Oxford

Neil Shephard
Nuffield College, University of Oxford

Carla Ysusi
Department of Statistics, University of Oxford

Abstract

We use high frequency financial data to proxy, via the realised variance, each day's financial variability. Based on a semiparametric stochastic volatility process, a limit theory shows you can represent the proxy as a true underlying variability plus some measurement noise with known characteristics. Hence filtering, smoothing and forecasting ideas can be used to improve our estimates of variability by exploiting the time series structure of the realised variances. This can be carried out based on a model or without a model. A comparison is made between these two methods.

10.1 Introduction

Neil Shephard was fortunate to have Jim Durbin as his supervisor and time series teacher during his first year of graduate studies at the London School of Economics (LSE) in 1986-7. It was just before Jim retired. Jim was very interested in state space models, having recently written the Harvey and Durbin (1986) influential seat-belt case study on structural time series models.

Ole Barndorff-Nielsen's main contact with the research work of Jim Durbin has been with his pathbreaking paper Durbin (1980). Together with the papers by Cox (1980) and Hinkley (1980), this was of key import for the discovery of the general form of the p^*-formula for the law of the maximum likelihood estimator and hence the development of the theory that has flown from that formula (see Barndorff-Nielsen and Cox (1994) and the survey paper by Skovgaard (2001)).

Jim's research has had a profound impact on statistics and econometrics. From modelling, estimating and testing time series models to instrumental variables and general estimating equations, through to modern distribution theory, his work has been characterised by energy and inventiveness. He has an original mind. His teaching at the LSE had a profound impact on the course of British econometrics for, with Denis Sargan, he revolutionised the technical standards expected of their students. The current high position of British econometrics is a legacy we largely owe to Denis and Jim.

This paper touches on a number of Jim's interests. It uses continuous time methods, discusses some asymptotic distributional theory and eventually builds towards what might be called a structural time series model.

We use high frequency financial data to proxy each day's financial variability. A limit theory shows you can represent the proxy as a signal, true underlying variability, plus some measurement noise with known characteristics. Hence time series filtering, smoothing and forecasting ideas can be used to improve our estimates of variability by exploiting the time series structure of the data.

In Section 10.2 we review the asymptotic distribution theory of realised variance, linking it to stochastic volatility and quadratic variation. Section 10.3 uses the distribution theory to derive an optimal filtering, smoothing and forecasting method for the signal in volatility models, the so-called integrated variance. We show that this can be implemented in a model free way or based on a parametric model. In Section 10.4 we discuss how to operationalise the model free approach, while Section 10.5 discusses the corresponding model based approach. In Section 10.6 we draw our conclusions. The Appendix contains a proof of a theorem we state in Section 10.3.

10.2 Every day is different: historical measures of variability

10.2.1 The continuous time framework

This paper looks at measuring and forecasting the level of variability of asset prices in a financial market. This theory assumes a flexible stochastic volatility (SV) model for log-prices y^* which follow

$$y^*(t) = \alpha^*(t) + \int_0^t \tau^{1/2}(u) \mathrm{d}w(u), \quad t \geq 0, \qquad (10.1)$$

where α^* has locally bounded variation paths and τ is a strictly positive process which is càdlàg. The processes $\tau^{1/2}$ and α^* are assumed to be stochastically independent of the standard Brownian motion w. We call $\tau^{1/2}$ the *instantaneous* or *spot volatility*, τ the corresponding *variance* and α^* the *drift* process. Also we define

$$\tau^*(t) = \int_0^t \tau(u) \mathrm{d}u,$$

the integrated variance. Throughout we will assume the following condition holds with probability 1:

(C) For all $\epsilon > 0$ $\tau^*(t) < \infty$ exists and α^* has the property that for $\delta \to 0$

$$\delta^{-3/4} \max_{1 \leq j \leq M} |\alpha^*(j\delta) - \alpha^*((j-1)\delta)| = o(1), \qquad (10.2)$$

where M is a positive integer and $M\delta = t$.

Condition (**C**) implies that the α^* process is continuous and so is predictable, while $\int_0^t \tau^{1/2}(u)\mathrm{d}w(u)$ is a continuous local martingale. Hence y^* is a rather flexible continuous semimartingale. Assumption (**C**) also allows the volatility to have, for example, deterministic diurnal effects, jumps, long memory, no unconditional mean or to be non-stationary.

Over an interval of time of length $\hbar > 0$, which we think of concretely as a day, returns on the ith day are defined as

$$y_i = y^*(\hbar i) - y^*\{(i-1)\hbar\}, \quad i = 1, 2, \ldots, T, \qquad (10.3)$$

which implies that

$$y_i | \alpha_i, \tau_i \sim N(\alpha_i, \tau_i), \quad \text{where} \quad \alpha_i = \alpha^*(i\hbar) - \alpha^*\{(i-1)\hbar\},$$

while

$$\tau_i = \tau^*(i\hbar) - \tau^*\{(i-1)\hbar\}.$$

Here τ_i is called the *actual variance* and α_i is the *actual mean*. Reviews of the literature on the SV topic are given in Taylor (1994), Shephard (1996) and Ghysels, Harvey and Renault (1996), while statistical and probabilistic aspects are studied in detail in Barndorff-Nielsen and Shephard (2001).

The focus of this paper will eventually be on filtering, smoothing and forecasting τ_i. For shorthand, we call filtering and smoothing 'measuring.'

10.2.2 Realised variance

Our econometric approach is motivated by the advent of complete records of quotes or transaction prices for many financial assets. Theoretical and empirical work suggests that the use of such high frequency data is both informative and simplifying for it brings us closer to the theoretical models based on continuous time. However, market microstructure effects (e.g. discreteness of prices, bid/ask bounce, irregular trading etc.) mean that there is a mismatch between asset pricing theory based on semimartingales and the data at very fine time intervals. This implies that we cannot simply rely on empirical computations based on literally infinitesimal returns. In practice we will use a large but not infinite number of high frequency returns in our empirical work.

We suppose there are M intra-$\hbar$ observations during each $\hbar > 0$ time period. Our approach is to think of M as large and increasing. Then high frequency observations will be defined as

$$y_{j,i} = y^*\left((i-1)\hbar + \frac{\hbar j}{M}\right) - y^*\left((i-1)\hbar + \frac{\hbar(j-1)}{M}\right), \qquad (10.4)$$

the jth intra-$\hbar$ return for the ith period (e.g. if $\hbar$ is a day, $M = 288$, then this is the jth five minute return on the ith day). This is illustrated in Figure 10.1 which displays $y^*(t)$ at five minute intervals for the first five days of the Olsen dollar/DM series. It starts on 1st December 1986 and ignores weekend breaks. This series is constructed every five minutes by the Olsen group from bid and ask quotes which appeared on the Reuters screen (see Dacorogna, Gencay, Müller, Olsen and Pictet (2001) for details). We have set it up so that $y^*(0) = 0$. Figure 10.1(b) displays the returns when $M = 1$, which correspond to daily price movements. (c) uses $M = 8$ and shows three hour returns. Finally (d) displays the case where $M = 48$, where we are using thirty minute returns.

The basis of our paper is to first work through the historical summary of variability, which can be thought of as estimators of past actual volatility τ_i. These are built using the M intra-$\hbar$ observations. The focus is on the

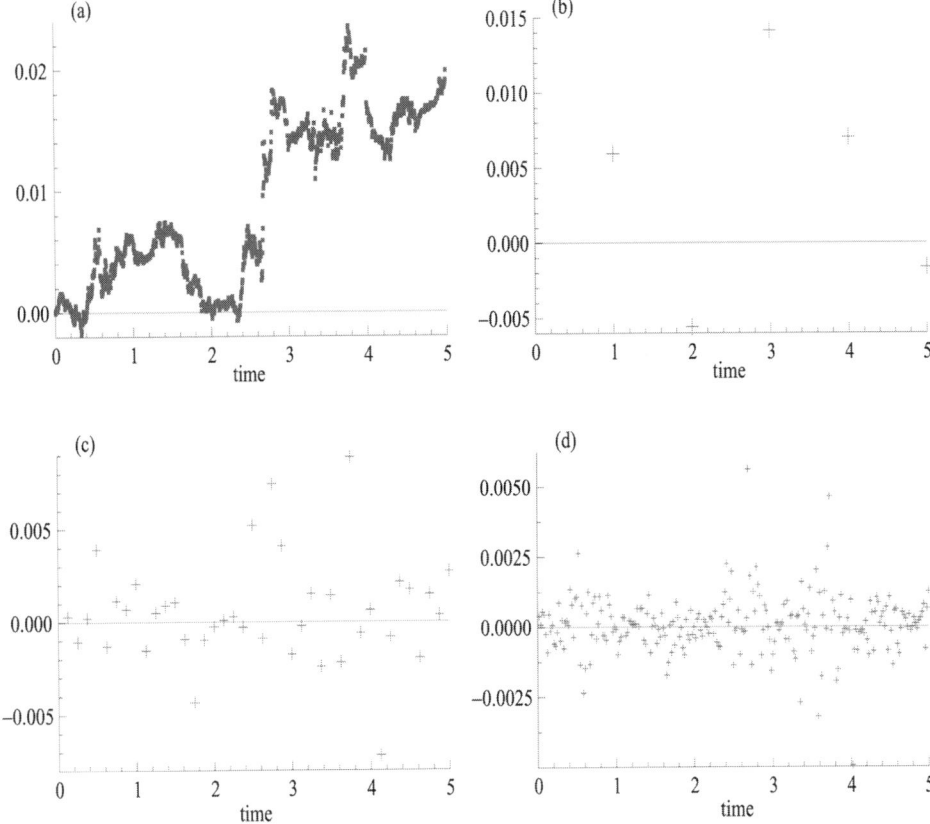

Fig. 10.1. Log-price and returns at different frequencies for the first five days of the Olsen data: (a) log-price $y^*(t)$ plotted every five minutes with $y^*(0) = 0$; (b) daily returns with $M = 1$; (c) three hour returns with $M = 8$; (d) thirty minute returns with $M = 48$.

realised variance[1]

$$[y_M^*]_i = \sum_{j=1}^M y_{j,i}^2. \quad (10.5)$$

Notice this estimator is entirely self-contained, that is, it only uses data from

---

[1] Sums of squared returns are often called realised volatility in econometrics, while we use the name realised variance for that term and realised volatility for the corresponding square root. The use of volatility to denote standard deviations rather than variances is standard in financial economics. We have chosen to follow this nomenclature rather than the one more familiar in econometrics. Confidence intervals for the realised volatility follow by square rooting the confidence intervals for the realised variance.

the ith time period to estimate τ_i. Its cousin *realised volatility*,

$$\sqrt{\sum_{j=1}^{M} y_{j,i}^2},$$

has been used in financial economics for many years by, for example, Schwert (1989), Taylor and Xu (1997), Christensen and Prabhala (1998), Andersen, Bollerslev, Diebold and Labys (2001) and Andersen, Bollerslev, Diebold and Ebens (2001). However, until recently little theory was known about realised variance outside the Brownian motion case. See the incisive review by Andersen, Bollerslev and Diebold (2003). Some other pieces on this work we would like to highlight are Meddahi (2002) and Andersen, Bollerslev and Meddahi (2004), although many other interesting papers exist which are discussed by Andersen, Bollerslev and Diebold (2003).

10.2.3 Properties of realised variance

It is very well known that the theory of quadratic variation (e.g. Jacod and Shiryaev (1987, p. 55), Protter (1990) and Back (1991)) implies that

$$[y_M^*]_i \xrightarrow{p} \tau_i,$$

as $M \to \infty$. This does not depend upon[2] the exact form of α^* or τ.

This consistency result is illustrated in Figure 10.2 which displays a simulated sample path of integrated variance τ_i from an Ornstein–Uhlenbeck (OU) process given by the solution to

$$\mathrm{d}\tau(t) = -\lambda \tau(t)\mathrm{d}t + \mathrm{d}z(\lambda t),$$

where z is a subordinator (a process with independent, stationary and nonnegative increments). In this example we construct the process so that $\tau(t)$ has a $\Gamma(4,8)$ stationary distribution, $\lambda = -\log(0.99)$ and $\hbar = 1$. Also drawn are the sample path of the realised variances $\sum_{j=1}^{M} y_{j,i}^2$ (depicted using crosses), where

$$y^*(t) = \beta \tau^*(t) + \int_0^t \tau^{1/2}(u)\mathrm{d}w(u)$$

and $\beta = 0.5$. The realised variances are computed using a variety of values of M. We see that as M increases the size of $\sum_{j=1}^{M} y_{j,i}^2 - \tau_i$ falls, illustrating the consistency of $\sum_{j=1}^{M} y_{j,i}^2$ for τ_i even though β is not zero.

In a recent paper Barndorff-Nielsen and Shephard (2002a), subsequently

---

[2] Indeed, the probability limit of realised variance is known under the even weaker assumption, that the price process is a semimartingale.

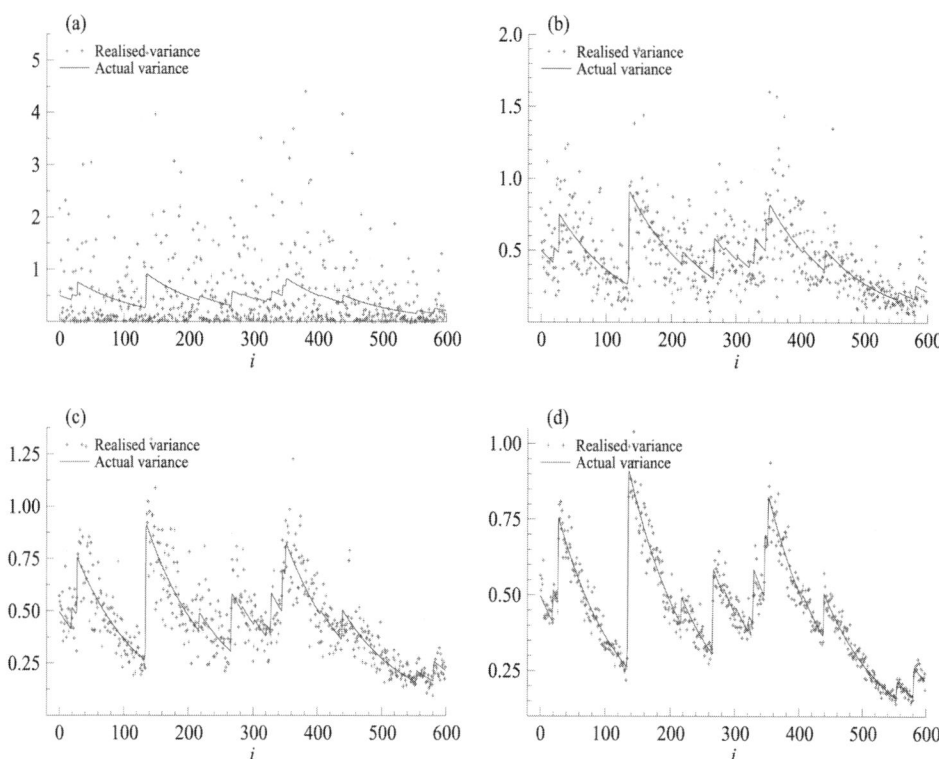

Fig. 10.2. Actual τ_i and realised $\sum_{j=1}^{M} y_{j,i}^2$ (with M varying) variance based upon a $\Gamma(4,8)$-OU process with $\lambda = -\log(0.99)$ and $\hbar = 1$. (a) $M = 1$; (b) $M = 12$; (c) $M = 48$; (d) $M = 288$.

extended in Barndorff-Nielsen and Shephard (2003, 2004), have strengthened the above result considerably. The main result is that:

Theorem 10.1 *Under assumption (C) for the SV model in (10.1), for any positive $\hbar$ and $M \to \infty$*

$$\frac{\sqrt{\frac{M}{\hbar}}\left([y_M^*]_i - \tau_i\right)}{\sqrt{2\tau_i^{[2]}}} \xrightarrow{d} N(0,1), \quad \text{where} \quad \tau_i^{[2]} = \int_{(i-1)\hbar}^{i\hbar} \tau^2(u)\mathrm{d}u. \qquad (10.6)$$

□

We call τ^2 and $\tau_i^{[2]}$ the *spot* and *actual quarticity*, respectively. Of course the problem with this theory is that $\tau_i^{[2]}$ is unknown. This is tackled by using the fact that

$$\frac{M}{\hbar}\sum_{j=1}^{M} y_{j,i}^4 \xrightarrow{p} 3\tau_i^{[2]}.$$

An implication of this is that we can use the feasible limit theory

$$\frac{[y_M^*]_i - \tau_i}{\sqrt{\frac{2}{3}\sum_{j=1}^{M} y_{j,i}^4}} \xrightarrow{\mathcal{L}} N(0,1), \qquad (10.7)$$

due to Barndorff-Nielsen and Shephard (2002a).

Of course in practice it may make sense to transform the above limit theorem to impose, *a priori*, positivity on the approximating distribution. In particular it seems natural to work with the logarithmic transformation of the realised variance ratio so that (see Barndorff-Nielsen and Shephard (2002b))

$$\frac{\sqrt{M/\hbar}\left\{\log[y_M^*]_i - \log\tau_i\right\}}{\sqrt{2\tau_i^{[2]}/(\tau_i)^2}} \xrightarrow{\mathcal{L}} N(0,1) \quad \text{or} \quad \frac{\log[y_M^*]_i - \log\tau_i}{\sqrt{\left(2/3\,[y_M^*]_i^2\right)\sum_{j=1}^{M} y_{j,i}^4}} \xrightarrow{\mathcal{L}} N(0,1).$$

The following remarks can be made about these results.

- $\sum_{j=1}^{M} y_{j,i}^2$ converges to $\int_{\hbar(i-1)}^{\hbar i} \tau(u)du$ at rate $\sqrt{M}$.
- The limit theorem is unaffected by the form of the drift process α^*, the smoothness assumption (**C**) is sufficient that its effect becomes negligible.
- Knowledge of the form of the volatility dynamics is not required in order to use this theory.
- The fourth moment of returns need not exist for the asymptotic normality to hold. In such heavy tailed situations, the stochastic denominator $\int_{(i-1)\hbar}^{i\hbar} \tau^2(u)du$ loses its unconditional mean.
- The volatility process τ can be nonstationary, exhibit long memory or include intra-day effects.
- $\sum_{j=1}^{M} y_{j,i}^2 - \int_{\hbar(i-1)}^{\hbar i} \tau(u)du$ has a mixed Gaussian limit implying that marginally it will have heavier tails than a normal.
- The magnitude of the error $\sum_{j=1}^{M} y_{j,i}^2 - \int_{\hbar(i-1)}^{\hbar i} \tau(u)du$ is likely to be large in times of high volatility.

10.3 Time series of realised variances

10.3.1 Motivation

So far we have analysed the asymptotics of $\sum_{j=1}^{M} y_{j,i}^2$ as $M \to \infty$ for a single i. In this section we will explicitly analyse a long time series of realised variances, trying to use the time series structure to construct more efficient estimators and forecasts of τ_i. This analysis will be based an assumption that integrated variance is covariance stationary. This is a much stronger assumption than the ones we have previously employed. To start out we

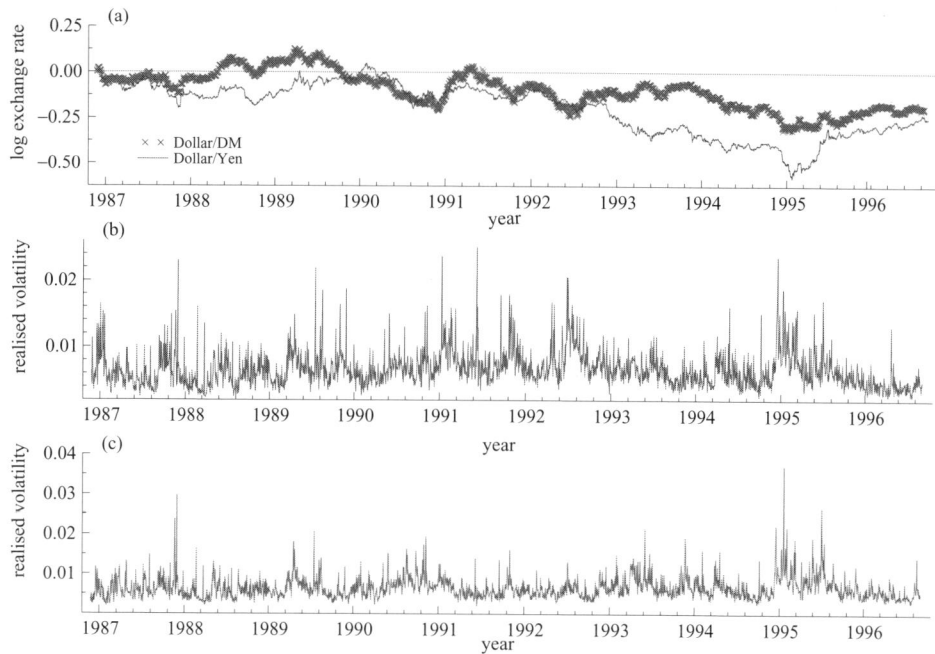

Fig. 10.3. Long time series of the daily movements in the dollar against the DM and yen: (a) the level of the log exchange rates compared to the rate at 1 December 1986; (b) realised volatility each day computed using $M = 144$ for the DM series; (c) realised volatility each day computed using $M = 144$ for the yen series.

have drawn Figure 10.3 which displays information on the Olsen data on the DM and yen against the US dollar. Figure 10.3(a) shows the movement of the log-prices from 1 December 1986 for ten years, with the log-prices transformed to be zero at the start of the sample. This is the same series as Figure 10.1(a) but now the graph is on a very long time scale. Figure 10.3(b) shows the daily realised volatility $\sqrt{\sum_{j=1}^{M} y_{j,i}^2}$ drawn against i, the day, for the DM series. It is computed using $M = 144$, corresponding to ten minute returns. It is quite a ragged series but with periods of increased volatility.

A similar picture emerges from the corresponding realised volatility for the yen given in Figure 10.3(c).

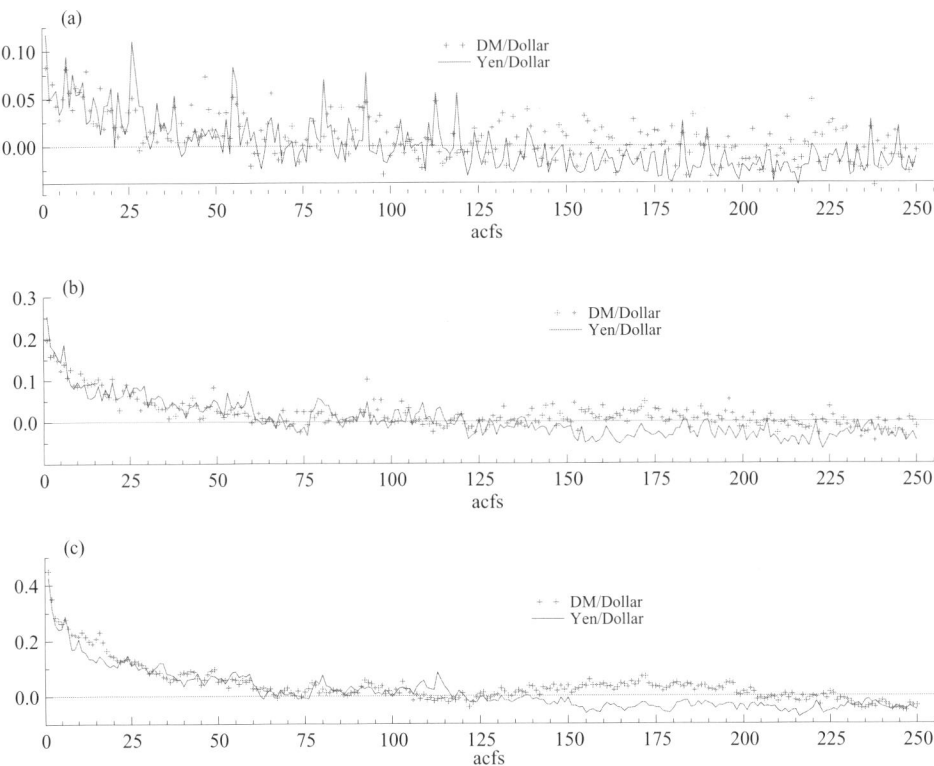

Fig. 10.4. Autocorrelations of realised variances using a long time series of the movements in the dollar against the DM and yen: (a) $M = 1$ case, which corresponds to daily returns; (b) $M = 8$ case; (c) $M = 72$.

10.3.2 Asymptotics

For each exchange rate we have computed realised variances each day. We can then regard the derived series as a daily time series

$$\sum_{j=1}^{M} y_{j,1}^2, \sum_{j=1}^{M} y_{j,2}^2, \ldots, \sum_{j=1}^{M} y_{j,T}^2.$$

This new series is of length T, the number of days in the sample.

The correlograms for the daily time series of realised volatilities of these quantities are displayed in Figure 10.4 for a variety of values of M. Throughout

250 lags are used in these figures which correspond to measuring correlations over a one year period. Figure 10.4(a) shows the results for $M = 1$. In this case the realised variances are simply squared daily returns. The correlogram has the well-known slow decay but starting at quite a low level. Figure 10.4(b) shows the effect of increasing M slightly to 8, now we are computing the realised quantities using 150 minute returns. Figure 10.4(c) shows the corresponding results for $M = 72$, which uses 20 minute returns. All the autocorrelations are boosted as M increases from 8, however, the broad story is the same. A clear observation is that the autocorrelations are becoming less jagged with the increase in M.

Having observed some of the empirical features of the realised variances we will now set out a theoretical framework for the study of the time series of realised variances.

We define sequences of realised and actual variances for the sth day to the pth day

$$[y^*_M]_{s:p} = \left(\sum_{j=1}^M y^2_{j,s}, \sum_{j=1}^M y^2_{j,s+1}, \ldots, \sum_{j=1}^M y^2_{j,p}\right)' \quad \text{and} \quad \tau_{s:p} = (\tau_s, \tau_{s+1}, \ldots, \tau_p)',$$

where we recall that $\tau_i = \int_{\hbar(i-1)}^{\hbar i} \tau(u)du$. The asymptotic theory of realised variance implies that

$$\sqrt{\frac{M}{\hbar}} ([y^*_M]_{s:p} - \tau_{s:p}) \xrightarrow{\mathcal{L}} N\left\{0, 2\text{diag}\left(\tau^{[2]}_{s:p}\right)\right\},$$

where $\tau^{[2]}_i = \int_{\hbar(i-1)}^{\hbar i} \tau^2(u)du$.

10.3.3 Linear estimators

Although estimating $\tau_{s:p}$ by $[y^*_M]_{s:p}$ has attractions, the variance of the error is typically quite large even when M is high. More precise estimators could be obtained by pooling neighbouring time series observations for realised variances tend to be highly correlated through time. This pooling will typically reduce the variance of the estimator, but will induce a bias.

To set up a formal framework for this discussion, abstractly write A as a matrix of nonstochastic weights. Then

$$\sqrt{\frac{M}{\hbar}} (A[y^*_M]_{s:p} - A\tau_{s:p}) |\tau^{[2]}_{[\nu]} \xrightarrow{\mathcal{L}} N\left\{0, 2A\text{diag}\left(\tau^{[2]}_{s:p}\right) A'\right\}.$$

Now consider the statistic

$$\widehat{\tau_{s:p}} = c\text{E}\left(\tau_{s:p}\right) + A[y^*_M]_{s:p}.$$

We assume that the realised variances constitute a covariance stationary process, which means that

$$\mathrm{E}\left(\tau_{s:p}\right) = \iota \mathrm{E}\left(\tau_t\right),$$

where $\iota = (1, 1, \ldots, 1)'$. Notice the stationarity is at the daily level, it does not need that the continuous time process τ is stationary.

The population weighted least squares estimator of $\tau_{s:p}$ sets

$$c = (I - A)\iota$$

and

$$\begin{aligned}
A &= \mathrm{Cov}\left(\tau_{s:p}, [y_M^*]_{s:p}\right) \left[\mathrm{Cov}\left([y_M^*]_{s:p}\right)\right]^{-1} \\
&= \mathrm{Cov}\left(\tau_{s:p}\right) \left[\mathrm{Cov}\left([y_M^*]_{s:p}\right)\right]^{-1} \\
&= \mathrm{Cov}\left(\tau_{s:p}\right) \left[\mathrm{Cov}\left(\tau_{s:p}\right) + \frac{2\hbar \mathrm{E}\left(\tau_i^{[2]}\right)}{M} I\right]^{-1}.
\end{aligned}$$

This follows as $\mathrm{Cov}\left(\tau_{s:p}, [y_M^*]_{s:p} - \tau_{s:p}\right) = 0$ and

$$\mathrm{E}\left\{\mathrm{diag}\left(\tau_{s:p}^{[2]}\right)\right\} = \mathrm{E}\left(\tau_i^{[2]}\right) I.$$

Notice that as $M \to \infty$ so $\widehat{A} \to I$ and $\widehat{\tau}_{s:p} \xrightarrow{p} \tau_{s:p}$. Unconditionally $\widehat{\tau}_{s:p}$ has a variance of

$$(2\hbar/M)\,\mathrm{E}\left(\tau_{s:p}^{[2]}\right) AA' + (I - A)\,\mathrm{Cov}\left(\tau_{s:p}\right)(I - A').$$

At the end of this section we will study conditions under which A is guaranteed to be nonnegative.

10.3.4 Implementation

In practice A has to be estimated from the data. Broadly this can be carried out in two ways:

(i) estimating A by using empirical averages from the data;
(ii) implying A from an estimated parametric model.

10.3.5 Positivity

Before going on to discuss the above issues of implementation we will take a moment to give conditions under which all the elements of

$$A = \mathrm{Cov}\,(\tau_{s:p}) \left[\mathrm{Cov}\,(\tau_{s:p}) + \frac{2\hbar \mathrm{E}\left(\tau_i^{[2]}\right)}{M} I \right]^{-1}$$

are non-negative. Such matrices are said to be totally nonnegative. The following example shows that A is not necessarily totally nonnegative.

Example 10.1 Suppose, $|a| < 1$ and we write $u_i = [y_M^*]_i - \tau_i$. Then

$$\mathrm{Cov}(\tau_{s:s+1}) = \begin{pmatrix} 1 & a \\ a & 1 \end{pmatrix}, \quad \mathrm{Cov}([y_M^*]_{s:s+1}) = \mathrm{Cov}(\tau_{s:s+1}) + I\mathrm{Var}(u_i),$$

and so

$$A = \frac{1}{\{\mathrm{Var}(u_i)+1\}^2 - a^2} \begin{pmatrix} \mathrm{Var}(u_i) + 1 - a^2 & a\mathrm{Var}(u_i) \\ a\mathrm{Var}(u_i) & \mathrm{Var}(u_i) + 1 - a^2 \end{pmatrix}.$$

Hence all weights are non-negative iff $a \geq 0$.

The next theorem gives conditions on $\mathrm{Cov}(\tau_{s:p})$ to ensure total non-negativity of A.

Theorem 10.2 *Assume that $\mathrm{Cov}(\tau_{s:p})$ is positive semi-definite. Then the necessary and sufficient condition for all the elements of A to be nonnegative for all values of $\sigma > 0$ is that $\mathrm{Cov}(\tau_{s:p})^{-1}$ has nonpositive off-diagonal elements.*

Proof. Given in the Appendix.

The condition that $\mathrm{Cov}(\tau_{s:p})^{-1}$ has to have non-positive off-diagonal elements has the following straightforward statistical interpretation.

Remark 10.1 Suppose X is a positive definite covariance matrix. We write the i, j element of X^{-1} as $x^{i,j}$. Then

$$\frac{-x^{i,j}}{\sqrt{x^{i,i} x^{j,j}}}$$

is the partial correlation between y_i and y_j. That is, it is the ordinary correlation between y_i and y_j conditioning on all the other elements of y (see, for example, Cox and Wermuth 1996, p. 69)).

10.4 Model free approach

Here we will discuss estimating A by using empirical averages from the data, delaying until the next section a discussion of a model based method.

If we have a large sample from a stationary process of realised variances and the daily process is ergodic then we have that

$$\widehat{\mathrm{E}\left(\tau_i^{[2]}\right)} = \left(\frac{1}{T}\sum_{i=1}^{T}\frac{M}{3\hbar}\sum_{j=1}^{M}y_{j,i}^4\right) \xrightarrow{p} \mathrm{E}\left(\tau_i^{[2]}\right),$$

as T and M go to infinity. Likewise $\mathrm{Cov}\left([y_M^*]_{s:p}\right)$ can be estimated by averages of the time series of realised variances. Hence

$$A = \mathrm{Cov}\left(\tau_{s:p}\right)\left[\mathrm{Cov}\left([y_M^*]_{s:p}\right)\right]^{-1}$$

can be replaced by

$$\widehat{A} = \left\{\widehat{\mathrm{Cov}\left([y_M^*]_{s:p}\right)} - \widehat{\mathrm{E}\left(\tau_i^{[2]}\right)}\frac{2\hbar}{M}I\right\}\left[\widehat{\mathrm{Cov}\left([y_M^*]_{s:p}\right)}\right]^{-1},$$

which is a feasible weighting matrix. This will imply $\widehat{c} = \left(I - \widehat{A}\right)\iota$ and

$$\widehat{\tau_{s:p}} = \widehat{c}\widehat{\mathrm{E}\left(\tau_{s:p}\right)} + \widehat{A}[y_M^*]_{s:p}.$$

This is a feasible, model free, optimal linear estimator of $\tau_{s:p}$ based on $[y_M^*]_{s:p}$.

10.4.1 Illustration

Table 10.1 contains the estimated weights for a single actual variance using a single realised variance sequence, so $s = p = i$, for the DM and Yen series. This is based on the entire time series sample of nearly 2500 days.

Table 10.1. *Estimated weights for $\widehat{\tau}_i$, the regression estimator of τ_i which uses only $[y_M^*]_i$ and an intercept. Results for the DM and yen series against the dollar. File:* `daily_timeseries.ox`

M	DM		Yen	
	$\widehat{A}$	$\widehat{c}$	$\widehat{A}$	$\widehat{c}$
1	0.182	0.817	0.229	0.770
8	0.449	0.550	0.513	0.486
72	0.778	0.221	0.789	0.210
288	0.877	0.122	0.906	0.093

We can see the results do not vary very much with the series being used.

In particular, for $M = 8$ the estimator of τ_i for the DM series would be

$$\widehat{\tau}_i = 0.550 \frac{1}{T} \sum_{j=1}^{T} [y_M^*]_j + 0.449 [y_M^*]_i. \qquad (10.8)$$

Thus for small values of M the regression estimator puts a moderate weight on the realised variance and more on the unconditional mean of the variances. As M increases this situation reverses, but even for large values of M the unconditional mean is still quite highly weighted. From now on we will solely focus on the DM series to make the exposition more compact.

In the dynamic case the results are more complicated to present. Here we start by considering estimating three actual variances using three contiguous realised variances – one lag, one lead and the contemporaneous realised variance. Thus

$$s : p = (i - 1, i, i + 1),$$

and so $\widehat{A}$ will be a 3×3 matrix and $\widehat{c}$ a 3×1 vector. In the case of $M = 8$ we have that

$$\{\text{Cov}([y_M^*]_{1:3})\}^{-1} = \frac{1}{100^2} \begin{pmatrix} 2.07 & -0.358 & -0.258 \\ -0.358 & 2.10 & -0.358 \\ -0.258 & -0.358 & 2.07 \end{pmatrix},$$

while

$$\widehat{A} = \begin{pmatrix} 0.418 & 0.100 & 0.072 \\ 0.100 & 0.409 & 0.100 \\ 0.072 & 0.100 & 0.418 \end{pmatrix}, \quad \widehat{c} = \begin{pmatrix} 0.408 \\ 0.388 \\ 0.408 \end{pmatrix}.$$

Thus the second row of $\widehat{A}$ implies the smoothed estimator of τ_i is

$$\widehat{\tau}_i = 0.100 [y_M^*]_{i-1} + 0.409 [y_M^*]_i + 0.100 [y_M^*]_{i+1} + 0.388 \frac{1}{T} \sum_{j=1}^{T} [y_M^*]_j.$$

The corresponding result for $M = 72$ is

$$\widehat{A} = \begin{pmatrix} 0.712 & 0.105 & 0.053 \\ 0.105 & 0.684 & 0.105 \\ 0.053 & 0.105 & 0.712 \end{pmatrix}, \quad \widehat{c} = \begin{pmatrix} 0.128 \\ 0.105 \\ 0.128 \end{pmatrix}.$$

This shows that the weighting on the diagonal elements of $\widehat{A}$ is much higher, while the size of $\widehat{c}$ has fallen by a factor of around 4. In both cases a lot of weight is put on neighbouring values of the realised variance and on

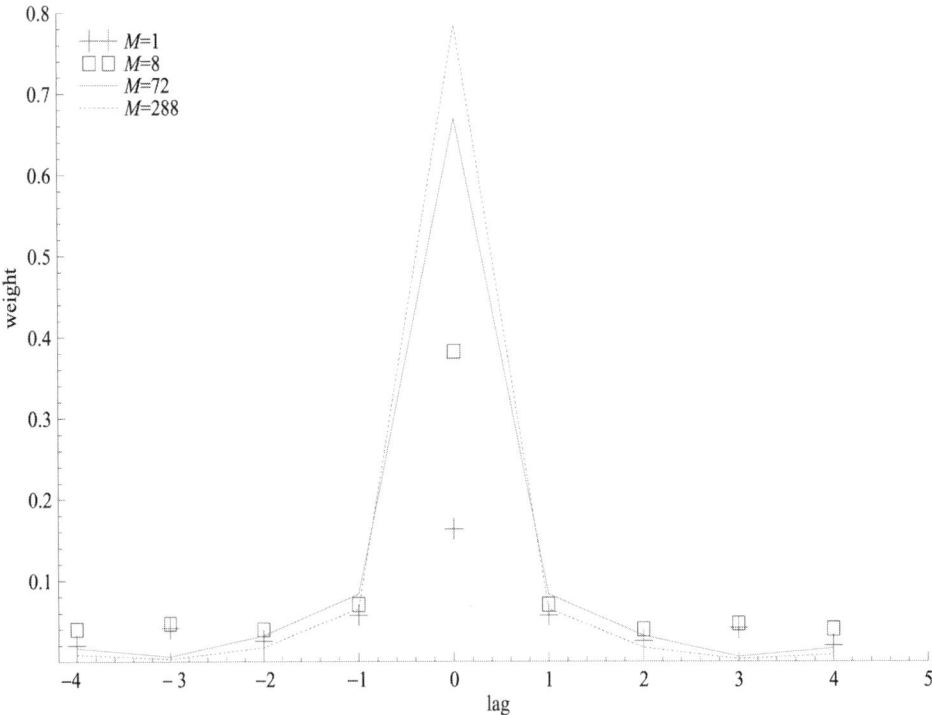

Fig. 10.5. Estimated weight vector for estimating τ_i using $[y^*_M]_{i-4}, [y^*_M]_{i-3}, \ldots,$ $[y^*_M]_{i+4}$ drawn against lag length. It is computed using the dollar against the DM. It shows that as M increases the weight on $[y^*_M]_i$ increases. Corresponding to these results is $\widehat{c}$, which moves from 0.548, 0.222, 0.0553, 0.026 as M increases through 1, 8, 72 to 288.

the intercept, although the weight on $[y^*_M]_i$ is not very much smaller than in the univariate case.

The corresponding filtered estimator (which seems a natural competitor to using the raw realised variance $[y^*_M]_i$) is obtained by using the last row of the $\widehat{A}$ matrix. Then we have, for $M = 8$,

$$\widehat{\tau}_i = 0.072[y^*_M]_{i-2} + 0.100[y^*_M]_{i-1} + 0.418[y^*_M]_i + 0.408\frac{1}{T}\sum_{j=1}^{T}[y^*_M]_j.$$

Here we see the usual decay in the weight as we go further back in time.

Figure 10.5 shows the middle row of $\widehat{A}$ for the case of estimating τ_i using nine realised variances, four lags and four leads together with $[y^*_M]_i$. It displays the weights as a function of M indicating how quickly the weights focus on $[y_M^*]_i$ as M increases. The legend of the figure also gives the value of the weight put on the unconditional mean of the realised variance. For $M = 72$ it is 0.0553, which is much lower than in the trivariate case of 0.105 and univariate case of 0.221.

Figure 10.6 shows a time series of realised variances for a number of values of M together with the corresponding estimator $\hat{\tau}_i$ based on nine observations, four leads, the current value and four lags. The smoothed estimator seems to deliver sensible answers, with the results being less sensitive to large values of the realised variances, in particular for small M.

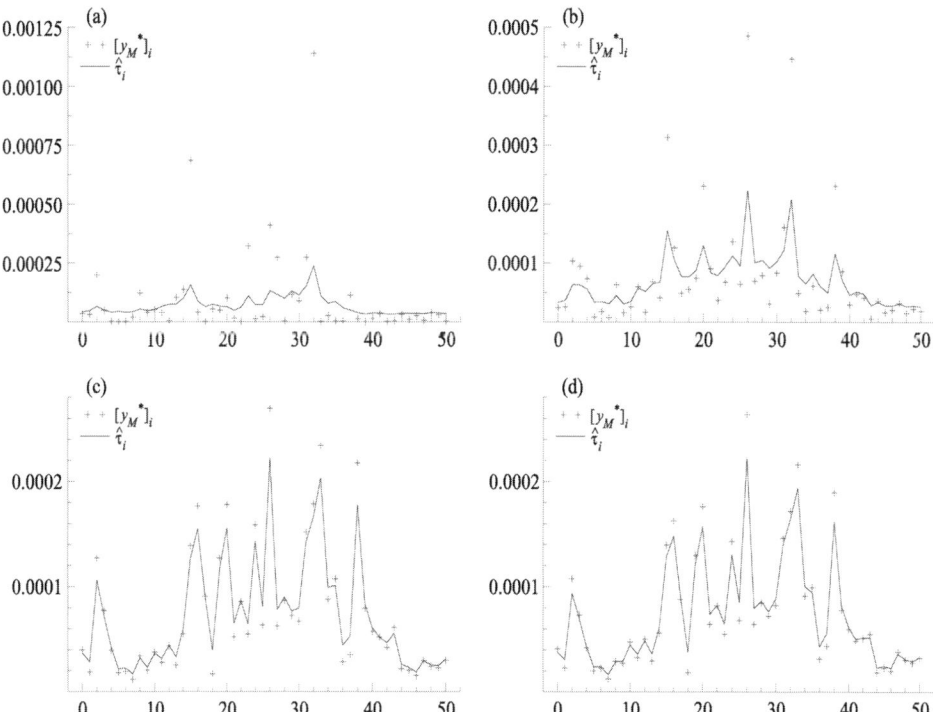

Fig. 10.6. The estimated τ_i using realised variance and weighted version of $[y_M^*]_{i-4}, [y_M^*]_{i-3}, \ldots, [y_M^*]_{i+4}$ computed using the dollar against the DM; (a) $M = 1$; (b) $M = 8$; (c) $M = 144$; and (d) $M = 288$.

Table 10.2 reports, using the DM data,

$$\frac{1}{T}\sum_{i=1}^{T}([y_M^*]_i - [y_{288}^*]_i)^2,$$

which is an empirical approximation to the mean square error of the realised variance estimator, using $[y^*_{288}]_i$ as a good proxy for τ_i. (The model based estimators would turn out to deliver even more accurate estimators, but this could be interpreted as biasing the results towards the model based approach and so here we use the raw realised variance.) The table shows a rapid decline in the mean square error with M. It also shows the corresponding results for the estimators based on just a regression on a constant and $[y^*_M]_i$, and $\widehat{\tau}_i$, which uses $[y^*_M]_{i-4}, [y^*_M]_{i-3}, \ldots, [y^*_M]_{i+3}, [y^*_M]_{i+4}$. The results reflect the fact that these adjusted estimators are much more efficient than the realised variance, although the difference between using the time series dynamics and the simple regression estimator is modest.

Table 10.2. *Mean square error of the realised variance and the regression estimator and the time series estimators $\widehat{\tau}_i$, which are based on $[y^*_M]_{i-4}$, $[y^*_M]_{i-3}, \ldots, [y^*_M]_{i+3}, [y^*_M]_{i+4}$. These are computed using $M = 1$, 8 and 72. The true value is taken as $[y^*_M]_i$ for 288.*

M	DM			yen		
	$[y^*_M]_i$	$(1-\widehat{A})E(\tau_i)+\widehat{A}[y^*_M]_i$	$\widehat{\tau}_i$	$[y^*_M]_i$	$(1-\widehat{A})E(\tau_i)+\widehat{A}[y^*_M]_i$	$\widehat{\tau}_i$
1	0.822	0.175	0.145	1.16	0.198	0.168
8	0.207	0.0989	0.0769	0.186	0.117	0.0985
72	0.0377	0.0345	0.0317	0.0424	0.0406	0.0378

10.4.2 Forecasting

Suppose we are interesting in forecasting τ_{p+1} based on the time series of realised variances $[y^*_M]_{s:p}$. Throughout we assume that the integrated and realised variances are second order stationary. The best linear forecast is given by

$$\widehat{\tau_{p+1|s:p}} = cE(\tau_{p+1}) + A[y^*_M]_{s:p},$$

where

$$c = 1 - A\iota$$

and

$$\begin{aligned}A &= \text{Cov}(\tau_{p+1}, [y^*_M]_{s:p})\left[\text{Cov}([y^*_M]_{s:p})\right]^{-1} \\ &= \text{Cov}([y^*_M]_{p+1}, [y^*_M]_{s:p})\left[\text{Cov}([y^*_M]_{s:p})\right]^{-1}.\end{aligned}$$

This is a somewhat surprising result for A can be computed without reference to the details of the asymptotic theory of error. It just falls out from the asymptotic relationship between the realised variances, which can be empirically determined. Hence $\widehat{\tau_{p+1|s:p}}$ is feasible. However, as $M \to \infty$ this is not consistent. Instead

$$A \to \mathrm{Cov}\left(\tau_{p+1}, \tau_{s:p}\right) \left[\mathrm{Cov}\left(\tau_{s:p}\right)\right]^{-1},$$

and so

$$\widehat{\tau_{p+1|s:p}} \xrightarrow{p} c\mathrm{E}\left(\tau_{p+1}\right) + \left\{\mathrm{Cov}\left(\tau_{p+1}, \tau_{s:p}\right) \left[\mathrm{Cov}\left(\tau_{s:p}\right)\right]^{-1}\right\} \tau_{s:p}.$$

Extensions to multistep ahead predictions are straightforward. Importantly the above forecasting framework means that the one-step ahead predictions are generated by a $p-s+1$ order autoregression plus intercept model, where the intercept follows a particularly simple constraint so that the weights on the lagged coefficients plus the intercept add to 1. Unconstrained autoregressive forecasting in the context of realised variances has been carried out by Andersen, Bollerslev, Diebold and Labys (2003).

The simplest interesting example of the above approach is where $s = p$. Then we are forecasting one-step ahead based on a single realised variance. This produces

$$\widehat{\tau_{p+1|p}} = \mathrm{E}\left(\tau_{p+1}\right) + \mathrm{Cor}\left([y_M^*]_{p+1}, [y_M^*]_p\right) \left\{[y_M^*]_p - \mathrm{E}\left(\tau_{p+1}\right)\right\}.$$

In practice we replace expectations by averages and correlations by empirical correlations. Table 10.3 provides empirical estimators of A and c for the DM and yen series for a variety of values of M.

Table 10.3. *Estimated weights for $\widehat{\tau_{p+1|p}}$, the regression estimator of τ_{p+1} which uses only $[y_M^*]_p$ and an intercept. Results for the DM and yen series against the dollar.*

M	DM		yen	
	$\widehat{A}$	$\widehat{c}$	$\widehat{A}$	$\widehat{c}$
1	0.083	0.917	0.117	0.883
8	0.197	0.803	0.254	0.746
72	0.471	0.529	0.428	0.572
288	0.540	0.460	0.517	0.483

We can see again that the results do not vary very much with the series being used. In particular, for $M = 8$ then the estimator of τ_i for the DM series would be

$$\widehat{\tau_{p+1|p}} = 0.803 \frac{1}{T} \sum_{j=1}^{T} [y_M^*]_j + 0.197 [y_M^*]_p.$$

Thus the forecast shrinks much more to the mean than does the corresponding smoother given in (10.8).

Table 10.4 provides the weights when we use six lags of realised variances to forecast τ_{p+1}. It shows again that quite a lot of weight is placed on the constant c, while the most recent realised variance is also highly weighted. This results from the fact that the autocorrelation function of the realised variances initially declines very rapidly, followed by a slower decay rate at higher lag lengths.

Table 10.4. *Estimated weights for one-step ahead forecast of integrated variance* τ_{p+1}

M	$[y_M^*]_{p-5}$	$[y_M^*]_{p-4}$	$[y_M^*]_{p-3}$	$[y_M^*]_{p-2}$	$[y_M^*]_{p-1}$	$[y_M^*]_p$	$\widehat{c}$
1	0.040	0.014	0.028	0.053	0.034	0.073	0.753
8	0.074	0.046	0.074	0.089	0.083	0.138	0.493
144	0.089	0.067	0.080	0.031	0.134	0.321	0.273
288	0.050	0.093	0.051	0.038	0.111	0.397	0.256

10.4.3 Log based theory

A similar style of argument could have been used based on the log realised variances. Here we will write

$$\log[y_M^*]_{s:p} = (\log[y_M^*]_s, \ldots, \log[y_M^*]_p)'$$

and

$$\log \tau_{s:p} = (\log \tau_s, \ldots, \log \tau_p)'.$$

The pooled estimator has the asymptotic distribution (see Barndorff-Nielsen and Shephard (2002c))

$$\sqrt{\frac{M}{\hbar}} \left(A \log[y_M^*]_{s:p} - A \log \tau_{s:p} \right) | \tau_{s:p}^{[2]}, \tau_{s:p}$$

$$\xrightarrow{\mathcal{L}} N \left\{ 0, 2AE \begin{pmatrix} \tau_s^{[2]}/(\tau_s^2) & 0 & 0 \\ 0 & \ddots & 0 \\ 0 & 0 & \tau_p^{[2]}/(\tau_p^2) \end{pmatrix} A' \right\},$$

which would allow us to choose A as a least squares estimator of $\log \tau_{s:p}$ repeating the above argument. Weighting based on the log realised variances has the advantage that the Monte Carlo evidence suggests that the asymptotics for the log realised variance is accurate with the errors being approximately homoskedastic which suggests the weighting will be more effective.

The important result that we need to use is that

$$\left\{ \frac{1}{T} \sum_{i=1}^{T} \frac{(M/3\hbar) \sum_{j=1}^{M} y_{j,i}^4}{\left(\sum_{j=1}^{M} y_{j,i}^2\right)^2} \right\} \xrightarrow{p} \mathrm{E}\left(\frac{\tau_i^{[2]}}{\tau_i^2}\right),$$

and hence

$$\widehat{\log \tau_{s:p}} = c \mathrm{E}\left(\log \tau_{s:p}\right) + A \log[y_M^*]_{s:p}.$$

Of course

$$\frac{1}{T} \sum_{i=1}^{T} \log[y_M^*]_i \xrightarrow{p} \mathrm{E}\left(\log \tau_i\right),$$

hence we are left with just determining $\widehat{c}$ and $\widehat{A}$. If we assume that the realised variances are a covariance stationary process then the weighted least squares statistic of $\log \tau_{s:p}$ sets

$$c = (I - A)\iota$$

and

$$\begin{aligned} A &= \left[\mathrm{Cov}\left(\log \tau_{s:p}\right) + \frac{2\hbar \mathrm{E}\left(\tau_i^{[2]}/\tau_i^2\right)}{M} I \right]^{-1} \mathrm{Cov}\left(\log \tau_{s:p}\right) \\ &= \left[\mathrm{Cov}\left(\log[y_M^*]_{s:p}\right)\right]^{-1} \mathrm{Cov}\left(\log \tau_{s:p}\right). \end{aligned}$$

Of course for this statistic

$$\widehat{\log \tau_{s:p}} \rightarrow \log \tau_{s:p}$$

as $M \rightarrow \infty$, as expected.

This style of approach extends to the multivariate case where the focus is on estimating the actual covariance matrix (see Barndorff-Nielsen and Shephard (2004)). Then it makes sense to use these regression approaches based on the logs of the realised variances and the Fisher transformation of the realised correlation. The asymptotic theory of the realised covariation allows this approach to be feasible without specifying a parametric model for the spot covariance matrix.

10.5 Model based approach

10.5.1 General discussion and example

Suppose we write (when they exist) ξ, ω^2 and r, respectively, as the mean, variance and the autocorrelation function of the continuous time stationary variance process τ. Here we recall the discussion of Barndorff-Nielsen and Shephard (2002a) on estimating and forecasting τ_i based upon a parametric models for τ and the time series of realised variances. Let us write $u_i = [y_M^*]_i - \tau_i$, then the asymptotic theory tells us that for large M the u_i are approximately uncorrelated with

$$\mathrm{Var}\left(\sqrt{M} u_i\right) \to 2\hbar^2 \left(\omega^2 + \xi^2\right)$$

as $M \to \infty$. Thus the second order properties of $[y_M^*]_i$ can be approximated. In particular $\mathrm{E}\left([y_M^*]_i\right) = \hbar\xi + o(1)$ and for $s > 0$

$$\begin{aligned}
\mathrm{Var}([y_M^*]_i) &= 2M^{-1}\hbar^2\left(\omega^2 + \xi^2\right) + \mathrm{Var}(\tau_i) + o(1), \\
\mathrm{Cov}([y_M^*]_i, [y_M^*]_{i+s}) &= \mathrm{Cov}(\tau_i, \tau_{i+s}) + o(1), \\
\mathrm{Cov}([y_M^*]_i, \tau_i) &= \mathrm{Var}(\tau_i) + o(1), \\
\mathrm{Cov}([y_M^*]_i, \tau_{i+s}) &= \mathrm{Cov}(\tau_i, \tau_{i+s}) + o(1).
\end{aligned}$$

$\mathrm{Var}(\tau_i)$ and $\mathrm{Cov}(\tau_i, \tau_{i+s})$ were given for all covariance stationary processes in Barndorff-Nielsen and Shephard (2001). In particular

$$\mathrm{Var}(\tau_i) = 2\omega^2 r^{**}(\hbar) \quad \text{and} \quad \mathrm{Cov}\{\tau_i, \tau_{i+s}\} = \omega^2 \diamond r^{**}(\hbar s), \qquad (10.9)$$

where

$$\diamond r^{**}(s) = r^{**}(s + \hbar) - 2r^{**}(s) + r^{**}(s - \hbar) \qquad (10.10)$$

and

$$r^{**}(t) = \int_0^t r^*(u) \mathrm{d}u \quad \text{where} \quad r^*(t) = \int_0^t r(u) \mathrm{d}u. \qquad (10.11)$$

Thus, for a given model for the covariance stationary process τ we can compute the approximate second order properties of the time series of $[y_M^*]_i$ and τ_i.

The above theory implies we can calculate asymptotically approximate best linear filtered, smoothed and forecast values of τ_i using standard regression theory. This has been independently and concurrently studied by Andersen, Bollerslev and Meddahi (2004) for some diffusion based models for τ. Their results are similar to those we present here.

Suppose we wish to estimate $\tau_{s:p}$ using $[y_M^*]_{s:p}$. Then the best linear estimator is

$$\widehat{\tau}_{s:p} = (I-A)\iota E(\tau_i) + A[y_M^*]_i$$
$$= A\{[y_M^*]_{s:p} - \hbar\xi\iota\} + \hbar\xi\iota,$$

where

$$A = \{\text{Cov}([y_M^*]_{s:p})\}^{-1}\text{Cov}(\tau_{s:p}, [y_M^*]_{s:p})$$
$$= \{\text{Cov}(\tau_{s:p}) + 2M^{-1}\hbar^2(\omega^2 + \xi^2)I\}^{-1}\text{Cov}(\tau_{s:p}).$$

The simplest special case of this is where $s = p = i$, that is, we use a single realised variance to estimate actual variance. Then the theory above suggests the efficient linear estimator is constructed using the scalar

$$A = \{r^{**}(\hbar) + M^{-1}\hbar^2(1 + \xi^2/\omega^2)\}^{-1} r^{**}(\hbar) \in [0,1], \quad (10.12)$$

which implies $\widehat{\tau}_i \geq 0$. Meddahi (2002) studied this particular regression, which we write as $\widehat{\tau}_i$ and call a Meddahi regression. It is always a consistent estimator of τ_i, but is more efficient than realised variance under the covariance stationarity assumptions.

In practice it is helpful to use the structure of the $\text{Cov}(\tau_{s:p})$ in order to carry out the required matrix inverse of $\text{Cov}([y_M^*]_{s:p})$.

10.5.2 Special case

Suppose τ has the autocorrelation function $r(t) = \exp(-\lambda|t|)$. This implies that

$$E(\tau_i) = \hbar\xi, \quad \text{Var}(\tau_i) = 2\omega^2\lambda^{-2}\left(e^{-\lambda\hbar} - 1 + \lambda\hbar\right),$$

and

$$\text{Cor}\{\tau_i, \tau_{i+s}\} = de^{-\lambda\hbar(s-1)}, \quad s = 1, 2, \ldots, \quad (10.13)$$

where

$$d = \frac{(1 - e^{-\lambda\hbar})^2}{2(e^{-\lambda\hbar} - 1 + \lambda\hbar)} \in [0,1].$$

In this case, in particular, the Meddahi regression has

$$\widehat{A} = \left\{\lambda^{-2}\left(e^{-\lambda\hbar} - 1 + \lambda\hbar\right) + M^{-1}\hbar^2(1+\xi^2/\omega^2)\right\}^{-1}\lambda^{-2}\left(e^{-\lambda\hbar} - 1 + \lambda\hbar\right)$$

The above structure implies τ_i has the autocorrelation function of an ARMA(1,1) model

$$\tau_i = \phi\tau_{i-1} + u_i + \theta u_{i-1}, \quad \phi = e^{-\lambda\hbar}.$$

The parameter θ was found numerically in Barndorff-Nielsen and Shephard (2001); however, it can be determined analytically as indicated by Meddahi (2002). In particular, write

$$c_i = \tau_i - \phi\tau_{i-1} = u_i + \theta u_{i-1},$$

then

$$\mathrm{Var}\,(c_i) = \left(1+\phi^2\right)\mathrm{Var}(\tau_i) - 2\phi\mathrm{Cov}(\tau_i, \tau_{i-1})$$

and

$$\begin{aligned}\mathrm{Cov}(c_i, c_{i-1}) &= \left(1+\phi^2\right)\mathrm{Cov}(\tau_i, \tau_{i-1}) - \phi\mathrm{Var}(\tau_i) - \phi\mathrm{Cov}(\tau_i, \tau_{i-2})\\ &= \mathrm{Cov}(\tau_i, \tau_{i-1}) - \phi\mathrm{Var}(\tau_i)\\ &= \mathrm{Var}(\tau_i)\left\{\mathrm{Cor}(\tau_i, \tau_{i-1}) - \phi\right\}.\end{aligned}$$

Note that $\mathrm{Cor}(\tau_i, \tau_{i-1}) \geq \phi$ as $\mathrm{e}^{\lambda\hbar} - \mathrm{e}^{-\lambda\hbar} \geq 2\lambda\hbar$. Write

$$\rho_1 = \frac{\mathrm{Cov}(c_i, c_{i-1})}{\mathrm{Var}\,(c_i)} \in \left[0, \frac{1}{2}\right], \quad \text{then} \quad \theta = \frac{1 - \sqrt{1-4\rho_1^2}}{2\rho_1} \in [0,1].$$

This argument extends to the case of a superposition where

$$r(t) = \sum_{j=1}^{J} w_j \exp(-\lambda_j |t|),$$

then τ_i can be represented as the sum of J uncorrelated ARMA(1,1) processes, with $\{w_j, \lambda_j\}$ determining the corresponding autoregressive and moving average roots $\{\phi_j, \theta_j\}$. This is important in practice as financial volatility tends to have components which have a great deal of memory in them.

In calculating $\widehat{\tau}_{s:p}$ Barndorff-Nielsen and Shephard (2001) conveniently placed $[y_M^*]_i$ into a linear state space representation so the filtering, smoothing and forecasting can be carried out using the Kalman filter (see, for example, Harvey (1989) and Durbin and Koopman (2001), Chapter 1)). In particular writing $\alpha_{1i} = (\tau_i - \hbar\xi)$ and $u_i = \sqrt{2M^{-1}\hbar^2(\omega^2+\xi^2)}v_{1i}$, then

$$\begin{aligned}[y_M^*]_i &= \hbar\xi + \begin{pmatrix}1 & 0\end{pmatrix}\alpha_i + u_i,\\ \alpha_{i+1} &= \begin{pmatrix}\phi & 1\\ 0 & 0\end{pmatrix}\alpha_i + \begin{pmatrix}\sigma_\sigma\\ \sigma_\sigma\theta\end{pmatrix}v_i,\end{aligned} \quad (10.14)$$

where v_i is a zero mean, unit variance, white noise sequence uncorrelated with u_i, which has a variance of $2M^{-1}\hbar^2\left(\omega^2+\xi^2\right)$. The parameters ϕ, θ and σ_σ^2 represent the autoregressive root, the moving average root and the variance of the innovation to the ARMA(1,1) representation of the τ_i process.

The extension to the superposition case is straightforward. In particular, in the case where $J = 2$ this becomes

$$[y_M^*]_i = \hbar\xi + \begin{pmatrix} 1 & 0 & 1 & 0 \end{pmatrix}\alpha_i + u_i,$$

$$\alpha_{i+1} = \begin{pmatrix} \phi_1 & 1 & 0 & 0 \\ 0 & 0 & 0 & 0 \\ 0 & 0 & \phi_2 & 1 \\ 0 & 0 & 0 & 0 \end{pmatrix}\alpha_i + \begin{pmatrix} \sigma_{\sigma 1} & 0 \\ \sigma_{\sigma 1}\theta_1 & 0 \\ 0 & \sigma_{\sigma 2} \\ 0 & \sigma_{\sigma 2}\theta_2 \end{pmatrix}v_i,$$

where again v_i is a zero mean, unit variance, white noise sequence.

Table 10.5. *Exact mean square error (steady state) of the estimators of actual volatility. The first two estimators are model based (smoother and one-step ahead predictor) and the third is $[y_M^*]_i$. These measures are calculated for different values of $\omega^2 = \mathrm{Var}(\tau(t))$ and λ, keeping $\xi = \mathrm{E}(\tau(t))$ fixed at 0.5. File:* `ssf_mse.ox`

M	$\xi = 0.5, \xi\omega^{-2} = 8$			$\xi = 0.5, \xi\omega^{-2} = 2$		
$e^{-\hbar\lambda} = 0.99$	Smooth	Predict	$[y_M^*]_i$	Smooth	Predict	$[y_M^*]_i$
1	0.0134	0.0226	0.624	0.0342	0.0625	0.998
12	0.00383	0.00792	0.0520	0.00945	0.0211	0.0833
48	0.00183	0.00430	0.0130	0.00440	0.0116	0.0208
288	0.000660	0.00206	0.00217	0.00149	0.00600	0.00347
$e^{-\hbar\lambda} = 0.9$	Smooth	Predict	$[y_M^*]_i$	Smooth	Predict	$[y_M^*]_i$
1	0.0345	0.0456	0.620	0.0954	0.148	0.982
12	0.0109	0.0233	0.0520	0.0259	0.0697	0.0832
48	0.00488	0.0150	0.0130	0.0108	0.0467	0.0208
288	0.00144	0.00966	0.00217	0.00280	0.0338	0.00347

Table 10.5 reports the mean square error of the model based one-step ahead predictor and smoother of actual variance, as well as the corresponding result for $[y_M^*]_i$. The results in the left hand block of the table correspond to the model which was simulated in Figure 10.2, while the other block represents other choices of the ratio of ξ to ω^2. The exercise is repeated for two values of λ.

The main conclusion from the results in Table 10.5 is that model based approaches can potentially lead to very significant reductions in mean square error, with the reductions being highest for persistent (low value of λ) variance processes with high values of $\xi\omega^{-2}$. Even for moderately large values of M the model based predictor can be more accurate than realised variance, sometimes by a considerable amount. This is an important result from a forecasting viewpoint. However, when there is not much persistence and M

is very large, this result is reversed and realised variance can be moderately more accurate. The smoother is always substantially more accurate than realised variance, even when M is very large and there is not much memory in variance.

Estimating the parameters of continuous time stochastic volatility models is known to be difficult due to our inability to compute the appropriate likelihood function. This has prompted the development of a sizable collection of methods to deal with this problem (e.g. Kim, Shephard and Chib (1998) and Gallant, Hsieh and Tauchen (1997)). Barndorff-Nielsen and Shephard (2002a) used quasi-likelihood estimation methods based on the time series of realised variance. The quasi-likelihood is constructed using the output of the Kalman filter. It is suboptimal for it does not exploit the non-Gaussian nature of the variance dynamics, but it provides a consistent and asymptotically normal set of estimators. Monte Carlo results reported in Barndorff-Nielsen and Shephard (2002a) indicate that the finite sample behaviour of this approach is quite good. Further, the estimation takes only a few seconds on a modern computer.

10.5.3 Empirical illustration

To illustrate some of these results we have fitted a set of superposition based models to the realised variance time series constructed from the five-minute US/DM exchange rate return data discussed above. Here we use the quasi-likelihood method to estimate the parameters of the model — ξ, ω^2, $\lambda_1, \ldots, \lambda_J$ and $w_1, \ldots, w_J$. We do this for a variety of values of M, starting with $M = 6$, which corresponds to working with four-hour returns. The resulting parameter estimates are given in Table 10.6. For the moment we will focus on this case.

The fitted parameters suggests a dramatic shift in the fitted model as we go from $J = 1$ to $J = 2$ or 3. The more flexible models allow for a factor which has quite a large degree of memory, as well as a more rapidly decaying component or two. A simple measure of fit of the model is the Box–Pierce statistic, which shows a large jump from a massive 302 when $J = 1$, down to a more acceptable number for a superposition model.

To provide a more detailed assessment of the fit of the model we have drawn a series of graphs in Figure 10.7 based on $M = 8$ and $M = 144$. Figure 10.7(a) shows the computed realised variance $[y_M^*]$, together with the corresponding smoothed estimate (based on $J = 3$) of actual variance using the model. These are based on the $M = 8$ case. We can see that realised variance is much more jagged than the smoothed quantity. These

Table 10.6. *Fit of the superposition of J volatility processes for an SV model based on realised variance computed using $M = 6$, $M = 18$ and $M = 144$. We do not record w_J as this is 1 minus the sum of the other weights. Estimation method: quasi-likelihood using output from a Kalman filter. BP denotes Box–Pierce statistic, based on 20 lags, which is a test of serial dependence in the scaled residuals. File:* `ssf_empirical.ox`

M	J	ξ	ω^2	λ_1	λ_2	λ_3	w_1	w_2	quasi-L	BP
6	3	0.4783	0.376	0.0370	1.61	246	0.212	0.180	−113,258	11.2
6	2	0.4785	0.310	0.0383	3.76		0.262		−113,261	11.3
6	1	0.4907	0.358	1.37					−117,397	302
18	3	0.460	0.373	0.0145	0.0587	3.27	0.0560	0.190	−101,864	26.4
18	2	0.460	0.533	0.0448	4.17		0.170		−101,876	26.5
18	1	0.465	0.497	1.83					−107,076	443
144	3	0.508	4.79	0.0331	0.973	268	0.0183	0.0180	−68,377	15.3
144	2	0.509	0.461	0.0429	3.74		0.212		−68,586	23.3
144	1	0.513	0.374	1.44					−76,953	765

are quite close to the semiparametric estimator given in Figure 10.6. Figure 10.7(b) shows the corresponding autocorrelation function for the realised variance series together with the corresponding empirical correlogram. We see from this figure that when $J = 1$ we are entirely unable to fit the data, as its autocorrelation function starts at around 0.6 and then decays to zero in a couple of days. A superposition of two processes is much better, picking up the longer-range dependence in the data. The superposition of two and three processes gives very similar fits, indeed in the graph they are indistinguishable.

We next ask how these results vary as M increases. We reanalyse the situation when $M = 144$, which corresponds to working with ten-minute returns. Figures 10.7(c) and (d) give the corresponding results. Broadly the smoother has not produced very different results, while the $J = 3$ case now gives a slightly different fit to the Acf than the $J = 2$. The latter result is of importance, for as M increases the correlogram becomes more informative, allowing us to discriminate between different models more easily.

10.5.4 Comparison

We can compare the fit of the smoothers from the model free and model based approaches. In Figure 10.8 we display, using crosses, the time series of the model free smoother, based on four leads and four lags. This is drawn, for a variety of values of M, as the square root of the estimate, so it is

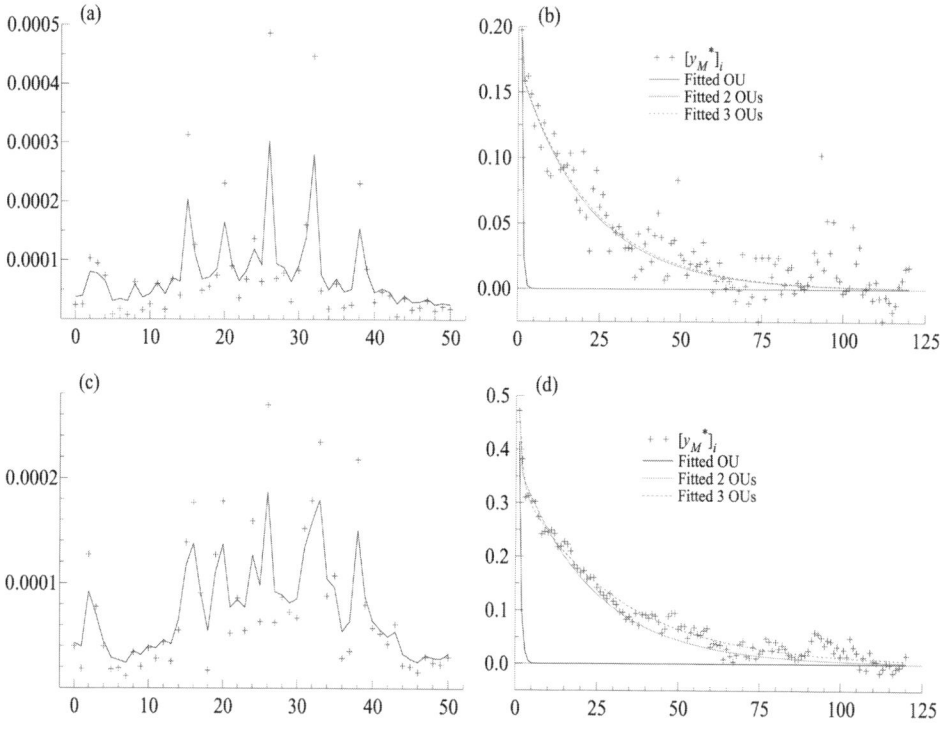

Fig. 10.7. Results from $M = 8$ and $M = 144$ using: (a) $M = 8$, first 50 observations of $[y_M^*]_i$ and a smoother; (b) $M = 8$, Acf of $[y_M^*]_i$ and the fitted version for various values of J; (c) $M = 144$, first 50 observations of $[y_M^*]_i$ and a smoother; (d) $M = 144$, Acf of $[y_M^*]_i$ and the fitted version for various values of J.

estimating the square root of integrated variance. The corresponding model based approach is drawn using a line and it shows a close connection with the model free estimator. Table 10.7 gives the correlations between the two estimators as a function of M and the number of leads and lags in the model free approach. As the number of leads and lags increases the connection between the two estimators becomes stronger. Likewise, as M increase the two estimators become more closely correlated.

10.6 Conclusion

In this paper we have shown how we can use a time series of realised variances to measure and forecast integrated variances. These high frequency financial data statistics allow either model based or model free approaches to the

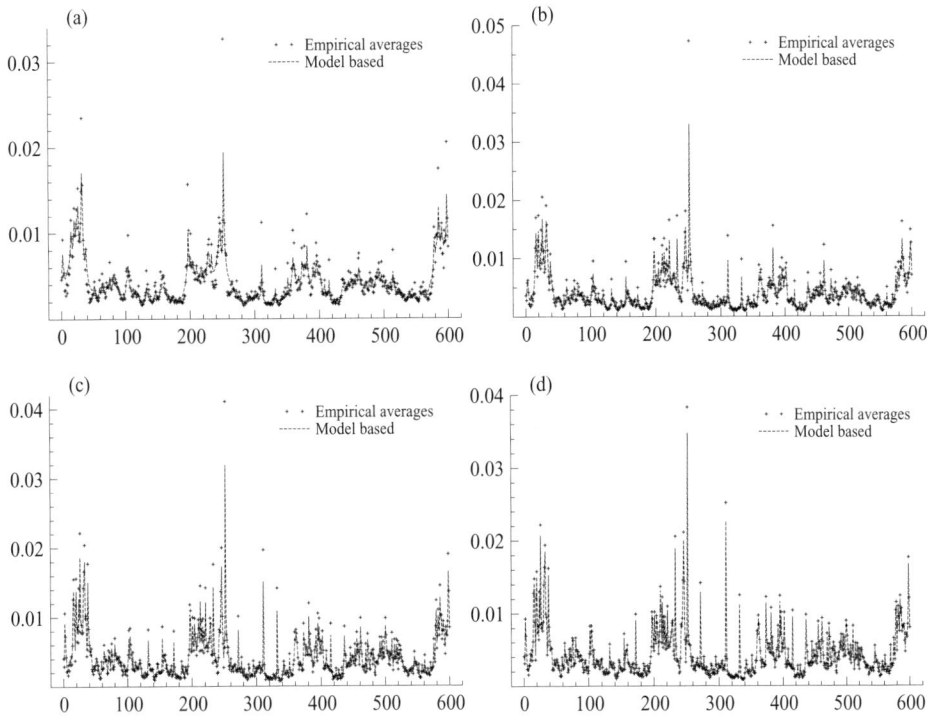

Fig. 10.8. Shows a comparison of the model free smoother based on four leads and lags and the model based approach. We show the estimators for the first 600 days in the sample, using a variety of values of M: (a) $M = 6$; (b) $M = 48$; (c) $M = 144$; (d) $M = 288$.

Table 10.7. *Correlations between the model free and model based smoothers based on the dollar/DM data. We vary M and the number of leads and lags.*

M	RV	1 lead, 1 lag	4 leads, 4 lags
6	0.702	0.849	0.929
48	0.903	0.924	0.932
144	0.961	0.985	0.989
288	0.984	0.997	0.998

problem. We have spent some time comparing the two smoothed estimators, which tend to be quite similar when M is large and we have employed quite a few leads and lags.

Appendix: Proof of theorem 10.2

We split the proof into two sections, dealing with the diagonal and non-diagonal elements of the matrix

$$A = (X + \sigma I)^{-1} X.$$

Here X is positive semi-definite and $\sigma > 0$.

(a) Diagonal elements of A

Since X and I commute then A is positive semi-definite, implying that A has nonnegative diagonal for all X.

To be more explicit write $X = V\Lambda V'$, where $I = VV'$ and Λ is diagonal. From

$$A = \{V(\Lambda + \sigma I)V'\}^{-1} V\Lambda V' = V\left\{(\Lambda + \sigma I)^{-1}\Lambda\right\}V',$$

it is seen that A is symmetric and positive definite since $(\Lambda + \sigma I)^{-1}\Lambda$ is diagonal with nonnegative diagonal elements.

(b) Off-diagonal elements of A

Rewrite

$$A = (X+\sigma I)^{-1} X = (I + \sigma X^{-1})^{-1} = \eta\left(\eta I + X^{-1}\right)^{-1}, \qquad \eta = 1/\sigma.$$

It suffices to consider off-diagonal elements of

$$N = \left(\eta I + X^{-1}\right)^{-1}.$$

The proof follows by induction. We use subscripts to denote the size of matrices, and superscripts to denote the elements of the inverse of a matrix.

Dimension 2. It holds that $N_2^{-1} = N_2^{\#}/\det(N_2)$, where

$$N_2^{\#} = \begin{pmatrix} \eta + X^{22} & -X^{12} \\ -X^{21} & \eta + X^{11} \end{pmatrix}.$$

Therefore the off-diagonal element is nonnegative, $N_2^{12} \geq 0$, for all σ if and only if $X^{12} \leq 0$.

Dimension $k+1$. Simultaneous permutation of the ith and jth columns and the ith and jth rows preserves the positive semi-definiteness of the matrix. Thus we can look at an arbitrary off-diagonal element to establish this result. Thus, look at the upper right element of N_{k+1}. This is given as

$$N_{k+1}^{1,k+1} = \frac{(-1)^k}{\det(N_{k+1})} \det \begin{pmatrix} X^{2,1} & \eta + X^{2,2} & \cdots & X^{2,k} \\ \vdots & & & \vdots \\ X^{k,1} & X^{k,2} & \cdots & \eta + X^{k,k} \\ X^{k+1,1} & X^{k+1,2} & \cdots & X^{k+1,k} \end{pmatrix}.$$

Expanding the latter determinant along the last row it follows that

$$N_{k+1}^{1,k+1} = \frac{(-1)^k}{\det(N_{k+1})} \sum_{j=1}^{k} (-1)^{k-j} X^{k+1,j} (-1)^{j+1} N_k^{1,j} \det(N_k)$$

$$= (-1) \frac{\det(N_k)}{\det(N_{k+1})} \sum_{j=1}^{k} X^{k+1,j} N_k^{1,j}.$$

By induction it holds that $N_k^{1,j} \geq 0$ for all j, and therefore a sufficient condition for $N_{k+1}^{1,k+1} \geq 0$ is that $X^{k+1,j} \leq 0$ for all $j \leq k$.

To prove necessity note that $N_k^{1,j} \det(N_k)$ is a polynomial in η of order $k-1$ if $j=1$ and of order $k-2$ if $j \neq 1$. Thus for large η

$$N_{k+1}^{1,k+1} \approx (-1) \frac{\det(N_k)}{\det(N_{k+1})} X^{k+1,1} N_k^{1,1},$$

so if $N_{k+1}^{1,k+1}$ is non-negative for large η then $X^{k+1,1}$ must be nonpositive.

This completes the proof.

Acknowledgments

Ole E. Barndorff-Nielsen's work is supported by CAF (www.caf.dk), which is funded by the Danish Social Science Research Council, and by MaPhySto (www.maphysto.dk), which is funded by the Danish National Research Foundation. Neil Shephard's research is supported by the UK's ESRC through the grant 'Econometrics of trade-by-trade price dynamics,' which is coded R00023839. All the calculations made in this paper are based on software written by the third author using the Ox language of Doornik (2001). We would like to thank Torben Andersen, Tim Bollerslev and Nour Meddahi for helpful conversations on this topic and Eric Zivot for his comments on our initial draft. We thank Michel M. Dacorogna for allowing us to use Olsen's high frequency exchange rate data in our study. Full details of this type of data are available in Dacorogna, Gencay, Müller, Olsen and Pictet (2001).

11
Practical filtering for stochastic volatility models

Jonathan R. Stroud
Department of Statistics, Wharton School, University of Pennsylvania

Nicholas G. Polson
Graduate School of Business, University of Chicago

Peter Müller
M. D. Anderson Cancer Center, University of Texas

Abstract

This paper provides a simulation-based approach to filtering and sequential parameter learning for stochastic volatility models. We develop a fast simulation-based approach using the practical filter of Polson, Stroud and Müller (2002). We compare our approach to sequential parameter learning and filtering with an auxiliary particle filtering algorithm based on Storvik (2002). For simulated data, there is close agreement between the two methods. For data on the S&P 500 market stock index from 1984–90, our algorithm agrees closely with a full MCMC analysis, whereas the auxiliary particle filter degenerates.

11.1 Introduction

Filtering and sequential parameter learning for stochastic volatility (SV) have many applications in financial decision making. SV models are commonly used in financial applications as their dynamics are flexible enough to model observed asset and derivative prices. However, many applied financial decision making problems are sequential in nature such as portfolio selection (e.g. Johannes, Polson and Stroud (2002b)) and option pricing. These applications require filtered estimates of spot volatility and sequential parameter estimates to account for estimation risk. In this paper, we provide a simulation-based approach for volatility state filtering that also incorporates sequential parameter learning. The methodology is based on the practical filter of Polson, Stroud and Müller (2002). Unlike previous simulation-based filtering methods, for example Kim, Shephard and Chib (1998) in the SV context, our algorithm incorporates sequential parameter learning within Markov chain Monte Carlo (MCMC).

Many authors have considered the problem of simulation-based filtering with known static parameters. A common approach uses particle filtering methods; see for example Gordon, Salmond and Smith (1993), Liu and Chen (1998), Carpenter, Clifford and Fearnhead (1999), Pitt and Shephard (1999b), or Doucet, Godsill and Andrieu (2000) for an excellent review of the subject. Liu and West (2001) provided an algorithm for joint state and parameter learning. Storvik (2002) provided an auxiliary particle filtering method that deals with sequential state and parameter learning. Other approaches include Chopin (2002) for static models, and Gilks and Berzuini (2001) and Fearnhead (2002) who provided useful approaches using MCMC, sufficient statistics and particle filters. For comparison purposes in our applications we use an auxiliary particle filter (APF) based on Pitt and Shephard (1999b) and Storvik (2002).

The goal of our analysis is to provide inference in sequential fashion about an unobserved state $\mathbf{x}_t$ given data $\mathbf{y}_{1:t}$ at each time t, for $t = 1, \ldots, T$, in a nonlinear, non-Gaussian state space model. In this paper we use the generic notation $\mathbf{y}_{1:t}$ to denote the vector $(\mathbf{y}_1, \ldots, \mathbf{y}_t)$. Our approach takes advantage of the mixture representation of the joint distribution of the states $\mathbf{x}_t$ and parameters θ given $\mathbf{y}_{1:t}$,

$$p(\mathbf{x}_t, \theta | \mathbf{y}_{1:t}) = \int p(\mathbf{x}_t, \theta | \mathbf{x}_{0:t-k}, \mathbf{y}_{1:t}) p(\mathbf{x}_{0:t-k} | \mathbf{y}_{1:t}) \mathrm{d}\mathbf{x}_{0:t-k}.$$

The practical filter of Polson, Stroud and Müller (2002) is a simulation-based filtering and sequential parameter learning algorithm. This approach has a number of computational advantages when parameter learning is also incorporated. The main advantage is that it does not degenerate as particle filtering methods can, see also Johannes, Polson and Stroud (2002a). Section 11.2 describes our approach to filtering and sequential parameter learning using the practical filter.

To illustrate our methodology, we use simulated data and daily data from the S&P 500. For the simulated data, we find that our methodology is comparable with the method of Storvik (2002), which builds on the APF, for both filtering and sequential parameter learning. However, for the S&P 500 data, we find that the APF degenerates for sequential parameter learning. This is partly due to the fact that the model is misspecified for these data. More specifically, the APF degenerates when sequentially estimating the volatility of volatility parameter. The practical filter avoids degeneracies and also compares favorably with a full MCMC analysis. However, for the volatility of volatility parameter, the practical filter does have some difficulties capturing the full posterior uncertainty.

The rest of the paper is outlined as follows. Section 11.2 describes the basic filtering problem and how the practical filter works. Section 11.3 describes the stochastic volatility model and shows how to implement the practical and particle filters for state and sequential parameter learning. An application to the S&P 500 is provided with the focus of providing a computationally fast algorithm. Finally, Section 11.4 concludes.

11.2 Practical filtering

The basic filtering problem for dynamic state space models requires that posterior distributions be recomputed at each time point. Let $\mathbf{y}_t$ denote the observation vector at time t, let $\mathbf{x}_t$ denote the state vector, θ the parameters, and $\mathbf{x}_{s:t} = (\mathbf{x}_s, \ldots, \mathbf{x}_t)$, $\mathbf{y}_{s:t} = (\mathbf{y}_s, \ldots, \mathbf{y}_t)$ the block of states and observations, respectively, from time s up to time t. We need to compute the joint conditional posterior distribution $p(\mathbf{x}_t, \theta | \mathbf{y}_{1:t})$. Filtering and sequential parameter learning are achieved by calculating the marginal state filtering distribution $p(\mathbf{x}_t | \mathbf{y}_{1:t})$ and the sequence of marginal parameter posteriors $p(\theta | \mathbf{y}_{1:t})$.

We now describe the practical filter which provides a sequential simulation-based approach to estimating these distributions. The key idea is to express the filtering distribution $p(\mathbf{x}_t, \theta | \mathbf{y}_{1:t})$ as a mixture of lag-filtering distributions. More specifically, we can write

$$p(\mathbf{x}_t, \theta | \mathbf{y}_{1:t}) = \int p(\mathbf{x}_t, \theta | \mathbf{x}_{0:t-k}, \mathbf{y}_{1:t}) p(\mathbf{x}_{0:t-k} | \mathbf{y}_{1:t}) d\mathbf{x}_{0:t-k}.$$

Here $p(\mathbf{x}_t, \theta|\mathbf{x}_{0:t-k}, \mathbf{y}_{1:t})$ is the lag-filtering distribution of the state and parameter vector given knowledge of $\mathbf{x}_{0:t-k}$ and data $\mathbf{y}_{1:t}$. The use of a block of k data points and updating has also been exploited in the particle filtering approach. Pitt and Shephard (2001) showed how to use it for the auxiliary particle filter. See also Clapp and Godsill (1999) and Godsill and Clapp (2001). Chopin (2002) provided a batch importance sampling procedure that extends to the parameter learning case and uses ideas in Gilks and Berzuini (2001).

Our algorithm proceeds in two stages. First, due to the sequential nature of the problem, we already have draws of the states $\mathbf{x}_{0:t-k}^{(g)}$ from $p(\mathbf{x}_{0:t-k}|\mathbf{y}_{1:t})$. Second, we use these to provide samples from the current filtering distribution $p(\mathbf{x}_t, \theta|\mathbf{y}_{1:t})$ by simulating from the filtering distribution

$$p\left(\mathbf{x}_{t-k+1:t}, \theta|\mathbf{x}_{0:t-k}^{(g)}, \mathbf{y}_{1:t}\right).$$

Samples from this distribution can be obtained by iteratively simulating from the complete conditionals

$$p\left(\mathbf{x}_{t-k+1:t}|\theta, \mathbf{x}_{t-k}^{(g)}, \mathbf{y}_{t-k+1:t}\right), \tag{11.1}$$

$$p\left(\theta|\mathbf{x}_{t-k+1:t}, \mathbf{x}_{0:t-k}^{(g)}, \mathbf{y}_{1:t}\right). \tag{11.2}$$

Notice that, due to the Markovian property of the states, the fixed-lag smoothing distribution, (11.1), depends on the history only through $\mathbf{x}_{t-k}^{(g)}$. Hence, we need only know the last stored state variable $\mathbf{x}_{t-k}^{(g)}$ to simulate from this conditional distribution. On the other hand, this is unfortunately not true for the parameter vector θ. Here we need the full history $\mathbf{x}_{0:t-k}^{(g)}$ to be able to draw from its conditional. However, a number of authors have pointed out that sufficient statistics, which are typically available, greatly simplify the problem (e.g. Storvik (2002), Fearnhead (2002) and Polson, Stroud and Müller (2002)). In such cases, one can exploit a sufficient statistic $\mathbf{s}_{t-k}^{(g)} = S\left(\mathbf{x}_{0:t-k}^{(g)}, \mathbf{y}_{1:t-k}^{(g)}\right)$ instead of saving the entire histories $\mathbf{x}_{0:t-k}^{(g)}$ and $\mathbf{y}_{1:t-k}^{(g)}$. This then simplifies the parameter draw (11.2) as

$$p\left(\theta|\mathbf{x}_{t-k+1:t}, \mathbf{s}_{t-k}^{(g)}, \mathbf{y}_{t-k+1:t}\right),$$

which implies that the computational cost for the θ update is fixed over time. Now our key assumption is that sequential draws $\mathbf{x}_{t-k}^{(g)}$ from the marginal posterior $p(\mathbf{x}_{t-k}|\mathbf{y}_{1:t})$ can reasonably be approximated by draws from $p(\mathbf{x}_{t-k}|\mathbf{y}_{1:t-1})$, hence our use of $\left(\mathbf{x}_{t-k}^{(g)}, \mathbf{s}_{t-k}^{(g)}\right)$ in the conditioning above.

Finally, to ensure a fast filtering algorithm, we need to efficiently simulate

$x_{t-k|1:t}$ from the fixed-lag smoothing distribution (11.1). This can be dealt with using a number of implementations. First, we have standard MCMC smoothing methods for state space models, for example Carlin, Polson and Stoffer (1992b). Moreover, in many instances with a careful choice of additional latent variables these can be implemented using the fast *forward-filtering backward-sampling* (FFBS) algorithm; see Carter and Kohn (1994), Frühwirth-Schnatter (1994c) and Shephard (1994) for conditionally Gaussian models, and Shephard and Pitt (1997) for non-Gaussian measurement models.

We now describe the practical filtering algorithm for state and sequential parameter learning for conditionally Gaussian models. Nonlinear or non-Gaussian models can be implemented by replacing the FFBS step with single-state updating or sub-blocking (see, Carlin, Polson and Stoffer (1992b) and Shephard and Pitt (1997), respectively).

Algorithm Filtering with sequential parameter learning
Initialisation: For $g = 1, \ldots, G$: generate $\theta^{(g)} \sim p(\theta)$.
Burn-in: For $t = 1, \ldots, k$:

    For $g = 1, \ldots, G$: initialise $\theta^0 = \theta^{(g)}$.

        For $i = 1, \ldots, I$:

            Generate $\mathbf{x}_{0:t}^i \sim p(\mathbf{x}_{0:t}|\theta^{i-1}, y_{1:t})$.
            Generate $\theta^i \sim p(\theta|\mathbf{x}_{0:t}^i, y_{1:t})$.

        Set $\left(\mathbf{x}_0^{(g)}, \theta^{(g)}\right) = (\mathbf{x}_0^I, \theta^I)$.

Sequential updating: For $t = k+1, \ldots, T$:

    For $g = 1, \ldots, G$: initialise $\theta^0 = \theta^{(g)}$.

        For $i = 1, \ldots, I$:

            Generate $\mathbf{x}_{t-k+1:t}^i \sim p(\mathbf{x}_{t-k+1:t}|\mathbf{x}_{t-k}^{(g)}, \theta^{i-1}, \mathbf{y}_{t-k+1:t})$
            Generate $\theta^i \sim p(\theta|\mathbf{x}_{0:t-k}^{(g)}, \mathbf{x}_{t-k+1:t}^i, \mathbf{y}_{1:t})$

        Set $\left(\mathbf{x}_{t-k+1}^{(g)}, \theta^{(g)}\right) = (\mathbf{x}_{t-k+1}^I, \theta^I)$ and leave $\mathbf{x}_{0:t-k}^{(g)}$ unchanged.

The speed and accuracy of the algorithm depends on the choice of the three parameters (G, I, k). First, the Monte Carlo sample size G relates to the desired accuracy for state and parameter estimation. In our stochastic volatility model, we find that $G = 250$ is sufficient. The number of MCMC iterations for fixed-lag smoothing is specified by I. In the stochastic volatility example we can use FFBS for the states and we can get a direct draw of the states given the parameters and hence $I = 1$ in this case. However, we also need to update the parameters given the states. With a small number of data points a larger value of I is necessary; however, as the data arrive this can be reduced as the previous parameter draw is already nearly a draw from the desired stationary distribution. The use of a reasonable burn-in period can also alleviate the problem with the choice of I. Finally, the choice of the lag length k depends on the memory of the process. For example, for a stationary AR(1) state process, the correlation decays geometrically in k. In many cases, a small value of k is sufficient; for example, we find that in the stochastic volatility model $k = 50$ usually suffices.

11.3 Filtering for stochastic volatility models

To illustrate our methodology we use a standard daily stock index return dataset from the Standard and Poor's S&P 500 market index from 1984–90. The stock market crash of October 1987 provides a negative return in excess of 20%. This can lead to filtered parameter estimates changing abruptly and provides a useful testing ground for the practical and particle filters. We now develop a fast practical filtering method and we compare this with the sequential particle filtering approach of Storvik (2002).

In the standard univariate log-stochastic volatility model (e.g. Jacquier, Polson and Rossi (1994), Ghysels, Harvey and Renault (1996)), the returns y_t and log-volatility states x_t follow a state space model of the form

$$y_t = \exp(x_t/2)\epsilon_t,$$
$$x_t = \alpha + \beta x_{t-1} + \sigma \eta_t$$

with initial log-volatility $x_0 \sim \mathcal{N}(m_0, C_0)$. The parameter vector θ consists of the volatility mean reversion parameters $\psi = (\alpha, \beta)$ and the volatility of volatility σ. Here we assume the errors ϵ_t and η_t are independent $\mathcal{N}(0,1)$ white noise sequences, although it is possible to include a cross correlation or leverage effect (e.g. Jacquier, Polson and Rossi (2003)).

We now describe how to implement the practical filter and the auxiliary particle filtering methods of Storvik (2002) for SV models. Previous work which has used particle filters on SV includes Kim, Shephard and Chib (1998), Pitt Shephard (1999b, 2001), but these papers only considered the problem of filtering the volatility with known parameters and not the case of sequential parameter learning.

11.3.1 Implementing the practical filter

In order to implement the practical filter we need to determine a fast algorithm for fixed-lag smoothing for the states. To do this, we transform the observations to $y_t^* = \log y_t^2$ and use the approximating mixture model of Kim, Shephard and Chib (1998):

$$\begin{aligned} y_t^* &= x_t + \epsilon_t^*, \\ x_t &= \alpha + \beta x_{t-1} + \sigma \eta_t, \end{aligned}$$

where ϵ_t^* is a discrete mixture of seven normals with known parameters and weights. To perform fixed-lag smoothing for $\mathbf{x}_{s:t} = (x_s, \ldots, x_t)$, we introduce a set of discrete mixture indicators $\mathbf{z}_{s:t} = (z_s, \ldots, z_t)$ with $z_t \in \{1, \ldots, 7\}$, and update the states and indicators in a two-block Gibbs sampler. First, we generate the states $\mathbf{x}$ given the indicators $\mathbf{z}$ using the FFBS algorithm. Then we generate the indicators given the states using independent multinomial draws.

For the parameters $\theta = (\psi, \sigma^2)$, the sufficient statistic structure $S(\cdot)$ is straightforward to determine. We assume a conjugate prior, $p(\theta) = p(\psi|\sigma^2) p(\sigma^2)$, where $\psi|\sigma^2 \sim \mathcal{N}(\psi_0, A_0^{-1}\sigma^2)$ and $\sigma^2 \sim \mathcal{IG}(a_0, b_0)$, which leads to closed-form posterior distributions at each time t,

$$\psi|\sigma^2, \mathbf{x}_{0:t}, \mathbf{y}_{1:t} \sim \mathcal{N}\left(\psi_t, A_t^{-1}\sigma^2\right) \quad \text{and} \quad \sigma^2|\mathbf{x}_{0:t}, \mathbf{y}_{1:t} \sim \mathcal{IG}(a_t, b_t).$$

The sufficient statistics for θ are $\mathbf{s}_t = \{\psi_t, A_t, a_t, b_t\}$, and the lag-$k$ updating recursions are given by $\psi_t = A_t^{-1}(A_{t-k}\psi_{t-k} + H'\mathbf{x})$, $A_t = A_{t-k} + H'H$, $a_t = a_{t-k} + 0.5k$, and $b_t = b_{t-k} + 0.5(\psi_{t-k}' A_{t-k}\psi_{t-k} + \mathbf{x}'\mathbf{x} - \psi_t' A_t \psi_t)$, where $\mathbf{x} = (x_{t-k+1}, \ldots, x_t)'$, $H = (H_{t-k+1}, \ldots, H_t)'$ and $H_t = (1, x_{t-1})'$.

The SV practical filtering algorithm then iterates between three blocks:

$$\begin{aligned} \text{States:} \quad & p(\mathbf{x}_{t-k+1:t}|\mathbf{x}_{t-k}, \mathbf{z}_{t-k+1:t}, \theta, \mathbf{y}_{t-k+1:t}), \\ \text{Indicators:} \quad & p(\mathbf{z}_{t-k+1:t}|\mathbf{x}_{t-k+1:t}, \mathbf{y}_{t-k+1:t}), \\ \text{Parameters:} \quad & p(\theta|\mathbf{s}_{t-k}, \mathbf{x}_{t-k+1:t}). \end{aligned}$$

This cycle is repeated I times for each sample path, $g = 1, \ldots, G$.

Our filtering approach also allows us to periodically refresh the states and parameters, if needed. More specifically, a full update of the state trajectories and parameters $(\mathbf{x}_{0:t}^{(g)}, \theta^{(g)}), g = 1, \ldots, G$, can be performed every T^* time steps. This improves the parameter learning problem dramatically although it slows down the algorithm. However, typically a few iterations are sufficient to obtain convergence to the smoothing distribution $p(\mathbf{x}_{0:t}, \mathbf{z}_{1:t}, \theta|\mathbf{y}_{1:t})$ due to our fast MCMC smoother. A similar approach has been used to improve particle filters using periodic MCMC moves (Doucet, Godsill and Andrieu 2000). However, this approach is more costly for particle filters since the number of particles, N, is typically much larger than the G paths required for our approach.

11.3.2 Auxiliary particle filtering with parameter learning

To compare our methodology to the particle filter we also implement the parameter learning algorithm of Storvik (2002). This is a sequential importance sampling procedure that assumes an initial set of N particles $(\mathbf{x}_{0:t-1}, \theta) \sim p(\mathbf{x}_{0:t-1}, \theta | \mathbf{y}_{1:t-1})$. Letting $\mathbf{s}_{t-1}$ denote the posterior sufficient statistics that are updated at each time step, the algorithm then draws

$$\theta \sim p(\theta | \mathbf{s}_{t-1}) \text{ and } \mathbf{x}_t \sim p(\mathbf{x}_t | \mathbf{x}_{t-1}, \theta)$$

and reweights $(\mathbf{x}_{0:t}, \theta)$ proportionally to the likelihood $p(\mathbf{y}_t | \mathbf{x}_t, \theta)$.

For the SV model, we implement an APF version of the Storvik algorithm using the sampling/importance resampling-based auxiliary proposal of Pitt and Shephard (1999b), Section 3.2). Given an initial particle set, the APF first selects particles with high 'likelihood', and propagates those forward to the next time step. In our implementation, we define the selection probabilities as $\lambda_t^j \propto p(y_t | \mu_t^j)$, where $\mu_t^j = \alpha^j + \beta^j x_{t-1}^j$ is the estimate of x_t for particle j. For each selected particle k_j, we sample (x_t^j, θ^j) from the dynamic prior $p(x_t, \theta | x_{0:t-1}^{k_j}, y_{1:t})$. We then reweight the particles by the likelihood $p(y_t | x_t^j)$ to obtain a sample from the posterior. The APF algorithm for joint state and parameter learning is given below.

Algorithm Auxiliary particle filter with parameter learning

Initialisation: For $j = 1, \ldots, N$:

> Generate $x_0^j \sim p(x_0)$, and set $\mathbf{s}_0^j = \mathbf{s}_0$.
> Initialise weights to $w_0^j = N^{-1}$.

Sequential updating: For $t = 1, \ldots, T$:

> Compute first-stage weights, $\lambda_t^j \propto p(y_t | \mu_t^j)$, for $j = 1, \ldots, N$.
> Sample the index k_j with probabilities λ_t^j, for $j = 1, \ldots, N$.
> For $j = 1, \ldots, N$:
>
>> Generate $\theta^j \sim p(\theta | \mathbf{s}_{t-1}^{k_j})$.
>> Generate $x_t^j \sim p(x_t | x_{t-1}^{k_j}, \theta^j)$.
>> Update $\mathbf{s}_t^j = S(x_t^j, \mathbf{s}_{t-1}^{k_j})$, and set $\tilde{\mathbf{x}}_t^j = (\theta^j, x_t^j, \mathbf{s}_t^j)$.
>> Compute the second-stage weights, $w_t^j = w_{t-1}^{k_j} p(y_t | x_t^j) / \lambda_t^{k_j}$.
>
> Normalise weights, $w_t^j = w_t^j / \sum_{i=1}^N w_t^i$.
> Compute effective sample size $N_{\text{eff}} = 1 / \sum_{i=1}^N (w_t^i)^2$.
> If $N_{\text{eff}} < 0.8N$, resample $\tilde{\mathbf{x}}_t^j$ with probability w_t^j, and set $w_t^j = N^{-1}$.

To improve efficiency of the algorithm, we use stratified sampling for the multinomial and resampling steps above (see, for example, Carpenter, Clifford and Fearnhead (1999)).

We now turn to our two applications.

11.3.3 Applications

We use the two filtering and sequential parameter learning algorithms, and compare the methods with estimates determined via full MCMC estimation. The two algorithms use the following specifications, chosen to make the running times similar:

(i) Storvik's algorithm with $N = 100\,000$ particles.
(ii) Practical filter with $G = 250, I = 50$ and $k = 50$.

Figure 11.1 displays the filtered volatility and sequential parameter learning plots for both algorithms on a simulated dataset. A time series of length 500 is generated using the parameter values

$$\theta = (\alpha, \beta, \sigma^2) = (-0.0084, 0.98, 0.04).$$

For both filtering algorithms, an initial prior of $x_0 \sim \mathcal{N}(0,1)$ is used along with the conjugate prior $p(\theta)$ described in Section 11.3.1 with hyperparameters $\psi_0 = (0, 0.95)'$, $A_0 = \text{diag}(10, 100)$, $a_0 = 2.25$ and $b_0 = 0.0625$. This implies a mildly informative prior which should give no advantage to either method. To implement the practical filter, refreshing is used every $T^* = 125$ time steps. For comparison purposes, we also run a full MCMC smoothing algorithm at each time point. The thick lines denote the full MCMC quantiles and the thin lines denote those from the two filters. As may be expected, there is close agreement between the practical filter and the true MCMC results for both the states x_t and the parameters θ in this simulated example. The auxiliary particle filter also performs quite well.

For daily data on the S&P 500, a different picture emerges. Figure 11.2 displays the filtered volatility and sequential parameter learning plots for both algorithms on the S&P 500 dataset. The priors used are the same as those in the simulation study. The 5, 50 and 95 percentiles are plotted along with the 'true' estimates obtained from a full MCMC analysis. For the practical filter, refreshing is used every $T^* = 250$ time steps. The main difference in the results comes in estimating the volatility of volatility parameter σ sequentially, and the differences are more pronounced at earlier time points. The advantage of practical filtering is that there are no degeneracies and at the end of the sample there is close agreement with the

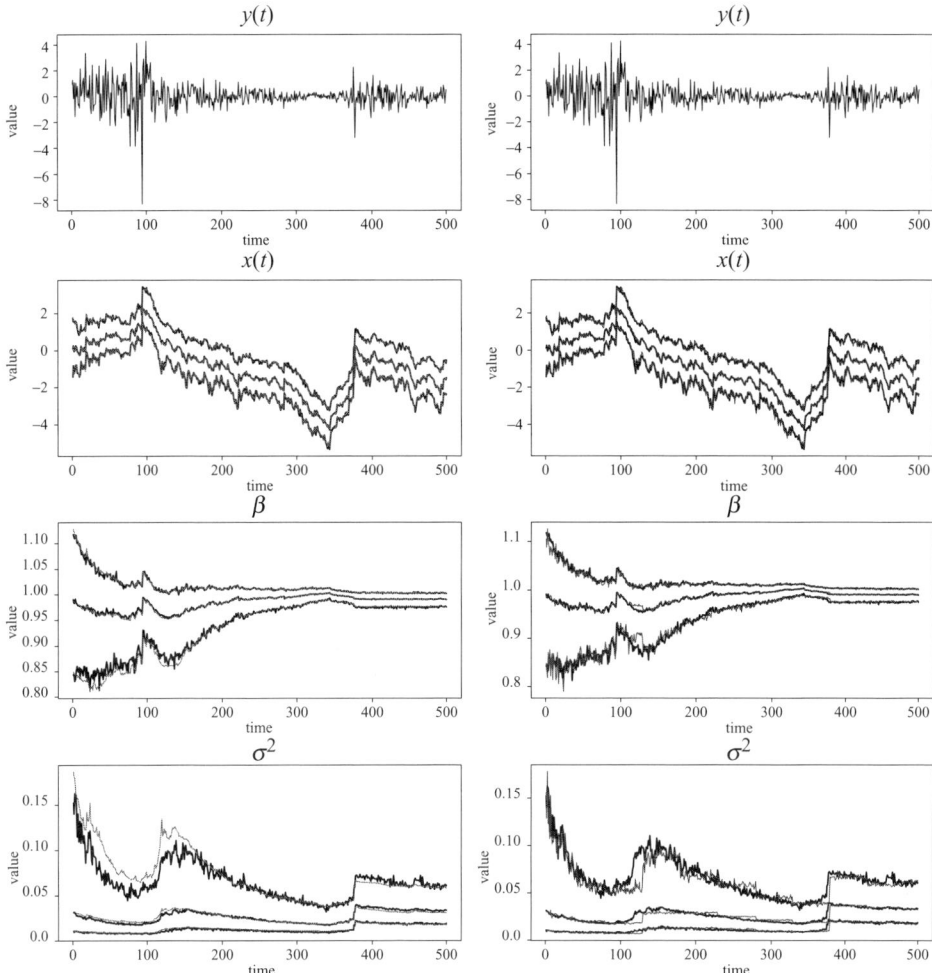

Fig. 11.1. SV model, simulated data. Filtered 5, 50, 95 percentiles. Left: Storvik's algorithm with $N = 100\,000$ particles. Right: Practical filter with $G = 250, I = 50, k = 50$. Thin lines are used to denote the filtering quantiles, and thick lines denote the quantiles from a full MCMC analysis.

full MCMC estimates after the Crash of 1987. The practical filter does a reasonably good job of tracking the shift in the filtered estimates of σ and eventually the quantiles match those obtained from the full analysis.

While the particle filter provides reliable inference about the state volatility vector similar to the practical filter, the difference lies in the sequential parameter estimates. The main problem with the particle filter is that it has great difficulty in handling the jump in σ during the stock market crash of October 1987. The sequential parameter posterior for σ nearly degenerates

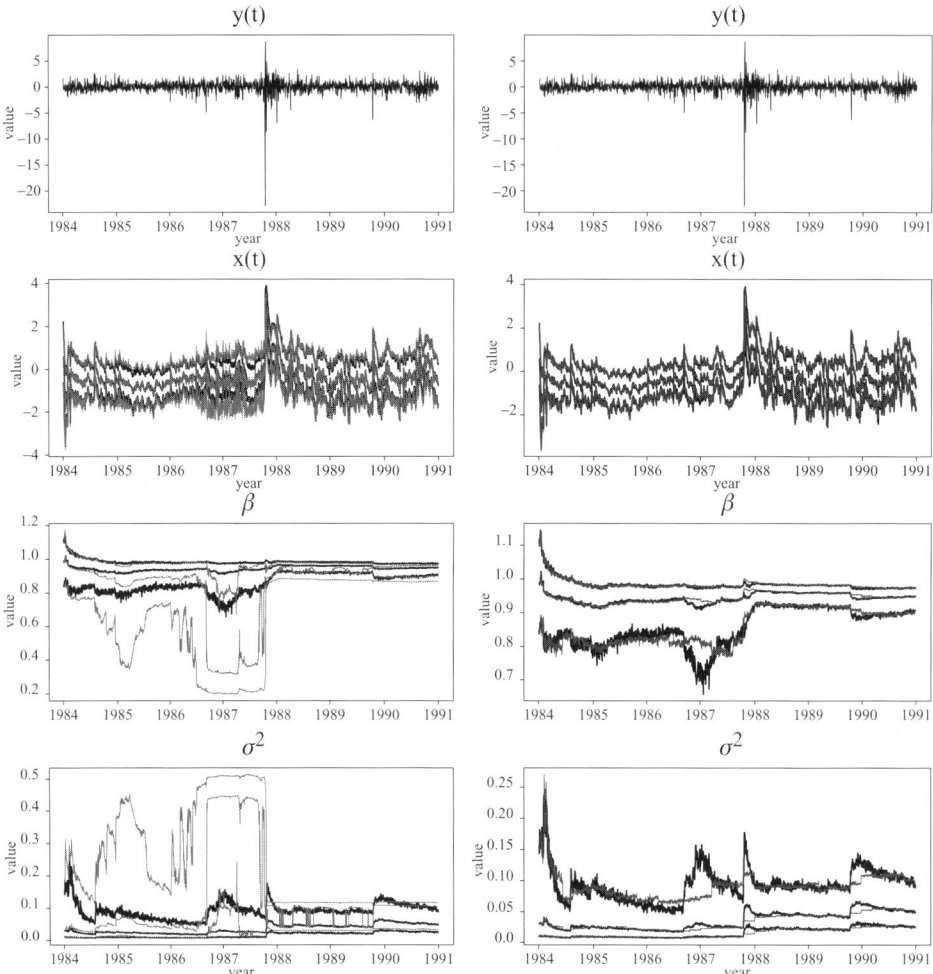

Fig. 11.2. SV model, S&P 500 data. Filtered 5, 50, 95 percentiles. Left: Storvik's algorithm with $N = 100\,000$ particles. Right: Practical filter with $G = 250, I = 50, k = 50$. Thin lines are used to denote the filtering quantiles, and thick lines denote the quantiles from a full MCMC analysis.

immediately after the crash as there are not enough particles to capture the appropriate posterior uncertainty in σ after the crash. Eventually it degenerates onto a value near the 'true' value learned from the full MCMC analysis but it totally misrepresents the appropriate posterior uncertainty.

11.4 Conclusions

In this paper we develop a sequential simulation-based methodology for filtering and sequential parameter learning in SV models. The method is easy to implement as it requires only simple modification of the smoothing algorithm. While filtering methods such as particle filtering can uncover states with known static parameters they have difficulty in also providing sequential parameter estimates. It is also not uncommon for these models to degenerate whereas our approach will not.

On a simulated dataset, we found that the filter of Storvik (2002) performed on a par with our methodology. For the S&P 500 daily stock return data, however, the auxiliary particle filter degenerates and has particular difficulty in estimating the volatility of volatility. Our procedure also avoids degeneracies but in dealing with the outlying crash it tends to oversmooth the uncertainty in our estimates for σ relative to a brute-force MCMC approach. This is also partly due to the fact that the learning parameters break the simple Markov property of updating and long-range dependence can reduce the accuracy of the particle filter approximation. We hope to study this further and provide other updating schemes.

12

On RegComponent time series models and their applications

William R. Bell

Statistical Research Divison, US Census Bureau

Abstract

We use the term 'RegComponent' model to refer to a regression model whose errors follow an ARIMA component time series model. This is a generalisation of 'RegARIMA' models for which the error term follows a standard ARIMA model. Specific forms of RegComponent models (e.g., 'structural models') have been used for some time in connection with modelling seasonal time series and for model-based seasonal adjustment, but this paper takes a broader view of RegComponent models. We discuss a general form for RegComponent models along with some theoretical considerations in their treatment regarding likelihood evaluation, likelihood maximisation, forecasting and signal extraction. We also illustrate the use of RegComponent models on some less familiar applications. These include: (i) modelling and seasonal adjustment of time series observed with sampling error; (ii) use of regression models with stochastically time-varying regression parameters and error terms that follow ARIMA or ARIMA component models; and (iii) using components present only occasionally to allow for seasonal or other types of heteroskedasticity. The examples presented illustrate both some of the general model capabilities and some of the capabilities of the US Census Bureau's REGCMPNT computer program, which handles general RegComponent models.

12.1 Introduction

This paper explores the use of RegComponent time series models. We use the term 'RegComponent' model to refer to a time series model with a regression mean function and an error term that follows an ARIMA component time series model. This is a generalisation of 'RegARIMA' models for which the error term follows a standard ARIMA (autoregressive-integrated-moving average) time series model (Box and Jenkins 1976). This paper provides an overview of RegComponent models and illustrates their use with three examples. The examples also illustrate capabilities of the REGCMPNT computer program developed by the time series staff of the US Census Bureau. This program was designed to do Gaussian maximum likelihood (ML) estimation and signal extraction for models of the general RegComponent form presented here.

RegComponent models of particular forms have been considered elsewhere, including Harvey (1989) and Durbin and Koopman (2001). These references consider so-called 'structural' time series models, and also consider inclusion of regression terms in the models. Structural models have seen considerable investigation for applications to seasonal time series, for which separate models are specified for the seasonal, trend and irregular components. The models used for the components can be specified in ARIMA form, though in some cases they are not initially formulated in ARIMA form.

In this paper we take a broader view of RegComponent models. The paper is essentially in two parts. In the first part, Section 12.2 discusses the general form of RegComponent models, and notes how some specific models, particularly the structural models, fit into this general form. We also note how RegComponent models can be put in state space form. Section 12.3 then discusses some theoretical considerations for RegComponent models regarding likelihood evaluation, likelihood maximisation, forecasting, and signal extraction. Section 12.3 presents these results as matrix expressions, though this is not to advocate that the computations actually be done this way. It is generally more convenient and more efficient to treat the RegComponent model in state space form and do computations using the Kalman filter and a suitable smoother. The REGCMPNT program, in fact, uses the state space approach. The matrix expressions, however, are useful for showing what is actually being computed, something that is often fairly obscure from discussions of recursive algorithms.

The second part of the paper, Sections 12.4, 12.5 and 12.6, presents the examples. Section 12.4 discusses how RegComponent models ideally can and should be used to account for sampling error when modelling time series from a repeated survey whose estimates are subject to significant amounts of sampling error. These considerations are illustrated by examining model-based seasonal adjustment allowing for sampling error in a time series of estimates of retail sales of drinking places from the US Monthly Retail Trade Survey. Section 12.5 then shows how a RegComponent model can accommodate stochastically time-varying regression parameters, illustrating this capability through its application to estimate time-varying trading-day effects in a monthly time series of retail sales of US department stores. Finally, Section 12.6 shows how a RegComponent model can allow for additional variability (noise) in some observations of a time series, illustrating the potential importance of this via modelling of a time series of regional US housing starts, a series for which extra variability in some winter months is appropriate.

A recurring theme in the three examples is that in such cases as these the use of a component model structure is essential to modelling the variation of interest; that is, a conventional RegARIMA model cannot account for the types of time series variation being modelled. Furthermore, the type of component model structure needed in all three examples is quite different from that of the structural time series models that have been so extensively studied.

12.2 RegComponent models

The general form of the RegComponent models considered here is

$$y_t = x_t'\beta + \sum_{i=1}^{m} h_{it} z_{it}. \tag{12.1}$$

In (12.1)

- y_t is the observed time series with observations at time points $t = 1, \ldots, n$.
- $x_t = (x_{1t}, \ldots, x_{rt})'$ is an $r \times 1$ row vector of known regression variables at time t and β is the corresponding vector of (fixed) regression parameters.
- h_{it} for $i = 1, \ldots, m$ are series of known constants that we call 'scale factors.' Often $h_{it} = 1$ for all i and t.
- z_{it} for $i = 1, \ldots, m$ are independent unobserved component series following ARIMA models.

A general notation for the ARIMA models for the z_{it} is

$$\phi_i(B)\delta_i(B)z_{it} = \theta_i(B)\epsilon_{it}, \tag{12.2}$$

where $\phi_i(B)$, $\delta_i(B)$ and $\theta_i(B)$ are the AR, differencing and MA operators, which are polynomials in the backshift operator B ($Bz_{it} = z_{i,t-1}$). We write

$$\phi_i(B) = 1 - \phi_{i1}B - \cdots - \phi_{i,p_i}B^{p_i},$$

where p_i is the order of $\phi_i(B)$, and similarly for $\delta_i(B)$ and $\theta_i(B)$. These polynomials can be multiplicative, as in seasonal models. The $\delta_i(B)$ typically have all their zeros on the unit circle, though we will not impose this restriction. We do require the $\phi_i(B)$ to have all their zeros outside the unit circle, and the $\theta_i(B)$ to have all their zeros on or outside the unit circle. The ϵ_{it} are iid $N(0, \sigma_i^2)$ (white noise) innovations, independent of one another (which implies $\text{cov}(\epsilon_{it}, \epsilon_{jt'}) = 0$ unless $i = j$ and $t = t'$). In the examples it will generally be more convenient to write out the specific models being used for that example than to relate the model to the general notation of (12.2).

Our use of the term 'differencing operators' for the $\delta_i(B)$ is rather loose. While $\delta_i(B)$ can be a nonseasonal difference of some order ($(1-B)^d$ for some $d > 0$), or a seasonal difference such as $1 - B^{12}$, it could also be a more general operator with unit roots, such as the (monthly) seasonal summation operator, $1 + B + \cdots + B^{11}$. We can also let the coefficients in $\delta_i(B)$ be parameters to be estimated, in which case the location of the zeros of $\delta_i(B)$ ultimately depends on these estimated parameters. For signal extraction we require that $\delta_i(B)$ and $\delta_j(B)$ for $i \neq j$ have no common unit roots or common explosive roots (zeros inside the unit circle), as dealing with such common roots in signal extraction presents problems (Bell 1984, 1991, Kohn and Ansley 1987). Such common roots do not present problems for model estimation or forecasting calculations, however.

We can easily see how the general RegComponent model (12.1) includes several important special cases. If $m = 1$ and $h_{1t} = 1$ for all t then (12.1) is the standard RegARIMA model. If $m = 3$ and $h_{it} = 1$ for all t and for all $i = 1, 2, 3$, and if the unobserved components z_{it} follow any of the seasonal, trend and irregular component models presented in Harvey (1989) or Durbin and Koopman (2001), then (12.1) is one of the structural time series models considered there (with regression terms). In particular, the basic structural model or BSM (Harvey 1989) is written (here shown without regression terms)

$$y_t = S_t + T_t + I_t, \tag{12.3}$$

where the unobserved components $z_{1t} = S_t$, $z_{2t} = T_t$ and $z_{3t} = I_t$ follow the models (for a monthly time series)

$$\begin{aligned}
(1 + B + \cdots + B^{11})S_t &= \epsilon_{1t}, \\
(1 - B)^2 T_t &= (1 - \eta B)\epsilon_{2t}, \\
I_t &= \epsilon_{3t},
\end{aligned} \quad (12.4)$$

where ϵ_{1t}, ϵ_{2t} and ϵ_{3t} are independent white noise series with variances σ_1^2, σ_2^2 and σ_3^2. The model formulation for T_t in Harvey (1989) is not directly in ARIMA form, but can be expressed in the ARIMA(0,2,1) form shown with the constraint $\eta \geq 0$. The general RegComponent form does not impose such a constraint, though if the constraint is violated by the fitted RegComponent model, this can only lead to an improved fit relative to the model with the constraint.

Gersch and Kitagawa (1983) and Kitagawa and Gersch (1984) proposed models similar to (12.4) but dropping the $(1 - \eta B)$ factor (that is, setting $\eta = 0$). They also considered different orders of differencing in the model for T_t and the addition of a stationary autoregressive component to the model.

Harvey (1989) proposed two modifications of the BSM that differ from (12.4) only in their models for the seasonal component, S_t. Bell (1993) noted that these two alternative models for S_t can be written in ARIMA component form as follows:

$$\text{TRIG-6:} \quad S_t = \sum_{j=1}^{6} S_{jt}, \quad \delta_j(B) S_{jt} = (1 - \alpha_j B)\varepsilon_{jt}, \quad (12.5)$$

and

$$\text{TRIG-1: TRIG-6 with the restriction} \quad \kappa_1^2 = \cdots = \kappa_6^2, \quad (12.6)$$

which is discussed below. Details of the TRIG-6 and TRIG-1 models are given in Tables 12.1 and 12.2. The 'differencing' operators $\delta_j(B)$ each correspond to a factor of $1 + B + \cdots + B^{11}$ at a different seasonal frequency $\lambda_j = 2\pi j/12$ $(j = 1, \ldots, 6)$. Thus,

$$\prod_{j=1}^{6} \delta_j(B) = 1 + B + \cdots + B^{11}.$$

In (12.6) the κ_j^2 denote the innovation variances in Harvey's formulation of the model for the S_{jt}. These determine the innovation variances $\sigma_j^2 = \text{var}(\varepsilon_{jt})$ in (12.5) through the relations $\kappa_6^2 = \sigma_6^2$ and $2\kappa_j^2 = (1 + \alpha_j^2)\sigma_j^2$ for $j = 1, \ldots, 5$. (Note: Bell (1993) incorrectly neglects to mention the relation

$\kappa_6^2 = \sigma_6^2$, but the results given there were obtained using this relation and are correct.) From the α_j values in Table 12.1, the TRIG-1 restriction $\kappa_1^2 = \cdots = \kappa_6^2$ restricts the σ_j^2 as follows: $\sigma_j^2 = 1.5\sigma_6^2$ for $j = 1, 5$; $\sigma_j^2 = (1 + \sqrt{3}/2)\sigma_6^2 \approx 1.866\sigma_6^2$ for $j = 2, 4$; and $\sigma_3^2 = 2\sigma_6^2$.

Table 12.1. *ARIMA representations for the individual TRIG-6 seasonal components*

$(1 - \sqrt{3}B + B^2)S_{1t}$	$= [1 - (\sqrt{3}/3)B]\varepsilon_{1t}$	$(\lambda_1 = \pi/6)$
$(1 - B + B^2)S_{2t}$	$= [1 - (2 - \sqrt{3})B]\varepsilon_{2t}$	$(\lambda_2 = \pi/3)$
$(1 + B^2)S_{3t}$	$= \varepsilon_{3t}$	$(\lambda_3 = \pi/2)$
$(1 + B + B^2)S_{4t}$	$= [1 + (2 - \sqrt{3})B]\varepsilon_{4t}$	$(\lambda_4 = 2\pi/3)$
$(1 + \sqrt{3}B + B^2)S_{5t}$	$= [1 + (\sqrt{3}/3)B]\varepsilon_{5t}$	$(\lambda_5 = 5\pi/6)$
$(1 + B)S_{6t}$	$= \varepsilon_{6t}$	$(\lambda_6 = \pi)$

Bell (1993) also obtained the ARIMA representation for the TRIG-1 model for S_t by applying $U(B) = 1 + B + \cdots + B^{11}$ to the TRIG-6 equation (12.5), giving

$$U(B)S_t = \sum_{j=1}^{6}[U(B)/\delta_j(B)](1 - \alpha_j B)\varepsilon_{jt}.$$

The right hand side of this equation is the sum of six independent MA(10) processes, which is itself then an MA(10) process. Using the TRIG-1 restrictions implied for the σ_j^2, the autocovariances through lag 10 of $U(B)S_t$ can be obtained up to a constant of proportionality. The resulting autocovariance generating function for $U(B)S_t$ can then be factored to give the MA(10) representation given in Table 12.2. The unknown constant of proportionality is absorbed into the innovation variance σ_c^2, which is the one seasonal parameter to be estimated in the TRIG-1 model.

Table 12.2. *ARIMA representation of the TRIG-1 seasonal component*

$(1 + B + \cdots + B^{11})S_t = (1 - \alpha_1 B - \cdots - \alpha_{10}B^{10})c_t$		$c_t \sim$ iid $N(0, \sigma_c^2)$	
$\alpha_1 = -0.737378$	$\alpha_2 = -0.627978$	$\alpha_3 = -0.430368$	$\alpha_4 = -0.360770$
$\alpha_5 = -0.219736$	$\alpha_6 = -0.180929$	$\alpha_7 = -0.088488$	$\alpha_8 = -0.071423$
$\alpha_9 = -0.020306$	$\alpha_{10} = -0.016083$		

If for some i the h_{it} in (12.1) are observed values of regression variables, then (12.1) is a model with stochastically time-varying regression parameters, these time-varying parameters being the z_{it} that correspond to the h_{it} that are regression variables. One example is the following seasonal component model proposed by Hannan 1970:

$$S_t = \sum_{k=1}^{5}\left[\cos\left(\frac{2\pi kt}{12}\right)\beta_{kt} + \sin\left(\frac{2\pi kt}{12}\right)\gamma_{kt}\right] + \cos(\pi t)\beta_{6t}, \quad (12.7)$$

where

$$(1-B)\beta_{kt} = \epsilon_{k1t} \sim \text{iid } N(0, \sigma_k^2), \quad k = 1, \ldots, 6,$$
$$(1-B)\gamma_{kt} = \epsilon_{k2t} \sim \text{iid } N(0, \sigma_k^2), \quad k = 1, \ldots, 5$$

with the ϵ_{k1t} and ϵ_{k2t} series all independent of one another, though for each k the pair ϵ_{k1t} and ϵ_{k2t} have a common variance σ_k^2. Though not immediately obvious, Bell (1993) noted that (12.7) is actually equivalent to the TRIG-6 model (12.5) under Gaussian assumptions (or in terms of second moments). Proietti (2000, p. 250) noted this connection for the case where the σ_k^2 in (12.7) are all equal. Note that the form in (12.7) uses the cosine and sine functions as scale factors (h_{kt}) that are trigonometric regression variables with the β_{kt} and γ_{kt} stochastically time-varying regression parameters. The form in (12.5) avoids the scale factors (all are 1) but has somewhat more complicated ARIMA models for the components.

The structural models just discussed are relatively well-known examples of RegComponent models. As noted in the Introduction, Sections 12.4–12.6 of this paper focus on some less-well-known examples. In the remainder of this section we show how RegComponent models can be put in state space form. This allows us to use the Kalman filter for likelihood evaluation and forecasting and a corresponding smoother for signal extraction. Section 12.3 gives matrix expressions that show what these calculations actually compute.

To put the general RegComponent model (12.1) in state space form we start with the state space representations of the ARIMA components z_{it} following (12.2), which are

$$z_{it} = [1, 0, \ldots, 0]\alpha_{it}, \quad (12.8)$$
$$\alpha_{it} = F_i \alpha_{i,t-1} + g_i \epsilon_{it}. \quad (12.9)$$

The state vector α_{it} for z_{it} following the ARIMA(p_i, d_i, q_i) model (12.2) can be defined in various ways. Akaike (1978) suggested one version. Here we use the following definition (Kohn and Ansley 1986, p. 754):

$$\alpha_{i,1t} = z_{it},$$
$$\alpha_{i,kt} = \sum_{j=k}^{r_i} \varphi_{ij} z_{i,t-1+k-j} - \sum_{j=k-1}^{q_i} \theta_{ij}\epsilon_{i,t-1+k-j}, \quad k = 2, \ldots, f_i,$$

where

$$\varphi_i(B) \equiv 1 - \varphi_{i1}B - \cdots - \varphi_{i,r_i}B^{r_i} = \phi_i(B)\delta_i(B)$$

is the product of the AR and differencing operators for component i (and we define $\sum_{j=k}^m = 0$ if $k > m$). For this definition of the state vector we have

$$F_i = \begin{bmatrix} \varphi_{i1} & 1 & & & \\ \varphi_{i2} & 0 & 1 & & \\ \vdots & & & \ddots & \\ & & & & 1 \\ \varphi_{i,f_i} & 0 & 0 & \cdots & 0 \end{bmatrix} \quad \text{and} \quad g_i = [1, -\theta_{i1}, \ldots, -\theta_{i,f_i-1}]',$$

where the state vector dimension $f_i = \max(r_i, q_i+1)$, and we define $\varphi_{ij} = 0$ for $j > r_i$ and $\theta_{ij} = 0$ for $j > q_i$. Note that $\text{var}(g_i\epsilon_{it}) = \sigma_i^2 g_i g_i'$.

Since the first element of α_{it} is z_{it}, the observation equation (12.8) simply picks out the first element of α_{it}. It is easy to see that the state space representation of $h_{it}z_{it}$ is obtained by replacing the observation equation (12.8) by

$$h_{it}z_{it} = [h_{it}, 0, \ldots, 0]\alpha_{it}$$

and keeping the same state equation (12.9). We then can see that the state space representation of the model for $y_t - x_t'\beta$ is

$$y_t - x_t'\beta = [h_{1t}, 0, \ldots, 0, h_{2t}, 0, \ldots 0, h_{mt}, 0, \ldots, 0]\alpha_t,$$
$$\alpha_t = F\alpha_{t-1} + \xi_{t-1},$$

where the state vector $\alpha_t = [\alpha_{1t}', \alpha_{2t}', \ldots, \alpha_{mt}']'$ is obtained by stacking the m state vectors for $z_{1t}, \ldots, z_{mt}$. Similarly $\xi_t = [\xi_{1t}', \ldots, \xi_{mt}']'$ is the stacked vector of the $\xi_{i,t} \equiv g_i\epsilon_{it}$, and F is a block diagonal matrix with the F_i as the diagonal blocks. Since the ϵ_{it}s are independent of one another so are the $\xi_{i,t}$s, thus ξ_t has a block diagonal covariance matrix with diagonal blocks given by the $\text{var}(\xi_{i,t}) = \sigma_i^2 g_i g_i'$.

One can also put the regression parameters β in the state vector α_t and thus obtain a complete state space representation of the model for y_t. Durbin and Koopman (2001, pp. 122–3) discuss both this approach and the approach of treating β separately in the Kalman filter. In Section 12.4.2 we give matrix results for the approach of treating β separately.

Initialisation of the Kalman filter is an important consideration for treating RegComponent models in state space form. Since $y_t - x_t'\beta$ is assumed to have mean zero ($x_t'\beta$ being the regression mean of y_t), the unconditional mean of the initial state vector is zero. The appropriate initial variance for stationary models uses the unconditional variance of the initial state vector. Akaike (1978) discussed this for stationary ARMA models for his choice of state space representation (see also Jones (1980)). For nonstationary models that involve differencing, the variance initialisations of Ansley and Kohn (1985), Bell and Hillmer (1991), de Jong (1991) and Koopman (1997) are appropriate, each providing a somewhat different way to calculate the same thing. Any of these initialisations, together with the Kalman filter and a corresponding smoother, provides a way to compute the same results as the matrix expressions shown in the next section. The REGCMPNT program uses the initialisation of Bell and Hillmer (1991).

12.3 Some theoretical considerations

The most efficient way to handle calculations with RegComponent models is to use the state space form with the Kalman filter (for likelihood evaluation and forecasting) and a suitable smoothing algorithm (for signal extraction). The REGCMPNT program takes this approach, using a fixed point smoother of reduced dimension (Anderson and Moore 1979) for signal extraction. As use of the Kalman filter and smoothers for ARIMA component models has been well covered in numerous references, such as Harvey (1989) and Durbin and Koopman (2001), we shall not repeat this material here. Instead, in this section we present results given as matrix expressions to discuss likelihood evaluation, likelihood maximisation, forecasting, and signal extraction with RegComponent models. Many of these results are taken from Bell and Hillmer (1988). Durbin and Koopman (2001, pp. 95–8) gave matrix expressions for estimation of the stacked state vectors and the innovations for all time points t.

The forecasts and signal extraction estimators shown are obtained by direct application of the 'transformation approach' of Ansley and Kohn (1985). The transformation approach estimators have minimum mean squared error (MMSE) among linear estimators that eliminate from the estimation error the effects of the nonstationary initial conditions. These estimators are MMSE among all linear estimators under Assumption A of Bell (1984) regarding these initial conditions. The forecasts are consistent with the forecasting approach proposed for ARIMA models in Box and Jenkins (1976). As noted earlier, the forecasts and signal extraction estimators are also what would be obtained from the Kalman filter and an associated smoother with one of the initialisations discussed at the end of Section 12.2.

While the matrix expressions that follow do not generally provide the best approach for doing the computations, they have four advantages. First, the matrix expressions show what is being computed, something that is fairly obscure from discussions of recursive filtering and smoothing algorithms. Second, the matrix expressions are easily programmed in mathematical programming languages, which makes the approach useful for checking results obtained when programming more efficient calculations via the Kalman filter and an associated smoother. Third, the matrix expressions give the complete covariance matrix of the forecasting and signal extraction errors for all time points. The standard Kalman filter and smoother calculations give error variances only for individual time points, and additional calculations are required to obtain error covariances between different time points (Durbin and Koopman 2001, pp. 77–81). Finally, the matrix expressions apply to models that cannot be put in state space form (e.g., long memory models), though we shall not illustrate this capability here.

For simplicity in the discussion of the following subsections we drop the scale factors from the model (12.1), that is, we assume $h_{it} = 1$ for all t and all i. Scale factors are difficult to treat in general terms with the matrix approach when they apply to components that require differencing ($\delta_i(B) \neq 1$). For concreteness we show results for the general two-component model

$$y_t = x_t'\beta + S_t + N_t \qquad (12.10)$$

though extensions to more than two components are straightforward. The components S_t and N_t can be thought of as 'signal' and 'noise,' though they could also represent 'seasonal' and 'nonseasonal.' We use the following notation for the ARIMA forms of the component models:

$$\phi_S(B)\delta_S(B)S_t = \theta_S(B)b_t, \qquad (12.11)$$
$$\phi_N(B)\delta_N(B)N_t = \theta_N(B)c_t, \qquad (12.12)$$

where b_t and c_t are independent white noise series with zero means and variances σ_b^2 and σ_c^2.

It is worth noting that when the scale factors h_{it} in (12.1) all equal 1, as in (12.10), the stochastic term $\sum_{i=1}^{m} z_{it}$ has an ARIMA representation, so that (12.10) can be expressed as a RegARIMA model. Harvey (1989) refers to this as the 'reduced form.' Conversely, many seasonal ARIMA models can be decomposed into ARIMA component form (Burman 1980, Hillmer and Tiao 1982). Despite this connection it will usually be more convenient to work with the model in whichever form it was originally specified, as the parameters of the alternative form will generally be subject to nonlinear constraints, something particularly inconvenient for model estimation.

12.3.1 Gaussian likelihood evaluation for RegComponent models

The likelihood based on observations $y = (y_1, \ldots, y_n)'$ is defined as the joint density of the differenced data. Box and Jenkins (1976) took this approach for ARIMA models, which is consistent with the transformation approach of Ansley and Kohn (1985). We shall assume that $\delta_S(B)$ and $\delta_N(B)$ have no common factors so that the differencing operator for y_t is

$$\delta(B) \equiv 1 - \delta_1 B - \cdots - \delta_d B^d = \delta_S(B)\delta_N(B).$$

As noted earlier, the assumption of no common factors in $\delta_S(B)$ and $\delta_N(B)$ is not necessary for model estimation or forecasting, but is important for signal extraction.

Let the differenced observed and component series be

$$w_t = \delta(B)y_t, \quad u_t = \delta_S(B)S_t \quad \text{and} \quad v_t = \delta_N(B)N_t.$$

Note that u_t and v_t follow the stationary ARMA models

$$\phi_S(B)u_t = \theta_S(B)b_t, \qquad (12.13)$$
$$\phi_N(B)v_t = \theta_N(B)c_t. \qquad (12.14)$$

Note also that

$$\begin{aligned} w_t &= \delta_S(B)\delta_N(B)\left(x_t'\beta + S_t + N_t\right) \\ &= [\delta(B)x_t']\beta + \delta_N(B)u_t + \delta_S(B)v_t. \end{aligned} \qquad (12.15)$$

We see that the regression mean function of w_t is $[\delta(B)x_t']\beta$, that is, it depends on the differenced regression variables

$$\delta(B)x_t' = [\delta(B)x_{1t}, \ldots, \delta(B)x_{rt}].$$

We let $S = (S_1, \ldots, S_n)'$, $N = (N_1, \ldots, N_n)'$ and define the vectors of differenced series $w = (w_{d+1}, \ldots, w_n)'$, $u = (u_{d_S+1}, \ldots, u_n)'$ and $v = (v_{d_N+1}, \ldots, v_n)'$. By analogy with (12.15) we can write

$$w = (\Delta X)\beta + \Delta_N u + \Delta_S v, \qquad (12.16)$$

where Δ, Δ_N and Δ_S are differencing matrices corresponding to $\delta(B)$, $\delta_N(B)$ and $\delta_S(B)$, respectively. For example

$$\Delta = \begin{bmatrix} -\delta_d & \cdots & -\delta_1 & 1 & & \\ & -\delta_d & \cdots & -\delta_1 & 1 & \\ & & \ddots & & \ddots & \ddots \\ & & & -\delta_d & \cdots & -\delta_1 & 1 \end{bmatrix}.$$

For (12.16) Δ, Δ_N and Δ_S must all be defined to have $n - d$ rows. (Note that we then cannot take $\Delta_N \Delta_S$, though if we instead defined Δ_S to have $n - d_S$ rows, we could take this product which would then yield Δ.) Finally, we let $\Sigma_w = \text{var}(w)$, $\Sigma_u = \text{var}(u)$, and $\Sigma_v = \text{var}(v)$, so from (12.16)

$$\Sigma_w = \Delta_N \Sigma_u \Delta_N' + \Delta_S \Sigma_v \Delta_S'. \tag{12.17}$$

The Gaussian likelihood function is the multivariate normal density of w. Let $X_\Delta = \Delta X$. The log-likelihood ℓ is (apart from the constant $-[(n-d)/2]\log(2\pi)$):

$$\ell = -\tfrac{1}{2}\log|\Sigma_w| - \tfrac{1}{2}(w - X_\Delta \beta)' \Sigma_w^{-1} (w - X_\Delta \beta). \tag{12.18}$$

The log-likelihood (12.18) depends in an obvious way on the regression parameters β, and also depends on the parameters $\phi_{Sj}, \theta_{Sj}, \sigma_b^2, \phi_{Nj}, \theta_{Nj}$ and σ_c^2 of (12.11) and (12.12) through the covariance matrix Σ_w via its dependence on the component covariance matrices Σ_u and Σ_v in (12.17). Autocovariances of the differenced components u_t and v_t for given parameter values in their ARMA models (12.13) and (12.14) can be computed by the algorithms of McLeod (1975) (see also McLeod (1977)) or Tunnicliffe-Wilson (1979). This determines Σ_u and Σ_v, and Σ_w then follows from (12.17). Alternatively, we can note that $\Delta_N \Sigma_u \Delta_N'$ is the covariance matrix of $n - d$ observations from the time series $\zeta_{St} \equiv \delta_N(B)u_t$, which follows the noninvertible ARMA model

$$\phi_S(B)\zeta_{St} = \delta_N(B)\theta_S(B)b_t.$$

We can compute autocovariances for this model to obtain $\Delta_N \Sigma_u \Delta_N'$. We can similarly compute $\Delta_S \Sigma_v \Delta_S'$.

We can use the algorithm in the Appendix to compute (12.18). Specifically, if $\Sigma_w = LDL'$ is the unit Cholesky decomposition of Σ_w, then (12.18) becomes

$$\ell = -\frac{1}{2}\sum_{t=d+1}^{n}\log(d_t) - \frac{1}{2}\sum_{t=d+1}^{n}\varepsilon_t^2/d_t, \tag{12.19}$$

where

$$\varepsilon \equiv (\varepsilon_{d+1},\ldots,\varepsilon_n)' = L^{-1}(w - X_\Delta \beta) = L^{-1}\Delta(y - X\beta) \tag{12.20}$$

and the d_t are the diagonal elements of D. Though not immediately obvious, the ε_t obtained from (12.20) are actually the Kalman filter innovations and the d_t the corresponding innovation variances. Harvey (1989, p. 131) and Durbin and Koopman (2001, pp. 14–15 and p. 97) noted this fact. The Kalman filter essentially provides an efficient way to compute the inverse

Cholesky decomposition of a covariance matrix. While the Kalman filter is generally more efficient numerically for ARIMA component models, the Cholesky decomposition approach is more general and has a role for models that cannot be put in state space form (so that the Kalman filter cannot be used).

12.3.2 Likelihood maximization for RegComponent models

The log-likelihood function, whether in the form (12.18) or (12.19), can be maximised by a nonlinear optimisation routine. In this subsection we discuss some alternative options in regard to how this is done. The alternative options relate to how we handle the regression parameters β, the innovation variances and constraints on the parameters.

In regard to β, for given values of the parameters of the ARMA models (12.13) and (12.14), which we generically denote by η, it follows from standard least squares results that the log-likelihood (12.18) is maximised by the generalised least squares (GLS) estimate:

$$\hat{\beta}(\eta) = \left(X'_\Delta \Sigma_w^{-1} X_\Delta\right)^{-1} X'_\Delta \Sigma_w^{-1} w = \left(X'\Delta'\Sigma_w^{-1}\Delta X\right)^{-1} X'\Delta'\Sigma_w^{-1}\Delta y. \tag{12.21}$$

For simplicity of notation we write just Σ_w though we could write $\Sigma_w(\eta)$ to more explicitly indicate the dependence of Σ_w on η. In addition, the asymptotic variance-covariance matrix of the MLE of β, given by (12.21) evaluated at the MLE $\hat{\eta}$ of η, is

$$\text{var}\left(\hat{\beta}\right) = \left(X'_\Delta \Sigma_w^{-1} X_\Delta\right)^{-1} = \left(X'\Delta'\Sigma_w^{-1}\Delta X\right)^{-1}. \tag{12.22}$$

Following the algorithm in the Appendix, with $\Sigma_w = LDL'$ as in the previous section, and defining

$$\tilde{X} = D^{-1/2} L^{-1} X_\Delta = D^{-1/2} L^{-1} \Delta X$$

and

$$\tilde{w} = D^{-1/2} L^{-1} w = D^{-1/2} L^{-1} \Delta y,$$

we can write (12.21) and (12.22) in terms of results of ordinary least squares regression of $\tilde{w}$ on $\tilde{X}$:

$$\hat{\beta}(\eta) = \left(\tilde{X}'\tilde{X}\right)^{-1} \tilde{X}'\tilde{w}$$

$$\text{var}\left(\hat{\beta}\right) = \left(\tilde{X}'\tilde{X}\right)^{-1}.$$

Kohn and Ansley (1985) essentially noted this result for models without differencing.

The result (12.21) raises two alternative possibilities for likelihood maximisation. One is to substitute $\hat{\beta}(\eta)$ back into the log-likelihood function (12.18), giving the log-likelihood concentrated over β:

$$\ell_c(\eta) = -\tfrac{1}{2}\log|\Sigma_w| - \tfrac{1}{2}\left(w - X_\Delta\hat{\beta}(\eta)\right)' \Sigma_w^{-1} \left(w - X_\Delta\hat{\beta}(\eta)\right). \quad (12.23)$$

We can then maximise $\ell_c(\eta)$ over η to get the MLE $\hat{\eta}$, and as noted $\hat{\beta}(\hat{\eta})$ then gives the MLE of β. The second approach, which we call iterative generalised least squares (IGLS), alternatively maximises the log-likelihood (12.18) or (12.19) over η for given β, and then maximises it over β for given η using the GLS result (12.21). The IGLS procedure can be started with some initial values for either η or β, such as setting the initial β to values obtained from the OLS regression of w on X_Δ.

Otto, Bell and Burman (1987), in a small study, compared the performance of these two alternative approaches to likelihood maximisation against simply including β in the nonlinear optimisation with η. Though this study was for RegARIMA models, the conclusions can be expected to hold more generally. They found that including β in the nonlinear optimisation was grossly inferior to the other two approaches, while the IGLS approach and a variant of the concentrated likelihood approach gave more comparable performance. It was pointed out by G. Tunnicliffe-Wilson (personal communication) that the variant of the concentrated likelihood approach that performed well is actually equivalent to an IGLS approach that only performs one iteration over η for each different value of β. That is, for each given β this approach does not iterate over η to convergence (maximising ℓ over η for that β). Instead, we take just one nonlinear step to improve the value of η before recalculating β. A more straightforward approach to maximising $\ell_c(\eta)$ gave significantly worse performance than IGLS. This observation shows the IGLS approach to be rather more general than it first appears, and also points out that the real question in tuning the optimisation is how refined to make the optimisation of $\ell(\eta, \beta)$ over η for given β before recalculating β? The two extreme choices are to optimise $\ell(\eta, \beta)$ over η to convergence for each given β (the IGLS approach), or to only take one nonlinear step in η for each given β (the successful variant of the concentrated likelihood approach).

Another optimisation choice concerns how one handles the innovation variances σ_b^2 and σ_c^2 assuming that neither is fixed, that is, both are to be estimated. (See later discussion for handling of fixed variances.) We can convert the parameterisation from (σ_b^2, σ_c^2) to $(\sigma_b^2, \tau^2 \sigma_b^2)$ for $\tau^2 \geq 0$ and estimate σ_b^2 and the variance ratio $\tau^2 = \sigma_c^2/\sigma_b^2$. This changes the expression (12.17) for Σ_w to

$$\Sigma_w = \sigma_b^2 \widetilde{\Sigma}_w = \sigma_b^2 \left(\Delta_N \widetilde{\Sigma}_u \Delta_N' + \tau^2 \Delta_S \widetilde{\Sigma}_v \Delta_S' \right), \qquad (12.24)$$

where $\widetilde{\Sigma}_w = \sigma_b^{-2} \Sigma_w$ and $\widetilde{\Sigma}_u = \sigma_b^{-2} \Sigma_u$ do not depend on σ_b^2, and $\widetilde{\Sigma}_v = \sigma_c^{-2} \Sigma_v$ does not depend on σ_c^2. Substituting for Σ_w from (12.24) into (12.18) and setting to zero the derivative of ℓ with respect to σ_b^2 gives

$$\hat{\sigma}_b^2(\eta^*) = \frac{1}{n-d} (w - X_\Delta \beta)' \widetilde{\Sigma}_w^{-1} (w - X_\Delta \beta), \qquad (12.25)$$

where η^* denotes η without σ_b^2. Substituting (12.25) back into (12.18) gives the log-likelihood concentrated over σ_b^2, which apart from constants is

$$\begin{aligned}\ell_2(\eta^*) &= -\frac{n-d}{2} \log(\hat{\sigma}_b^2(\eta^*)) - \frac{1}{2} \log \left| \widetilde{\Sigma}_w \right| \\ &= -\frac{n-d}{2} \log \left[\hat{\sigma}_b^2(\eta^*) \left| \widetilde{\Sigma}_w \right|^{1/(n-d)} \right].\end{aligned}$$

Letting $\rho = |\widetilde{\Sigma}_w|^{1/2(n-d)} > 0$ we note from (12.18)–(12.20) and (12.25) that

$$\ell_2(\eta^*) = -\frac{n-d}{2} \log \left(\frac{\rho^2}{n-d} \sum_{t=d+1}^{n} \tilde{\varepsilon}_t^2 / \tilde{d}_t \right), \qquad (12.26)$$

where the $\tilde{\varepsilon}_t$ and $\tilde{d}_t$ are obtained using the Cholesky decomposition of $\widetilde{\Sigma}_w = \tilde{L}\tilde{D}\tilde{L}'$. Thus, analogous to (12.20), $\tilde{\varepsilon} = \tilde{L}^{-1}(w - X_\Delta \beta)$. We see from (12.26) that maximising $\ell_2(\eta^*)$ is equivalent to minimising the sum of squares

$$\sum_{t=d+1}^{n} \left(\rho \tilde{\varepsilon}_t / \tilde{d}_t^{1/2} \right)^2.$$

This turns the optimisation problem into a nonlinear least squares problem so that nonlinear least squares software can be used. The REGCMPNT program does this and uses the Minpack software of More, Garbow and Hillstrom (1980).

The REGCMPNT program allows one or more of the η parameters to be fixed, something useful for modelling time series with sampling error as discussed in Section 12.4. If one of the variances, say σ_c^2, is fixed then we cannot concentrate out σ_b^2 as just described. We must instead maximise the log-likelihood (12.18) over all free (not fixed) parameters. However, we can express (12.19) as

$$\ell = -\frac{1}{2}\left\{\left(\sqrt{\sum_{t=d+1}^{n} \log(d_t)}\right)^2 + \sum_{t=d+1}^{n} \varepsilon_t^2/d_t\right\} \quad (12.27)$$

as long as

$$\sum_{t=d+1}^{n} \log(d_t) = \log|\Sigma_w| \geq 0,$$

which holds if $|\Sigma_w| \geq 1$. Since Σ_w depends on the scale of the data, this condition can always be satisfied by rescaling the data as necessary. Therefore, (12.27) again turns the optimisation problem into a least squares problem.

In maximising the log-likelihood ℓ or ℓ_2, the parameters of (12.11) and (12.12) are subject to various constraints: the AR operators must have all their zeros outside the unit circle, the MA operators must have all their zeros on or outside the unit circle, and the innovation variances σ_b^2 and σ_c^2 (and the variance ratio τ^2 for ℓ_2) must be nonnegative. The constraints on the AR and MA parameters can be enforced by appropriately shrinking the step in the nonlinear search whenever the search attempts a step that would violate these constraints. REGCMPNT enforces the nonnegativity constraints on the variances by doing the nonlinear search over the standard deviations σ_b and σ_c. The standard deviations are allowed to go negative in the nonlinear search but upon convergence are squared to get the MLEs of the variances. This causes no problems since the log-likelihood as a function of $\sigma_b < 0$ is a mirror image of that for $\sigma_b > 0$, and also $\partial \ell / \partial \sigma_b = \left(\partial \ell / \partial \sigma_b^2\right) 2\sigma_b$ is zero at $\sigma_b = 0$ (and similarly for σ_c). When a variance is concentrated out so ℓ_2 is used REGCMPNT maximises the concentrated likelihood over τ rather than τ^2.

One final point concerns model identifiability – ensuring that the likelihood function is not constant over certain combinations of parameters. Hotta (1989) discussed conditions for model identifiability in stationary ARMA component models, and Harvey (1989, pp. 205–9) discussed such conditions for structural models. The situation becomes more complex when the models may include scale factors, or have some parameters held fixed as being known *a priori*. For example, while two white noise components in a model are not identified if both their variances are unknown, they are both identified if either of their variances is known. They are also identified if the components are multiplied by different sets of scale factors. Because the flexibility of models allowed by the REGCMPNT program complicates the identifiability issue, the program makes no checks of model identifiability, but leaves it up to the program user to avoid specifying models that are not identified.

12.3.3 Forecasting in RegComponent models

Forecasting in nonstationary RegComponent models can be performed by forecasting the differenced series from the differenced data and then translating these results into forecasts of the original series y_t. To be more specific, let y_f, w_f, X_f and $X_{\Delta f}$ denote vectors and matrices, respectively, of values of y_t, w_t, x_t and $\delta(B)x_t$ for some forecast period $t = n+1, \ldots, n+K$. Denote the covariance matrix of the extended differenced series as

$$\mathrm{var}\begin{bmatrix} w \\ w_f \end{bmatrix} = \begin{bmatrix} \Sigma_w & \Sigma_{wf} \\ \Sigma_{fw} & \Sigma_{ff} \end{bmatrix}.$$

Then, if the model parameters η and β were known, the forecast of w_f from standard linear projection results would be given by

$$\hat{w}_f = X_{\Delta f}\beta + \Sigma_{fw}\Sigma_w^{-1}(w - X_\Delta \beta). \tag{12.28}$$

In practice we substitute the MLEs of η (to determine Σ_{fw} and Σ_w) and β into (12.28). We can use the algorithm in the Appendix to compute

$$\Sigma_{fw}\Sigma_w^{-1}(w - X_\Delta \beta)$$

in (12.28). To translate $\hat{w}_f$ into forecasts of y_f we can simply use the recursive relation

$$\hat{y}_t = \delta_1 \hat{y}_{t-1} + \cdots + \delta_d \hat{y}_{t-d} + \hat{w}_t, \quad t = n+1, \ldots, n+K, \tag{12.29}$$

where $\hat{y}_j = y_j$ for $1 \leq j \leq n$. If there is no differencing then (12.29) simply says $\hat{y}_t = \hat{w}_t$.

The covariance matrix of the forecast errors $w_f - \hat{w}_f$ is, again from standard linear projection results (assuming the model parameters are known),

$$\mathrm{var}(w_f - \hat{w}_f) = \Sigma_{ff} - \Sigma_{fw}\Sigma_w^{-1}\Sigma_{wf}. \tag{12.30}$$

We can use the algorithm in the Appendix to compute $\Sigma_{fw}\Sigma_w^{-1}\Sigma_{wf}$ in (12.30). To translate these results into the covariance matrix of the forecast errors $y_f - \hat{y}_f$ we define a $(K+d) \times (K+d)$ nonsingular matrix $\tilde{\Delta}_f$ as follows:

$$\tilde{\Delta}_f = \left[\begin{array}{ccc|ccc} & I_d & & & 0_{d \times K} & \\ \hline -\delta_d & \cdots & -\delta_1 & 1 & & \\ & \ddots & & \ddots & \ddots & \\ & & -\delta_d & \cdots & -\delta_1 & 1 \end{array}\right].$$

We then have, noting (12.29)

$$\tilde{\Delta}_f \begin{bmatrix} y_{n+1-d} \\ \vdots \\ y_n \\ \hat{y}_f \end{bmatrix} = \begin{bmatrix} y_{n+1-d} \\ \vdots \\ y_n \\ \hat{w}_f \end{bmatrix}. \qquad (12.31)$$

The analogous relation holds, of course, if we replace $\hat{y}_f$ and $\hat{w}_f$ by y_f and w_f. Noting the structure of $\tilde{\Delta}_f$ (lower triangular with I_d in the upper left hand $d \times d$ corner) its partitioned inverse can be written as

$$\tilde{\Delta}_f^{-1} = \begin{bmatrix} I_d & 0 \\ A_K & C_K \end{bmatrix}. \qquad (12.32)$$

The form of A_K and C_K is given in Bell (1984, Section 2). In particular,

$$C_K = \begin{bmatrix} 1 & & & \\ \xi_1 & 1 & & \\ \vdots & \ddots & \ddots & \\ \xi_{K-1} & \cdots & \xi_1 & 1 \end{bmatrix},$$

where ξ_k is the coefficient of B^k in the expansion of

$$\delta(B)^{-1} = 1 + \xi_1 B + \xi_2 B^2 + \cdots.$$

From (12.31) and (12.32) we see that

$$y_f = A_K y^\dagger + C_K w_f, \qquad (12.33)$$
$$\hat{y}_f = A_K y^\dagger + C_K \hat{w}_f, \qquad (12.34)$$

where $y^\dagger = (y_{n+1-d}, \ldots, y_n)'$, so that

$$y_f - \hat{y}_f = C_K(w_f - \hat{w}_f). \qquad (12.35)$$

Hence

$$\text{var}(y_f - \hat{y}_f) = C_K \text{var}(w_f - \hat{w}_f) C_K', \qquad (12.36)$$

where $\text{var}(w_f - \hat{w}_f)$ is given by (12.30).

To account for the error in estimating β, but still assuming that η is known, we let

$$\tilde{w}_f = X_{\Delta f} \hat{\beta} + \Sigma_{fw} \Sigma_w^{-1} \left(w - X_\Delta \hat{\beta} \right) \qquad (12.37)$$

be the vector of forecasts of w_f obtained using the estimated regression parameters $\hat{\beta}$ from (12.21). Then the corresponding forecasts of y_f can be written as for (12.34):

$$\tilde{y}_f = A_K y^\dagger + C_K \tilde{w}_f. \tag{12.38}$$

We now write the forecast error in $\tilde{y}_f$ as the sum of two parts:

$$y_f - \tilde{y}_f = (y_f - \hat{y}_f) + (\hat{y}_f - \tilde{y}_f). \tag{12.39}$$

From (12.34) and (12.38) we see that

$$\hat{y}_f - \tilde{y}_f = C_K(\hat{w}_f - \tilde{w}_f). \tag{12.40}$$

By the optimality of $\hat{y}_f$ we know that $y_f - \hat{y}_f$ is orthogonal to the differenced data w, while from (12.28), (12.37) and (12.40) $\hat{y}_f - \tilde{y}_f$ is a linear function of w. Hence, the two parts in (12.39) are orthogonal and

$$\text{var}(y_f - \tilde{y}_f) = \text{var}(y_f - \hat{y}_f) + \text{var}(\hat{y}_f - \tilde{y}_f).$$

Furthermore, from (12.28) and (12.37)

$$\hat{w}_f - \tilde{w}_f = X_{\Delta f}(\beta - \hat{\beta}) - \Sigma_{fw}\Sigma_w^{-1}X_\Delta(\beta - \hat{\beta}).$$

If we let $P_w = \Sigma_{fw}\Sigma_w^{-1}$ be the matrix for projecting w_f on w we can write this as

$$\hat{w}_f - \tilde{w}_f = (X_{\Delta f} - P_w X_\Delta)(\beta - \hat{\beta}). \tag{12.41}$$

Putting together (12.36), (12.40) and (12.41) we compute the covariance matrix of (12.39) as

$$\begin{aligned}&\text{var}(y_f - \tilde{y}_f) \\ &= C_K \text{var}(w_f - \hat{w}_f)C_K' + C_K \text{var}(\hat{w}_f - \tilde{w}_f)C_K' \\ &= C_K \left\{ \text{var}(w_f - \hat{w}_f) + (X_{\Delta f} - P_w X_\Delta) \text{var}(\beta - \hat{\beta})(X_{\Delta f} - P_w X_\Delta)\right\}' C_K'.\end{aligned} \tag{12.42}$$

Note that to compute $P_w X_\Delta$ we apply the forecasting calculations for w to each column of $X_{\Delta f}$. The general approach leading to (12.42) was suggested by Kohn and Ansley (1985).

Also note that analogous to (12.33) we could write $X_f = A_K X^\dagger + C_K X_{\Delta f}$ and can thus see that, analogous to (12.35),

$$X_f - P_y X = C_K (X_{\Delta f} - P_w X_\Delta),$$

where $P_y X$ projects X_f on X column by column in the same way that we forecast y_f from y. This result applies whether we do the direct matrix calculations or the same thing using the Kalman filter. Therefore, we can alternatively write (12.42) as

$$\text{var}(y_f - \tilde{y}_f) = \text{var}(y_f - \hat{y}_f) + (X_f - P_y X)\,\text{var}(\beta - \hat{\beta})\,(X_f - P_y X)', \tag{12.43}$$

where $\text{var}(y_f - \hat{y}_f)$ is given by (12.36) and $\text{var}(\beta - \hat{\beta})$ by (12.22).

12.3.4 Signal extraction for RegComponent models

Bell and Hillmer (1988) gave matrix results for finite sample signal extraction for nonstationary RegComponent models. We only summarise their results here. (Kohn and Ansley (1987) discussed finite sample nonstationary signal extraction from a general perspective, not focusing on particular model forms.) The focus in this section is on models for which one or more of the components is nonstationary requiring differencing ($\delta_i(B) \neq 1$.) Signal extraction results for stationary models follow from standard linear projection (Gaussian conditional expectation) results. They are an easy special case of the results presented here.

The simplest nonstationary results obtain for the case when only one of the components requires differencing. Suppose S_t is nonstationary and N_t is stationary ($\delta_N(B) = 1$). (Actually, for this case we can let N_t be nonstationary in certain ways, such as having nonconstant scale factors so that its variances change over time, as long as N_t does not require differencing.) Then the signal extraction estimate of the vector S is obtained by projecting N on the differenced data w and subtracting this estimate of N from the data y. Allowing for the regression effects this gives

$$\begin{aligned}\hat{S} &= (y - X\beta) - \text{cov}(N, w)\Sigma_w^{-1}(w - X_\Delta \beta) \\ &= (y - X\beta) - \Sigma_N \Delta_S' \Sigma_w^{-1}(w - X_\Delta \beta),\end{aligned} \tag{12.44}$$

where the last relation is obtained using (12.16), noting that when N_t does not require differencing we have $v = N$ in (12.16). The vector of signal extraction errors is $S - \hat{S} = \hat{N} - N$, and its covariance matrix is

$$\text{var}(S - \hat{S}) = \Sigma_N - \Sigma_N \Delta_S' \Sigma_w^{-1} \Delta_S \Sigma_N. \tag{12.45}$$

We can use the algorithm in the Appendix to compute (12.44) and (12.45). The results above assume β is known. As with forecasting, we substitute $\hat{\beta}$ for β in (12.44). Accounting for uncertainty about β in the signal extraction error variances is discussed briefly at the end of this section.

When both components are nonstationary requiring differencing the results become more complicated. We assume the 'differencing' operators $\delta_S(B)$ and $\delta_N(B)$ have no common factors, otherwise explicit assumptions are needed about initial conditions (Bell 1984, 1991, Kohn and Ansley 1987). To estimate the vector S we first develop estimates of the vector u of differences of S_t, then develop estimates of the starting values for S_t using the transformation approach of Ansley and Kohn (1985), and then combine the two pieces to develop the signal extraction estimates of the remaining S_t.

In signal extraction, it is most convenient to let the starting values occur at the beginning of the series (in contrast to forecasting, where we placed the 'starting values' $y^\dagger$ at the end of the series). The results, however, are actually invariant to the location of the 'starting values' (Bell and Hillmer 1991). For S_t we need d_S starting values, which we label $S_* = (S_1, \ldots, S_{d_S})'$. The starting values for N_t are $N_* = (N_1, \ldots, N_{d_N})'$. Let $z_t = y_t - x_t'\beta$. From Bell (1984) the starting values for z_t, $z_* = (z_1, \ldots, z_d)'$, are related to S_* and N_* by

$$z_* = [H_1 \ H_2] \begin{bmatrix} S_* \\ N_* \end{bmatrix} + \tilde{C}_1 u_{(d)} + \tilde{C}_2 v_{(d)}, \tag{12.46}$$

where $u_{(d)} = (u_{d_S+1}, \ldots, u_d)$ and $v_{(d)} = (v_{d_N+1}, \ldots, v_d)$ are the elements of the vectors u and v up to time point d. The matrices H_1 $(d \times d_S)$ and $\tilde{C}_1$ $(d \times d_N)$, and similarly H_2 $(d \times d_N)$ and $\tilde{C}_2$ $(d \times d_S)$, are obtained from the differencing operators $\delta_S(B)$ and $\delta_N(B)$, respectively, via analogous relations to (12.31) and (12.32). That is, if we define a $d \times d$ matrix $\tilde{\Delta}_{S,d}$ from $\delta_S(B)$ so that

$$\tilde{\Delta}_{S,d} \begin{bmatrix} S_1 \\ \vdots \\ S_d \end{bmatrix} \equiv \begin{bmatrix} I_{d_S} & | & 0_{d_S \times d_N} \\ \hline -\delta_{S,d_S} & \cdots & -\delta_{S,1} & 1 \\ & \ddots & & \ddots & \ddots \\ & & -\delta_{S,d_S} & \cdots & -\delta_{S,1} & 1 \end{bmatrix} \begin{bmatrix} S_* \\ S_{d_S+1} \\ \vdots \\ S_d \end{bmatrix}$$

$$= \begin{bmatrix} S_* \\ u_d \end{bmatrix}$$

then H_1 and $\tilde{C}_1$ are, respectively, the first d_S and last d_N columns of $\tilde{\Delta}_{S,d}^{-1}$:

$$\tilde{\Delta}_{S,d}^{-1} = \begin{bmatrix} H_1 & \tilde{C}_1 \end{bmatrix} = \begin{bmatrix} I_{d_S} & 0_{d_S \times d_N} \\ A_{S,d_N} & C_{S,d_N} \end{bmatrix}, \tag{12.47}$$

where (12.47) defines A_{S,d_N} and C_{S,d_N} (see (12.32) and the surrounding discussion). We analogously define H_2 and $\tilde{C}_2$ from $\delta_N(B)$. Note that $\tilde{C}_1$ augments C_{S,d_N} with d_S rows of zeros, and that $\tilde{C}_2$ would include d_N rows of zeros. More explicitly

$$\tilde{C}_1 = \begin{bmatrix} 0_{d_S \times d_N} \\ \hline 1 \\ \xi_{S1} & 1 \\ \vdots & \ddots & \ddots \\ \xi_{S,d-d_S-1} & \cdots & \xi_{S1} & 1 \end{bmatrix},$$

$$\tilde{C}_2 = \begin{bmatrix} 0_{d_N \times d_S} \\ \hline 1 \\ \xi_{N1} & 1 \\ \vdots & \ddots & \ddots \\ \xi_{N,d-d_N-1} & \cdots & \xi_{N1} & 1 \end{bmatrix},$$

where

$$\delta_S(B)^{-1} = 1 + \xi_{S1}B + \xi_{S2}B^2 + \cdots$$

and

$$\delta_N(B)^{-1} = 1 + \xi_{N1}B + \xi_{N2}B^2 + \cdots.$$

Bell (1984) noted that the matrix $[H_1 \ H_2]$ is nonsingular. Letting $G = [H_1 \ H_2]^{-1}$, we see from (12.46) that

$$\begin{bmatrix} S_* \\ N_* \end{bmatrix} = G\left(z_* - \tilde{C}_1 u_{(d)} - \tilde{C}_2 v_{(d)}\right). \tag{12.48}$$

The transformation approach estimates of S_* and N_* are obtained by estimating u and v from the differenced data w and then substituting the elements of these vectors up to $t = d$ for $u_{(d)}$ and $v_{(d)}$ in (12.48). The signal extraction estimates of u and v based on w are

$$\hat{u} = \Sigma_u \Delta_N' \Sigma_w^{-1}(w - X_\Delta \beta), \tag{12.49}$$
$$\hat{v} = \Sigma_v \Delta_S' \Sigma_w^{-1}(w - X_\Delta \beta). \tag{12.50}$$

These can be computed using the algorithm in the Appendix. We then have the transformation approach estimates of S_* and N_*:

$$\begin{bmatrix} \hat{S}_* \\ \hat{N}_* \end{bmatrix} = G\left(z_* - \tilde{C}_1 \hat{u}_{(d)} - \tilde{C}_2 \hat{v}_{(d)}\right). \tag{12.51}$$

With the results from (12.49) and (12.51) we can compute the remaining estimates $\hat{S}_t$ for $l = d_S + 1, \ldots, n$ recursively via

$$\hat{S}_t = \delta_{S1}\hat{S}_{t-1} + \cdots + \delta_{S,d_S}\hat{S}_{t-d_S} + \hat{u}_t. \qquad (12.52)$$

Bell and Hillmer (1988) gave expressions for $\text{var}(S - \hat{S})$, but these are somewhat complicated when there is more than one component requiring differencing. They also show how to account for error in estimating the regression parameters in the covariance matrix calculations, and in doing so consider the complication that we may want to add part or all of the estimated regression effects $X\hat{\beta}$ together with $\hat{S}$. For example, in a seasonal adjustment application we may add estimated trading-day regression effects to the estimated stochastic seasonal to get the estimated combined seasonal-trading-day component.

12.4 Example: seasonal adjustment of a time series with sampling error

Many economic time series are estimates from repeated sample surveys and, as such, are subject to sampling error. It is worth keeping in mind that the sampling error component of such time series is real and not just a modelling construct. This contrasts with the seasonal, trend and irregular components in seasonal adjustment, which are typically artificial constructs used for modelling or interpretation. There are several very good reasons for explicitly accounting for the sampling error component of time series from repeated surveys:

(i) Sampling error can be a very important source of variation in time series, particularly in very disaggregated time series.
(ii) Sampling error can have unusual time series properties not well captured by applying conventional time series models to the observed series. This includes variances that change significantly over time due to: (a) dependence of sampling variances on the level of the series, (b) sample size changes, or (c) more major survey redesigns. Another possible complication is unusual patterns of autocorrelation due to sample overlap at unusual lags or to independent redrawings of the sample that result in a break in autocorrelation at some point.
(iii) Theoretical considerations about how the series should behave (e.g., from economic theory) surely apply to the underlying true series and do not take into account the sampling error. Failure to account for the sampling error component when fitting the model may compromise attempts to test theory with the observed time series data. For example, Bell and Wilcox (1993) showed how attempts to validate

economic theories of consumption with data on retail sales could be adversely affected by sampling error in the estimates of retail sales.

Scott and Smith (1974) and Scott, Smith and Jones (1977) proposed using time series signal extraction techniques to improve estimates from repeated sample surveys. These ideas were somewhat ahead of theoretical techniques and computational results for RegComponent models. More recently, this approach has been taken up by a number of researchers at or working with statistical agencies. Relevant papers include Binder and Dick (1989, 1990), Bell and Hillmer (1990), Pfeffermann (1991), Tiller (1992), and Harvey and Chung (2000). Seasonal adjustment accounting for a sampling error component has seen somewhat less attention, though Wolter and Monsour (1981), Pfeffermann (1994) and Bell and Kramer (1999) all considered accounting for sampling error in attempts to derive variances for seasonal adjustments obtained with X-11 filters. Bell and Otto (1992) and Bell and Pugh (1990) gave examples of model-based seasonal adjustment allowing for a sampling error component in the models. Another such example is presented next.

We examine here a monthly time series of estimates of US retail sales of drinking places from September 1977 to October 1989. The series is plotted in Figure 12.1(a). The estimates derive from the US monthly retail trade survey, but, unlike the published estimates, the estimates used here were not benchmarked to estimates from the corresponding annual retail trade survey. We use the unbenchmarked series for illustration since the properties of the sampling error are simpler in the unbenchmarked series. Bell and Hillmer (1990) developed a model for this series, composed of models for the underlying true series and the sampling error component. They focused on use of the model to investigate possible improvements in the estimates from signal extraction. The series was also analysed by Bell and Wilcox (1993) who kept the same sampling error model, but used an airline model for the true series. Here we use the Bell and Wilcox model. Letting y_t represent the logarithms of the survey estimates, Y_t the corresponding true population quantities and e_t the sampling error in y_t as an estimate of Y_t, the model has $y_t = Y_t + e_t$ with component models

$$(1-B)(1-B^{12})\left(Y_t - x_t'\beta\right) = (1-\theta_1 B)(1-\theta_{12}B^{12})b_t \qquad (12.53)$$

and

$$(1-0.75B)(1-0.66B^3)(1-0.71B^{12})e_t = (1+0.13B)c_t,$$
$$\sigma_c^2 = 0.93 \times 10^{-4}, \qquad (12.54)$$

where b_t and c_t are mutually independent Gaussian white noise series with variances σ_b^2 and σ_c^2. The parameter values for the sampling error model are shown in (12.54) to emphasise that they are estimated separately using estimated autocovariances of the sampling errors and then held fixed when the RegComponent model is fitted to the series y_t to estimate β, θ_1, θ_{12} and σ_b^2. The factor $(1 - 0.75B)$ in the model for the sampling error e_t, however, arises from the composite estimation procedure used to produce the survey estimates (see Bell and Hillmer (1989)), that is, the 0.75 is known and not an estimated model parameter. Also, because the retail trade survey sample was redrawn independently in September 1977, January 1982 and January 1987, the sampling errors from the three different samples are uncorrelated. This means that the covariance matrix of the vector e is block diagonal with three blocks corresponding to the three independent samples, and with a covariance matrix within each diagonal block corresponding to the model for e_t in (12.54). Bell and Hillmer (1989) noted how iterations of the Kalman filter and smoother can be modified at the times of the covariance breaks in e_t to handle these appropriately. It should be noted that the form of the model for e_t was developed for the survey design that was in effect during the time frame of the series analysed. The retail trade survey has since been redesigned (Cantwell and Black 1998) and the sampling error model in (12.54) would no longer be appropriate.

Something worth noting about (12.54) is the high value (0.71) of the seasonal AR parameter. This implies substantial seasonal autocorrelation in the sampling error, which we will see strongly affects seasonal adjustment results if the sampling error is not accounted for.

The term $x_t'\beta$ in (12.53) is used to model trading-day effects as in Bell and Hillmer 1983, though a length-of-month effect is also included in x_t. (The next section notes that we could instead account for length-of-month effects by dividing the original estimates by the length-of-month variable. We include it as a regressor here for comparability with the results of Bell and Wilcox (1993).) ML estimates of the parameters of (12.53) from REGCMPNT, leaving out $\hat\beta$ for brevity, are $\hat\theta_1 = 0.23$, $\hat\theta_{12} = 0.88$, and $\hat\sigma_b^2 = 3.97 \times 10^{-4}$. Given this estimated model, we can decompose Y_t into seasonal and nonseasonal components as in Burman (1980) and Hillmer and Tiao (1982). Thus we have

$$\begin{aligned} y_t &= Y_t + e_t \\ &= (x_t'\beta + S_t + N_t) + e_t. \end{aligned} \quad (12.55)$$

REGCMPNT also provides signal extraction estimates and their variances for the components of (12.55) as discussed in Section 12.3.4 (calculating

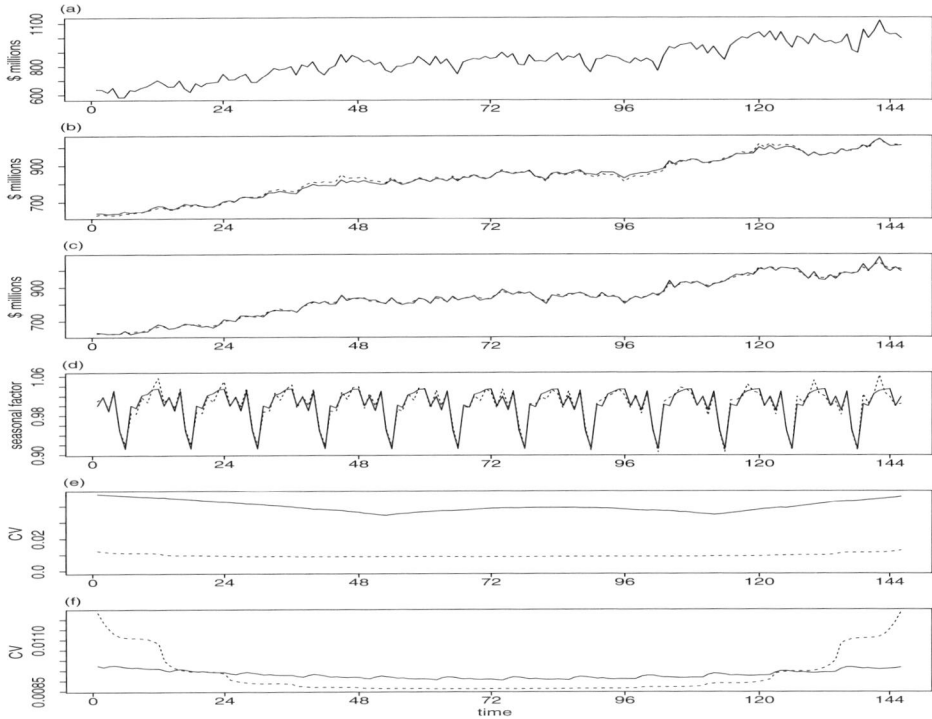

Fig. 12.1. US retail sales of drinking places July 1977 to October 1989: (a) original series (not benchmarked), (b) estimated nonseasonals, (c) seasonally adjusted series, (d) estimated seasonal factors, (e) CV of estimated nonseasonals, (f) CV of estimated seasonal factors. In (b)–(f) the solid lines are for models with a sampling error component and the dashed lines are for models without.

these results via a Kalman smoother rather than from the matrix expressions of Section 12.3.4). These signal extraction estimates are plotted in Figures 12.1(b)–(d) and are discussed subsequently. Since logarithms were taken of the original time series the signal extraction point estimates of the components of (12.55) are exponentiated before being plotted. This puts the estimates of N_t on the original scale of the time series, and estimates of S_t show up as multiplicative seasonal factors that vary around 1.0. Signal extraction standard deviations for the decomposition (12.55) of the log survey estimates can be interpreted as approximate coefficients of variation (CVs) for the multiplicative components on the original scale. The CVs are plotted in Figures 12.1(e)–(f).

For comparison Figures 12.1(b) (f) also show signal extraction results when the airline model with trading-day effects as in (12.53) is used for the series y_t rather than for the component Y_t, that is, when we ignore the sampling error component e_t. The parameter estimates when fitting this model are then (again omitting $\hat{\beta}$) $\hat{\theta}_1 = 0.29$, $\hat{\theta}_{12} = 0.56$, and $\hat{\sigma}^2 = 6.37 \times 10^{-4}$. Using this model we decompose y_t into two components as

$$y_t = \tilde{S}_t + \tilde{N}_t. \tag{12.56}$$

Omitting the sampling error component of course affects the properties of the model and the decomposition, so that $\tilde{S}_t$ and $\tilde{N}_t$ of (12.56) are different from S_t and N_t of (12.55). Notice in particular how the estimate of θ_{12} drops from 0.88 for the model with the sampling error component to 0.56 for the model without the sampling error component. The value of θ_{12} reflects the stability of the seasonal component implied by the model decomposition: values of θ_{12} closer to 1 correspond to more stable seasonal components, with $\theta_{12} = 1$ implying a fixed seasonal, that is, a component that follows a deterministic seasonal pattern (Bell 1987).

Figure 12.1(b) shows signal extraction estimates of $\exp(N_t)$ from (12.55) and $\exp(\tilde{N}_t)$ from (12.56). We see more variation over time in the estimates of $\exp(\tilde{N}_t)$ (the dashed curve) because $\tilde{N}_t$ and $\tilde{S}_t$ have to absorb the sampling error component from (12.55). Figure 12.1(c), in contrast, shows estimates of the two seasonally adjusted series: $\exp(y_t - x_t'\beta - S_t)$ from (12.55) and $\exp(y_t - x_t'\beta - \tilde{S}_t)$ from (12.56). Note that the former leaves all the sampling error in the adjusted series, while the latter indirectly removes some of the sampling error to the extent it is absorbed by the estimate of $\tilde{S}_t$. As a result, the seasonally adjusted series from (12.55) (solid line) shows somewhat larger fluctuations over time than that from (12.56) (dashed line). Figure 12.1(d) shows the estimates of the seasonal factors, $\exp(S_t)$ from (12.55) and $\exp(\tilde{S}_t)$ from (12.56), and this shows more clearly what is going on. The model (12.55) assumes that the component $\exp(S_t)$ evolves gradually over time due to its large estimate of $\hat{\theta}_{12} = 0.88$. In contrast, the model (12.56) assumes that the component $\exp(\tilde{S}_t)$ varies much more rapidly over time due to its smaller estimate of $\hat{\theta}_{12} = 0.56$. This behavior is reflected in the signal extraction estimates. Effectively, the estimates of $\exp(\tilde{S}_t)$ from (12.56) pick up some rapidly varying seasonality from the sampling error in the series that is missing from the estimates of $\exp(S_t)$.

Figure 12.1(e) shows standard deviations for the signal extraction estimates of N_t and $\tilde{N}_t$, which, as noted above, are also approximate CVs of the exponentiated estimates plotted in Figure 12.1(b). The standard deviations computed for $\tilde{N}_t$ are much lower than those for N_t reflecting the fact that estimating and removing all the sampling error (in estimating N_t) has a cost in terms of variance of the estimates. Notice also the pattern in the standard deviations for N_t caused by the breaks in autocorrelation of e_t in January 1982 and January 1987, which correspond (nearly) to the two minimum points on the solid curve. (The standard deviation for December 1986 is actually very slightly lower than that for January 1987.) Keep in mind that both sets of signal extraction variances are computed assuming the respective models used are true. While models are never exactly true, we know the airline model applied directly to y_t for use in (12.56) fails to properly account for the sampling error in the series. Even if it manages to approximate to some extent the variation due to the sampling error, it clearly has no means of modelling the effects of the autocovariance breaks in the sampling error.

Figure 12.1(f) shows standard deviations for the signal extraction estimates of S_t and $\tilde{S}_t$, which are also standard deviations of the seasonally adjusted series. The signal extraction standard deviations for $\tilde{S}_t$ (dashed line) are actually the same as those for $\tilde{N}_t$ in Figure 12.1(e), since in a two-component decomposition the adjusted series and estimated nonseasonal are the same. (We are ignoring, in both Figure 12.1(e) and (f), for simplicity of illustration, variance contributions from estimating β.) The dashed lines in Figures 12.1(e) and (f) appear different because the vertical scale is so different. In the dashed line of Figure 12.1(f) we see the characteristic pattern of seasonal adjustment standard deviations that rise near either end of the series and show a seasonal pattern in between. Again, these assume the airline model without sampling error is actually correct, and thus do not appropriately reflect the influence of the sampling error. The standard deviations for the estimates of S_t (solid curve), in contrast, are lower than those from the dashed curve near the ends of the series but higher in the middle. This reflects the greater stability of the seasonal component from (12.55), which results in less difference between the centre and ends of the series in the solid curve than in the dashed curve. Also, the autocovariance breaks in the sampling error component here generate some irregularities in the seasonal pattern of the signal extraction standard deviations for S_t.

The main point of these results is that application of the airline model, or any standard ARIMA or ARIMA component model for that matter, directly to the series y_t cannot capture certain aspects of the sampling error variation in this series. While opinions could differ as to the most appropriate model for the true series Y_t and sampling error component e_t for this series, if the true model for e_t is at all like the model (12.54) used here, standard ARIMA

12.5 Example: modeling time-varying trading-day effects

Trading-day effects (Bell and Hillmer 1983) reflect variations in a monthly flow time series (sales, shipments, etc.) due to the changing composition of months with respect to the numbers of times each day of the week occurs in the month. Fixed trading-day regression effects were included in the model for retail sales of drinking places used in the previous section. In detail, the trading-day model used was as follows:

$$TD_t = \sum_{i=1}^{6} \beta_i T_{it}, \qquad (12.57)$$

where, for month t, T_{1t} is the number of Mondays minus the number of Sundays, ..., T_{6t} is the number of Saturdays minus the number of Sundays. From Bell and Hillmer (1983) the β_i represent the deviations of the Monday, ..., Saturday effects from the average daily effect. The corresponding deviation of the Sunday effect from the average daily effect is $\beta_7 = -(\beta_1 + \cdots + \beta_6)$. It is also important to augment (12.57) to account for length-of-month effects. In the example of the previous section we did this by including as an additional regressor $m_t =$ number of days in month t. Here we instead divide the series by m_t (before taking logarithms). This latter approach to handling length-of-month effects is the default option in the trading-day model of the X-12-ARIMA seasonal adjustment program, and is generally preferred for modelling logged series. (Findley, Monsell, Bell, Otto and Chen (1998, pp. 132–4 and 142) noted that X-12 actually divides by a descasonalised and detrended version of m_t, but for models with seasonal differencing division by m_t is equivalent.)

A relevant question regarding trading-day effects is whether they remain constant over time? This is very pertinent for retail sales time series in which trading-day effects presumably depend on consumers' shopping patterns and on the hours that retail stores are open, two things that have changed over time in the USA. Seasonal adjustment practitioners sometimes deal with this issue by restricting the length of the series to which the trading-day model is fit. In this section we investigate possible time variation in trading-day effects in the time series of retail sales of US department stores from January 1967 to December 1993, which is shown in the first plot of Figure 12.2. To allow for time variation in the trading-day effects we generalise (12.57) by converting the fixed effects β_i to stochastic components following random walk models:

$$TD_t = \sum_{i=1}^{6} \beta_{it} T_{it}, \quad (1-B)\beta_{it} = \epsilon_{it} \quad i = 1, \ldots, 6, \qquad (12.58)$$

where the ϵ_{it} are mutually independent white noise series with variances σ_i^2. As noted in Section 12.2 this sort of model can be accommodated in a RegComponent model by letting the regression variables T_{it} be scale factors for the ARIMA components β_{it}.

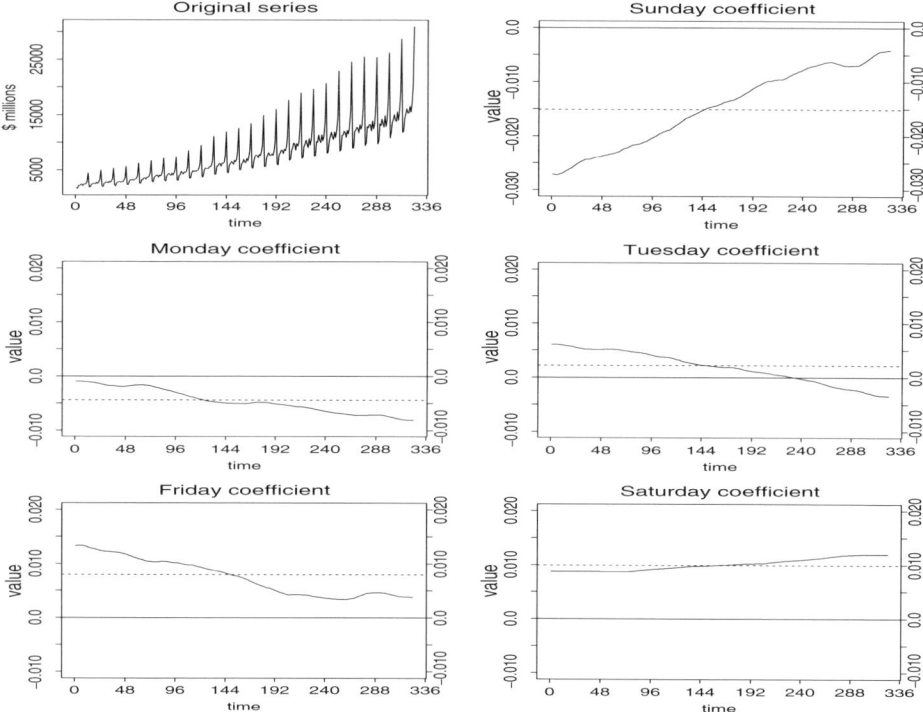

Fig. 12.2. US retail sales of department stores, January 1967 to December 1993. Original series and estimates of time-varying trading-day coefficients.

Let y_t denote the series obtained by taking logarithms of the retail department store sales after dividing the figures by the length of month. Apart from the time-varying trading day effects, the model we use includes an Easter effect as in Bell and Hillmer (1983) with a ten-day window, and an airline model for the remainder:

$$(1-B)(1-B^{12})(y_t - TD_t - \alpha E_t) = (1-\theta_1 B)(1-\theta_{12} B^{12}) a_t, \qquad (12.59)$$

where E_t is the Easter effect variable and α the associated parameter, and a_t is white noise with variance σ_a^2. For comparison we shall fit (12.59) both with TD_t following the fixed effects model (12.57) and the time-varying coefficients model (12.58). Although the retail sales of department stores figures are estimates from the monthly retail trade survey that also yielded the estimates of retail sales of drinking places modelled earlier, the model (12.59) contains no sampling error component. This is because essentially all the major department stores in the US are included in the sample, hence the estimates contain essentially no sampling error. (As with all surveys the estimates do contain nonsampling errors, but limited information on these does not permit modelling them.)

ML fitting of (12.59) with TD_t following (12.58) gives $\hat{\alpha} = 0.032$, $\hat{\theta}_1 = 0.49$, $\hat{\theta}_{12} = 0.52$, and $\hat{\sigma}_a^2 = 0.00039$. Note the large estimated Easter effect of over 3%. When the model is fitted with the fixed trading-day effects model (12.57), the estimates of these four parameters are about the same. It is worth noting that estimating the model with time-varying trading-day is more difficult than estimating the model with fixed trading-day, taking considerably more iterations to converge (48 versus 7), due to the large number of variance components. We turn now to the estimates of the trading-day effects.

Table 12.3. *Trading-day estimation for US retail sales of department stores. Note that $\hat{\beta}_7$ is minus the sum of the other six trading-day coefficients. For the fixed effects model (12.58) $\hat{\beta}_7$ is -0.0150.*

i	1 (Mon.)	2 (Tues.)	3 (Wed.)	4 (Thurs.)	5 (Fri.)	6 (Sat.)
$\hat{\sigma}_i$ eq. (12.58)	0.00042	0.00038	0	0	0.00046	0.00025
$2n^{0.5}\hat{\sigma}_i$	0.015	0.014	0	0	0.017	0.009
$\hat{\beta}_i$ eq. (12.57)	−0.0044	0.0022	−0.0073	0.0068	0.0079	0.0099

Table 12.3 shows estimation results for both the fixed and time-varying trading-day effect models. The estimated fixed effect coefficients $\hat{\beta}_i$ show moderate negative effects for Monday and Wednesday, a small positive effect for Tuesday, moderate positive effects for Thursday and Friday, a somewhat larger positive effect for Saturday, and a large negative effect for Sunday ($\hat{\beta}_7 = -0.015$). Only the Tuesday coefficient is statistically insignificant at the 0.05 level. The estimated standard deviations, $\hat{\sigma}_i$, of the innovations ϵ_{it} from (12.58) reflect how much month-to-month variation is allowed in the time-varying coefficients β_{it}. Those for Wednesday and Thursday are zero, so

the estimated model (12.58) allows no time variation in these coefficients. All the other $\hat{\sigma}_i$ look small, but the month-to-month variation in the coefficients should be small. Translating these to estimates of the standard deviations of the change over the entire series, which from the random walk model in (12.58) is $[n\hat{\sigma}_i^2]^{0.5}$, gives an indication of the amount of potential variation in the coefficients over the full length of the series. Considering $2n^{0.5}\hat{\sigma}_i$ as the half width of a 95% confidence interval for the change in the coefficients, we see that apart from Wednesday and Thursday these range from 0.9% to 1.7%, indicating a potential for practically significant change in the coefficients over the course of the series.

Figure 12.2 plots the signal extraction estimates of the time-varying trading-day coefficients β_{it}. The estimated fixed coefficients are also plotted as horizontal dotted lines, and the horizontal axis at zero is shown as a solid line. The plots of the estimated Wednesday and Thursday coefficients are omitted since they are constant over time and very close to the estimates from the fixed effects model. The coefficient plots shown have the same vertical axes except that for Sunday, which is shifted down relative to the others. All have the same vertical axis range (about 0.03), however, which allows comparisons across all plots of the amount of change in the estimated coefficients over time. The plots for Tuesday and Friday show decreases over time in the trading-day coefficients of about 1%, and the plot for Monday shows a decrease of about 0.7%. At the end of the series the estimates suggest Monday, Tuesday and Wednesday had small or moderate negative impacts on sales, while Thursday and Friday had small positive impacts. The estimated Saturday coefficients reflect somewhat larger positive impacts on sales (about 1% on average), with relatively little time variation, increasing about 0.3% over the series. The big story from these plots, though, is the large change over time in the estimated Sunday coefficients, which increase about 2.5% over the series. At the start of the series the estimated Sunday effect is negative and considerably larger in magnitude than the effect for any of the other days, but by the end of the series the estimated Sunday effect, while still negative, is relatively small.

Note that the estimated fixed effect coefficients appear near the middle of the path of the estimated time-varying coefficients. Signal extraction standard deviations for the estimated β_{it} are about 0.0025 near the middle and 0.0035 near the ends of the series. Corresponding 95% confidence intervals around the $\hat{\beta}_{it}$ nearly all include the corresponding estimated fixed coefficients except for Sunday near the ends of the series. Still, given the plausibility of change over time in the daily effects, the patterns of the estimates shown in Figure 12.2 are intriguing.

12.6 Example: modelling seasonal heteroskedasticity in a housing starts time series

We consider the time series of monthly total housing starts for the northeast region of the US from August 1972 to March 1989. Standard ARIMA modelling suggested the airline model for logarithms of the series. Letting y_t denote the logged series, the fitted model is

$$(1 - B)(1 - B^{12})y_t = (1 - 0.64B)(1 - 0.58B^{12})a_t, \quad \hat{\sigma}_a^2 = 0.050. \quad (12.60)$$

However, outlier detection with this model in the X-12-ARIMA program (Findley, Monsell, Bell, Otto and Chen 1998), which uses a modification of the approach described in Chang, Tiao and Chen (1988), suggested several possible additive outliers (AOs) in January and February. Table 12.4 shows the outliers detected via AO detection with a critical value of 3.0. Note that all the outliers detected are in January and February, and many are quite large – both in statistical significance as measured by the AO t-statistics, and in practical significance as measured by the estimated AO effects, which are the estimated regression coefficients of the AO indicator variables. Since logs were taken of the series, the coefficients can be exponentiated to see their estimated multiplicative effects on the untransformed series. These multiplicative effects range from 0.35 for February 1978 to 1.82 for January 1989. The large number of outliers detected in January and February suggest that there may be extra variability in these months, which could be due to occasional effects of unusually bad or unusually good weather on construction activity.

Table 12.4. *Additive outlier detection for total housing starts, northeast region of the US, 8/72–3/89 – Automatically identified additive outliers (critical value = 3.0)*

time point	AO effect	AO t-statistic
Feb. 1975	0.44	3.08
Jan. 1977	−0.68	−4.65
Feb. 1977	−0.51	−3.46
Jan. 1978	−0.64	−4.40
Feb. 1978	−1.05	−7.17
Feb. 1980	−0.77	−5.36
Feb. 1981	−0.61	−4.27
Jan. 1982	−0.52	−3.68
Jan. 1989	0.60	4.16

To model extra variability in January and February we augment (12.60) as follows, with parameter estimates from the fitted model shown:

$$y_t = z_{1t} + h_{2t}z_{2t}$$
$$(1-B)(1-B^{12})z_{1t} = (1-0.54B)(1-1.00B^{12})b_t, \quad (12.61)$$
$$z_{2t} = c_t, \quad \hat{\sigma}_c^2 = 0.12, \quad \hat{\sigma}_b^2 = 0.023,$$

where b_t and c_t are mutually independent white noise series, and where

$$h_{2t} = \begin{cases} 1 & t \sim \text{January or February} \\ 0 & \text{all other months} \end{cases}.$$

This use of the scale factors h_{2t} in (12.61) allows the additional additive white noise c_t to affect the series only in January and February.

Notice that $\hat{\sigma}_b^2$ is less than half $\hat{\sigma}_a^2$, showing how much the variance estimate in (12.60) is affected by the winter months. The large value of $\hat{\sigma}_c^2$, though not directly comparable with the estimated innovations variances $\hat{\sigma}_a^2$ and $\hat{\sigma}_b^2$ of the airline models, clearly shows the importance of the additional variance in January and February. Notice also how the estimate of the seasonal moving average parameter increases from 0.58 for model (12.60) to 1.00 for model (12.61), the latter estimate implying fixed seasonality in the series (Bell 1987). Thus, when we allow for the additional noise in the winter months the seasonality in the series appears fixed. Although the REGCMPNT program does not perform AO detection, examination of standardised residuals from (12.61) shows no extreme values, suggesting outliers may not be a problem for (12.61). Finally, a comparison of Akaike information criterion (AIC) values for the two models yields a very large difference of approximately 50 in favor of (12.61), confirming the value of allowing for the additional winter variance.

Appendix: Computations with the Cholesky decomposition

The following algorithm, based on use of the Cholesky decomposition, can be used for several of the computations discussed in this paper.

Algorithm: Let Σ be a positive definite $n \times n$ covariance matrix with Cholesky decomposition $\Sigma = \bar{L}\bar{L}'$, where $\bar{L}$ is lower triangular. To compute $A\Sigma^{-1}B$, where A and B are vectors or matrices conformable with Σ^{-1}, we can proceed as follows:

1. Solve $\bar{L}R_1 = B$ for $R_1 = \bar{L}^{-1}B$.
2. Solve $\bar{L}R_2 = A'$ for $R_2 = \bar{L}^{-1}A'$.
3. Compute $A\Sigma^{-1}B = R_2'R_1$.

Note that if $A = B'$ then $R_2 = R_1$, we can skip Step 2, and $A\Sigma^{-1}B = R_1'R_1$. Note also that

$$\log(\det(\Sigma)) = 2\sum_{i=1}^{n} \log(\ell_{ii}),$$

where the ℓ_{ii} are the diagonal elements of $\bar{L}$.

Another approach that is more suitable for some applications is to replace Steps 2 and 3 by

2(a) Solve $\bar{L}'R_3 = R_1$ for $R_3 = (\bar{L}^{-1})'R_1$.
3(a) Compute $A\Sigma^{-1}B = AR_3$.

Note that in Steps 1 and 2 we solve lower triangular systems of linear equations, and in Step 2(a) we solve an upper triangular system. Efficient implementation of these algorithms requires computer routines for computing the Cholesky decomposition *and* for solving triangular systems of linear equations.

We could instead use the unit Cholesky decomposition, that is, $\Sigma = LDL'$, where $D = \mathrm{diag}(d_1, \ldots, d_n)$ and $L = D^{-1/2}\bar{L}$ is lower triangular with ones on its diagonal. We can substitute L for $\bar{L}$ in Steps 1 and 2 above, but Step 3 then becomes $A\Sigma^{-1}B = R_2'D^{-1}R_1$. Note that then

$$\log(\det(\Sigma)) = \sum_{i=1}^{n} \log(d_i).$$

If Σ is a band matrix we can use the more efficient band Cholesky decomposition and corresponding linear equation solving routines. To use this approach with ARIMA component models (without scale factors) we can extend the ideas of Ansley (1979) by taking $\phi(B)w_t$, which follows a moving average model and so has a band covariance matrix. Here $\phi(B)$ is the autoregressive operator from the model for z_t; in the context of Section 12.3, $\phi(B) = \phi_S(B)\phi_N(B)$.

Computer routines for the regular and band Cholesky decomposition, and corresponding routines for solving triangular systems of linear equations, are available from the Linpack package (Dongarra, Moler, Bunch and Stewart 1979), and also from its successor package LAPACK (Anderson et al., 1992).

Acknowledgments

This paper reports the results of research and analysis undertaken by US Census Bureau staff. The views expressed are those of the author and do not necessarily reflect those of the US Census Bureau.

I wish to thank John Aston, David Findley, Andrew Harvey, Siem Jan Koopman and Neil Shephard for useful comments that improved the paper. Steve Hillmer deserves special mention for his collaboration in the development of RegComponent models for application to time series observed with sampling error, as well as for his part in developing the first version of the REGCMPNT software for this application. Mark Otto wrote the original REGARIMA modelling program (now the RegARIMA modelling part of the X-12 program) which provided the framework around which the REGCMPNT program was built. Mark also incorporated into this framework pieces of the REGCMPNT code that were written by me and by other current and former members of the time series staff of the Statistical Research Division of the US Census Bureau. Those other staff members who developed key parts of the REGCMPNT program include Larry Bobbitt, Jim Bozik and Marian Pugh. In addition, Matthew Kramer and Brian Monsell provided valuable assistance in development and testing of the program.

13

State space modelling in macroeconomics and finance using `SsfPack` in `S+Finmetrics`

Eric Zivot
Department of Economics, University of Washington

Jeffrey Wang
Ronin Capital LLC

Siem Jan Koopman
Department of Econometrics and Operations Research, Vrije University, Amsterdam

Abstract

This article surveys some common state space models used in macroeconomics and finance and shows how to specify and estimate these models using the `SsfPack` library of algorithms implemented in the `S-PLUS` module `S+FinMetrics`. Examples include recursive regression models, time varying parameter models, exact autoregressive moving average models and calculation of the Beveridge–Nelson decomposition, unobserved components models, stochastic volatility models, and term structure models.

State Space and Unobserved Component Models: Theory and Applications, eds. Andrew C. Harvey, Siem Jan Koopman and Neil Shephard. Published by Cambridge University Press. © Cambridge University Press 2004

13.1 Introduction

The first version of SsfPack[1] appeared in 1998 and was developed further during the years that the last author was working with Jim Durbin on their 2001 textbook on state space methods. The fact that SsfPack functions are now a part of the S-PLUS software is partly due to Jim Durbin. He convinced Doug Martin that SsfPack would be very beneficial to S-PLUS. Indeed the persuasive arguments of Jim Durbin has initiated the development of SsfPack functions for S-PLUS as part of the S+FinMetrics module. It is therefore an honour for us, the developers of SsfPack for S+FinMetrics, to contribute to this volume with the presentation of various applications in economics and finance that require the use of SsfPack for S+FinMetrics in empirical research.

State space modelling in economics and finance has become widespread over the last decade. Textbook treatments of state space models are given in Harvey (1989, 1993), Hamilton (1994), West and Harrison (1997), Kim and Nelson (1999), Shumway and Stoffer (2000), Durbin and Koopman (2001) and Chan (2002). However, until recently there has not been much flexible software for the statistical analysis of general models in state space form. A modern set of state space modelling tools are available in SsfPack which is a suite of C routines for carrying out computations involving the statistical analysis of univariate and multivariate models in state space form. The routines allow for a variety of state space forms from simple time invariant models to complicated time varying models. Functions are available to put standard models like autoregressive moving average and spline models in state space form. General routines are available for filtering, smoothing, simulation smoothing, likelihood evaluation, forecasting and signal extraction. Full details of the statistical analysis are provided in Durbin and Koopman (2001), and the reader is referred to the papers by Koopman, Shephard and Doornik (1999, 2001) for technical details on the algorithms used in the SsfPack functions.

The SsfPack 2.3 routines are implemented in Ox and the SsfPack 3.0 routines are in Insightful's new S-PLUS module S+FinMetrics[2]. The implementation of the SsfPack functions in Ox is described in Koopman, Shephard and Doornik (1999). Its implementation in S+FinMetrics is described in

---

[1] Information about SsfPack can be found at http://www.ssfpack.com
[2] Ox is an object-oriented matrix programming language developed by Doornik (1999); more information is available at http://www.nuff.ox.ac.uk/users/doornik. S+FinMetrics is an S-PLUS module for the analysis of economic and financial time series. It was conceived by the first two authors and Doug Martin, and developed at Insightful Corporation. Its use and functionality is described in detail in Zivot and Wang (2003); more information is available at http://www.insightful.com/products/default.asp

Zivot and Wang (2003, Chapter 14). This paper gives a selected overview of state space modelling with some economic and financial applications utilising the S+FinMetrics/SsfPack functions.

This article is organised as follows. Section 13.2 deals with: (i) the general state space model and its specific `SsfPack` state space representation, (ii) descriptions of some functions for putting common time series models into state space form and (iii) the process of simulating observations from a given state space model. Section 13.3 summarises the main algorithms used for the analysis of state space models. These include the Kalman filter, smoothing and forecasting. Further, the estimation of the unknown parameters in a state space model is described. The remaining part of the paper presents several applications of state space modelling in economics and finance. These include recursive estimation of the capital asset pricing model (CAPM) with fixed and time varying parameters (Section 13.4), maximum likelihood estimation of autoregressive moving average models and unobserved component models together with trend-cycle decompositions based on these models including the Beveridge–Nelson decomposition (Section 13.5), estimation of a stochastic volatility (SV) model using Monte Carlo simulation techniques (Section 13.6) and the estimation and analysis of a simple affine term structure model (Section 13.7).

13.2 Linear state space representation

Many dynamic time series models in economics and finance may be represented in state space form. Some common examples are autoregressive moving average (ARMA) models, time varying regression models, dynamic linear models with unobserved components, discrete versions of continuous time diffusion processes, SV models, nonparametric and spline regressions. The linear Gaussian state space model is represented as the system of equations

$$\boldsymbol{\alpha}_{t+1} = \mathbf{d}_t + \mathbf{T}_t \boldsymbol{\alpha}_t + \mathbf{H}_t \boldsymbol{\eta}_t, \tag{13.1}$$

$$\boldsymbol{\theta}_t = \mathbf{c}_t + \mathbf{Z}_t \boldsymbol{\alpha}_t, \tag{13.2}$$

$$\mathbf{y}_t = \boldsymbol{\theta}_t + \mathbf{G}_t \boldsymbol{\varepsilon}_t, \tag{13.3}$$

where $\boldsymbol{\alpha}_{t+1}$, $\boldsymbol{\eta}_t$ and $\boldsymbol{\theta}_t$ are $m \times 1$, $r \times 1$ and $N \times 1$ vectors respectively, $t = 1, \ldots, n$ and

$$\boldsymbol{\alpha}_1 \sim N(\mathbf{a}, \mathbf{P}), \quad \boldsymbol{\eta}_t \sim \text{iid } N(\mathbf{0}, \mathbf{I}_r), \quad \boldsymbol{\varepsilon}_t \sim \text{iid } N(\mathbf{0}, \mathbf{I}_N), \tag{13.4}$$

with the assumption that $E[\boldsymbol{\varepsilon}_t \boldsymbol{\eta}_t'] = \mathbf{0}$. The initial mean vector $\mathbf{a}$ and initial variance matrix $\mathbf{P}$ are fixed and known but that can be generalised. The state vector $\boldsymbol{\alpha}_t$ contains unobserved stochastic processes and unknown fixed effects and the transition equation (13.1) describes the evolution of the state vector over time using a first-order Markov structure. The measurement equation (13.3) describes the vector of observations $\mathbf{y}_t$ in terms of the state vector $\boldsymbol{\alpha}_t$ through the signal $\boldsymbol{\theta}_t$ and a vector of disturbances $\boldsymbol{\varepsilon}_t$. It is assumed that the innovations in the transition equation and the innovations in the measurement equation are independent, but this assumption can be relaxed. The time varying deterministic matrices $\mathbf{T}_t, \mathbf{Z}_t, \mathbf{H}_t, \mathbf{G}_t$ are called system matrices and are usually sparse selection matrices. The vectors $\mathbf{d}_t$ and $\mathbf{c}_t$ contain fixed components and may be used to incorporate known effects or known patterns into the model; otherwise they are equal to zero.

The state space model (13.1)–(13.3) may be compactly expressed as

$$\begin{pmatrix} \boldsymbol{\alpha}_{t+1} \\ \mathbf{y}_t \end{pmatrix} = \boldsymbol{\delta}_t + \boldsymbol{\Phi}_t \boldsymbol{\alpha}_t + \mathbf{u}_t, \tag{13.5}$$

$$\boldsymbol{\alpha}_1 \sim N(\mathbf{a}, \mathbf{P}), \quad \mathbf{u}_t \sim \text{iid } N(\mathbf{0}, \boldsymbol{\Omega}_t), \tag{13.6}$$

where

$$\boldsymbol{\delta}_t = \begin{pmatrix} \mathbf{d}_t \\ \mathbf{c}_t \end{pmatrix}, \quad \boldsymbol{\Phi}_t = \begin{pmatrix} \mathbf{T}_t \\ \mathbf{Z}_t \end{pmatrix}$$

and

$$\mathbf{u}_t = \begin{pmatrix} \mathbf{H}_t \boldsymbol{\eta}_t \\ \mathbf{G}_t \boldsymbol{\varepsilon}_t \end{pmatrix}, \quad \boldsymbol{\Omega}_t = \begin{pmatrix} \mathbf{H}_t \mathbf{H}_t' & \mathbf{0} \\ \mathbf{0} & \mathbf{G}_t \mathbf{G}_t' \end{pmatrix}.$$

The initial value parameters are summarised in the $(m+1) \times m$ matrix

$$\Sigma = \begin{pmatrix} \mathbf{P} \\ \mathbf{a}' \end{pmatrix}. \tag{13.7}$$

For multivariate models, that is, $N > 1$, it is assumed that the $N \times N$ matrix $\mathbf{G}_t \mathbf{G}_t'$ is diagonal. This restriction can be circumvented by including the disturbance vector ε_t in the state vector α_t.

13.2.1 Initial conditions

The variance matrix $\mathbf{P}$ of the initial state vector α_1 is assumed to be of the form

$$\mathbf{P} = \mathbf{P}_* + \kappa \mathbf{P}_\infty, \tag{13.8}$$

where $\mathbf{P}_\infty$ and $\mathbf{P}_*$ are symmetric $m \times m$ matrices with ranks k_∞ and k_*, respectively, and κ represents the diffuse prior that is typically taken as a large scalar value, say, 10^7. The matrix $\mathbf{P}_*$ captures the covariance structure of the stationary components in the initial state vector, and the matrix $\mathbf{P}_\infty$ is used to specify the initial variance matrix for nonstationary components and fixed unknown effects. When the ith diagonal element of $\mathbf{P}_\infty$ is negative, the corresponding ith column and row of $\mathbf{P}_*$ are assumed to be zero, and the corresponding row and column of $\mathbf{P}_\infty$ will be taken into consideration. Further, the algorithms implement an 'exact diffuse prior' approach as described in Durbin and Koopman (2001, Chapter 5) and, with more algorithmic detail, in Koopman, Shephard and Doornik (2001).

13.2.2 State space representation in S+FinMetrics/SsfPack

State space models in S+FinMetrics/SsfPack utilise the compact representation (13.5) with initial value information (13.7). The following two examples describe its functionality.

Example 13.1 State space representation of the local level model

Consider the simple local level model for the stochastic evolution of the logarithm of an asset price y_t

$$\alpha_{t+1} = \alpha_t + \eta_t^*, \qquad \eta_t^* \sim \text{iid } N(0, \sigma_\eta^2), \tag{13.9}$$

$$y_t = \alpha_t + \varepsilon_t^*, \qquad \varepsilon_t^* \sim \text{iid } N(0, \sigma_\varepsilon^2), \tag{13.10}$$

$$\alpha_1 \sim N(a, P), \qquad E[\varepsilon_t^* \eta_t^*] = 0. \tag{13.11}$$

In this model, the observed value y_t is the sum of the unobservables α_t and ε_t^*. The level α_t is the state variable and represents the underlying signal. The transition equation (13.9) shows that the state evolves according to a random walk. The component ε_t^* represents random deviations (noise) from the signal that are assumed to be independent from the innovations to α_t. The strength of the signal relative to the noise is measured by $q = \sigma_\eta^2/\sigma_\varepsilon^2$.

The state space form (13.5) of the local level model has time invariant parameters

$$\boldsymbol{\delta} = \begin{pmatrix} 0 \\ 0 \end{pmatrix}, \quad \boldsymbol{\Phi} = \begin{pmatrix} 1 \\ 1 \end{pmatrix}, \quad \boldsymbol{\Omega} = \begin{pmatrix} \sigma_\eta^2 & 0 \\ 0 & \sigma_\varepsilon^2 \end{pmatrix} \tag{13.12}$$

with errors $\sigma_\eta \eta_t = \eta_t^*$ and $\sigma_\varepsilon \varepsilon_t = \varepsilon_t^*$. Since the state variable α_t is $I(1)$, the unconditional distribution of the initial state α_1 does not have a finite variance[3]. In this case, it is customary to set $a = E[\alpha_1] = 0$ and $P = \text{var}(\alpha_1) = \kappa$ in (13.11) with $\kappa \to \infty$ to reflect that no prior information is available. Using (13.8), the initial variance is specified with $a = 0$, $P_* = 0$

---
[3] The short-hand notation $I(1)$ is for a nonstationary variable that needs to be differenced once to become stationary, see Hamilton (1994) for further details.

and $P_\infty = 1$. Therefore, the initial state matrix (13.7) for the local level model has the form

$$\Sigma = \begin{pmatrix} -1 \\ 0 \end{pmatrix}, \qquad (13.13)$$

where -1 implies that $P_\infty = 1$.

```
> sigma.e = 1
> sigma.n = 0.5
> a1 = 0
> P1 = -1
> ssf.ll.list = list(mPhi=as.matrix(c(1,1)),
+ mOmega=diag(c(sigma.n^2,sigma.e^2)),
+ mSigma=as.matrix(c(P1,a1)))
> ssf.ll.list
$mPhi:        $mOmega:             $mSigma:
     [,1]         [,1] [,2]             [,1]
[1,] 1       [1,] 0.25 0          [1,] -1
[2,] 1       [2,] 0.00 1          [2,]  0

> ssf.ll = CheckSsf(ssf.ll.list)
> class(ssf.ll)
[1] "ssf"
> names(ssf.ll)
 [1] "mDelta" "mPhi" "mOmega" "mSigma" "mJPhi"
 [6] "mJOmega" "mJDelta" "mX" "cT" "cX"
[11] "cY" "cSt"
> ssf.ll
     $mPhi:   $mOmega:              $mSigma:     $mDelta:
        [,1]        [,1] [,2]          [,1]         [,1]
[1,] 1        [1,] 0.25 0         [1,] -1       [1,] 0
[2,] 1        [2,] 0.00 1         [2,]  0       [2,] 0
$mJPhi: $mJOmega: $mJDelta:
[1] 0   [1] 0     [1] 0
$mX:   $cT:   $cX:   $cY:   $cSt:
[1] 0  [1] 0  [1] 0  [1] 1  [1] 1
attr(, "class"):
[1] "ssf"
```

Listing 13.1. Local level model in SsfPack.

In S+FinMetrics/SsfPack, a state space model is specified by creating either a list variable with components giving the minimum components necessary for describing a particular state space form or by creating an 'ssf' object. To illustrate, consider Listing 13.1 where a list variable is created that contains the state space parameters in (13.12)–(13.13), with $\sigma_\eta = 0.5$ and $\sigma_\varepsilon = 1$. In the list variable ssf.ll.list, the component names match the state space form parameters in (13.5) and (13.7). This naming convention, summarised in Table 13.1, must be used for the specification of any valid

Table 13.1. `SsfPack` state space form list components

state space parameter	list component name
δ	mDelta
Φ	mPhi
Ω	mOmega
Σ	mSigma

state space model. An 'ssf' object may be created from the list variable `ssf.ll.list` using the function `CheckSsf` as done in Listing 13.1. The function `CheckSsf` takes a list variable with a minimum state space form, coerces the components to matrix objects and returns the full parameterisation of a state space model used in many of the `S+FinMetrics/SsfPack` state space modelling functions.

Example 13.2 State space representation of a time varying parameter regression model

Consider a CAPM with time varying intercept and slope

$$y_t = \alpha_t + \beta_{M,t} x_{M,t} + \nu_t, \quad \nu_t \sim \text{iid } N(0, \sigma_\nu^2),$$
$$\alpha_{t+1} = \alpha_t + \xi_t, \quad \xi_t \sim \text{iid } N(0, \sigma_\xi^2), \quad (13.14)$$
$$\beta_{M,t+1} = \beta_{M,t} + \varsigma_t, \quad \varsigma_t \sim \text{iid } N(0, \sigma_\varsigma^2),$$

where y_t denotes the return on an asset in excess of the risk free rate, and $x_{M,t}$ denotes the excess return on a market index. In this model, both the abnormal excess return α_t and asset risk $\beta_{M,t}$ are allowed to vary over time following a random walk specification. Let $\boldsymbol{\alpha}_t = (\alpha_t, \beta_{M,t})'$, $\mathbf{x}_t = (1, x_{M,t})'$, $\mathbf{H}_t = \text{diag}(\sigma_\xi, \sigma_\varsigma)'$ and $G_t = \sigma_\nu$. Then the state space form (13.5) of (13.14) is

$$\begin{pmatrix} \boldsymbol{\alpha}_{t+1} \\ y_t \end{pmatrix} = \begin{pmatrix} \mathbf{I}_2 \\ \mathbf{x}_t' \end{pmatrix} \boldsymbol{\alpha}_t + \begin{pmatrix} \mathbf{H}_t \eta_t \\ G_t \varepsilon_t \end{pmatrix}$$

and has parameters

$$\Phi_t = \begin{pmatrix} \mathbf{I}_2 \\ \mathbf{x}_t' \end{pmatrix}, \quad \Omega = \begin{pmatrix} \sigma_\xi^2 & 0 & 0 \\ 0 & \sigma_\varsigma^2 & 0 \\ 0 & 0 & \sigma_\nu^2 \end{pmatrix}. \quad (13.15)$$

Since α_t is $I(1)$ the initial state vector α_1 requires an infinite variance so it is customary to set $\mathbf{a} = \mathbf{0}$ and $\mathbf{P} = \kappa \mathbf{I}_2$ with $\kappa \to \infty$. Using (13.8), the initial variance is specified with $\mathbf{P}_* = \mathbf{0}$ and $\mathbf{P}_\infty = \mathbf{I}_2$. Therefore, the initial state matrix (13.7) for the time varying CAPM has the form

$$\Sigma = \begin{pmatrix} -1 & 0 \\ 0 & -1 \\ 0 & 0 \end{pmatrix}.$$

The state space parameter matrix Φ_t in (13.15) has a time varying system element $\mathbf{Z}_t = \mathbf{x}'_t$. In S+FinMetrics/SsfPack, the specification of this time varying element in Φ_t requires an index matrix $\mathbf{J}_\Phi$ and a data matrix $\mathbf{X}$ to which the indices in $\mathbf{J}_\Phi$ refer. The index matrix $\mathbf{J}_\Phi$ must have the same dimension as Φ_t. The elements of $\mathbf{J}_\Phi$ are all set to -1 except the elements for which the corresponding elements of Φ_t are time varying. The nonnegative index value indicates the column of the data matrix $\mathbf{X}$ which contains the time varying values.[4] For example, in the time varying CAPM, the index matrix $\mathbf{J}_\Phi$ has the form

$$\mathbf{J}_\Phi = \begin{pmatrix} -1 & -1 \\ -1 & -1 \\ 1 & 2 \end{pmatrix}.$$

The specification of the state space form for the time varying CAPM requires values for the variances σ^2_ξ, σ^2_ς and σ^2_ν as well as a data matrix $\mathbf{X}$ whose rows correspond with $\mathbf{Z}_t = \mathbf{x}'_t = (1, r_{M,t})$. For example, let $\sigma^2_\xi = (0.01)^2$, $\sigma^2_\varsigma = (0.05)^2$ and $\sigma^2_\nu = (0.1)^2$ and construct the data matrix $\mathbf{X}$ using the excess return data in the S+FinMetrics 'timeSeries' excessReturns.ts. The state space form is then created as in Listing 13.2. Notice in the specification of Φ_t the values associated with $\mathbf{x}'_t$ in the third row are set to zero. In the index matrix $\mathbf{J}_\Phi$, the (3,1) element is 1 and the (3,2) element is 2 indicating that the data for the first and second columns of $\mathbf{x}'_t$ come from the first and second columns of the component mX, respectively.

In the general state space model (13.5), it is possible that all of the system matrices δ_t, Φ_t and Ω_t have time varying elements. The corresponding index matrices $\mathbf{J}_\delta$, $\mathbf{J}_\Phi$ and $\mathbf{J}_\Omega$ indicate which elements of the matrices δ_t, Φ_t and Ω_t are time varying and the data matrix $\mathbf{X}$ contains the time varying components. The naming convention for these components is summarised in Table 13.2.

---

[4] When there are time varying elements in $\mathbf{T}_t$, the initial values of these elements in the specification of Φ_t should not be set to $-1, 0$ or 1 due to the way SsfPack handles sparse matrices. We suggest setting these elements equal to 0.5 so that the sparse matrix operations know that there should be a nontrivial number there.

```
> X.mat = cbind(1,as.matrix(seriesData(excessReturns.ts[,"SP500"])))
> msft.ret = excessReturns.ts[,"MSFT"]
> Phi.t = rbind(diag(2),rep(0,2))
> Omega = diag(c((.01)^2,(.05)^2,(.1)^2))
> J.Phi = matrix(-1,3,2)
> J.Phi[3,1] = 1
> J.Phi[3,2] = 2
> Sigma = -Phi.t
> ssf.tvp.capm = list(mPhi=Phi.t,
+ mOmega=Omega,
+ mJPhi=J.Phi,
+ mSigma=Sigma,
+ mX=X.mat)
> ssf.tvp.capm
$mPhi:                    $mOmega:
     [,1] [,2]                  [,1]   [,2]   [,3]
[1,]  1    0              [1,]  0.0001 0.0000 0.00
[2,]  0    1              [2,]  0.0000 0.0025 0.00
[3,]  0    0              [3,]  0.0000 0.0000 0.01
$mJPhi:                   $mSigma:
     [,1] [,2]                  [,1] [,2]
[1,] -1   -1              [1,]  -1    0
[2,] -1   -1              [2,]   0   -1
[3,]  1    2              [3,]   0    0
$mX:
numeric matrix: 131 rows, 2 columns.
        SP500
1   1   0.002839
...
131 1  -0.0007466
```

Listing 13.2. The CAPM model in SsfPack.

Table 13.2. *SsfPack* time varying state space components

parameter index matrix	list component name
$\mathbf{J}_\delta$	mJDelta
$\mathbf{J}_\Phi$	mJPhi
$\mathbf{J}_\Omega$	mJOmega

time varying component data matrix	list component name
X	mX

13.2.3 Model specification

SsfPack has functions for the creation of the state space representation of some common time series models. These functions and models are summarised in Table 13.3. For other models, the system matrices can be created within S-PLUS and the function CheckSsf.

Table 13.3. *SsfPack* functions for creating common state space models

function	description
GetSsfReg	linear regression model
GetSsfArma	stationary and invertible ARMA model
GetSsfRegArma	linear regression model with ARMA errors
GetSsfStsm	structural time series model
GetSsfSpline	nonparametric cubic spline model

A complete description of the underlying statistical models and use of these functions is given in Zivot and Wang (2003, Chapter 14). The use of some of these functions will be illustrated in the applications to follow.

13.2.4 Simulating observations

Once a state space model has been specified, it is often interesting to draw simulated values from the model. Simulation from a given state space model is also necessary for Monte Carlo and bootstrap exercises. The SsfPack function SsfSim may be used for such a purpose. The arguments expected from SsfSim are as illustrated in Listing 13.3. The variable ssf represents either a list with components giving a minimal state space form or a valid 'ssf' object, n is the number of simulated observations, mRan is user-specified matrix of disturbances, and a1 is the initial state vector.

Example 13.3 Simulating observations from the local level model

The code in Listing 13.3 generates 250 observations on the state variable α_{t+1} and observations y_t in the local level model (13.9)–(13.11). The function SsfSim returns a matrix containing the simulated state variables α_{t+1} and observations y_t. These values are illustrated in Figure 13.1.

```
> set.seed(112)
> ll.sim = SsfSim(ssf.ll.list,n=250)
> class(ll.sim)
[1] "matrix"
> colIds(ll.sim)
[1] "state" "response"
```

Listing 13.3. Simulating observations and states in SsfPack.

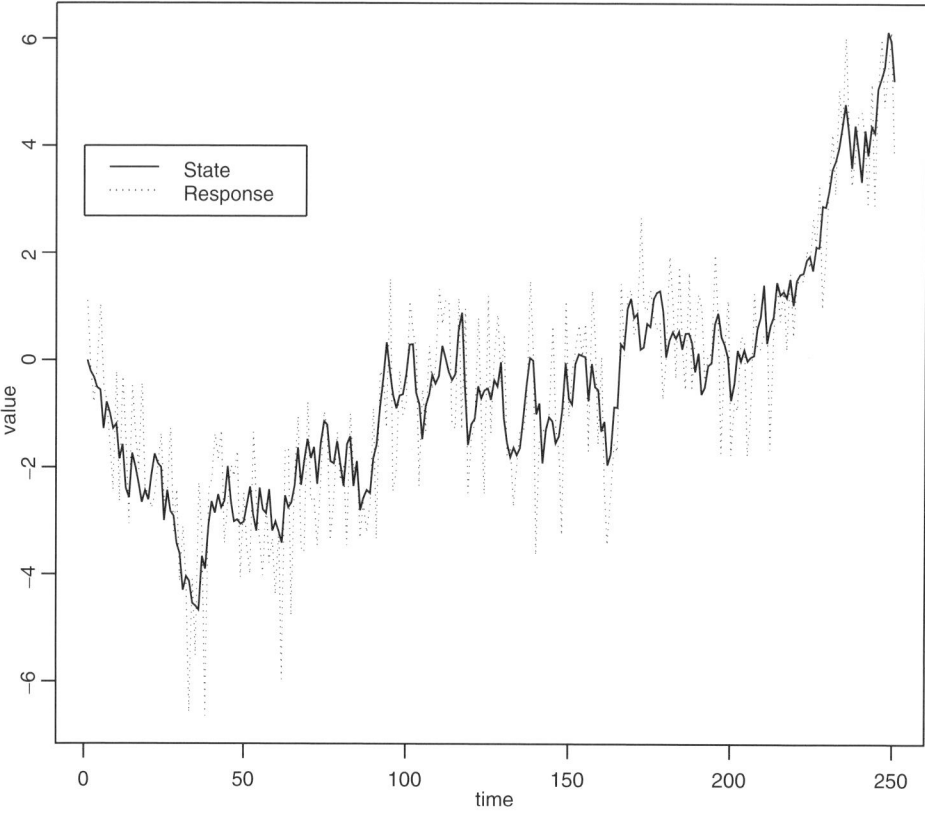

Fig. 13.1. Simulated values from local level model created using the `SsfPack` function `SsfSim`.

13.3 Overview of algorithms

In this section we review the algorithms that are considered in `SsfPack`. The most relevant equations are reproduced but the review does not aim to be complete and to provide the detail of implementation. The derivations of the algorithms can be found in, for example, Harvey (1989), Shumway and Stoffer (2000) and Durbin and Koopman (2001). Details of numerical implementation are reported in Koopman, Shephard and Doornik (1999) and Koopman, Shephard and Doornik (2001) unless indicated otherwise.

The `SsfPack` functions for computing the algorithms described below are summarised in Table 13.4. All of the functions except `KalmanSmo` have an optional argument `task` which controls the task to be performed by the function. The values of the argument `task` with brief descriptions are given in Table 13.5.

13.3.1 Kalman filter

The Kalman filter is a recursive algorithm for the evaluation of moments of the normally distributed state vector α_{t+1} conditional on the observed data $\mathbf{Y}_t = (y_1, \ldots, y_t)$. Let $\mathbf{a}_t = E[\alpha_t | \mathbf{Y}_{t-1}]$ denote the conditional mean of α_t based on information available at time $t-1$ and let $\mathbf{P}_t = \text{var}(\alpha_t | \mathbf{Y}_{t-1})$ denote the conditional variance of α_t. The filtering or updating equations of the Kalman filter compute $\mathbf{a}_{t|t} = E[\alpha_t | \mathbf{Y}_t]$ and $\mathbf{P}_{t|t} = \text{var}(\alpha_t | \mathbf{Y}_t)$, as well as the measurement equation innovation or one-step ahead prediction error $\mathbf{v}_t = \mathbf{y}_t - \mathbf{c}_t - \mathbf{Z}_t \mathbf{a}_t$ and prediction error variance $\mathbf{F}_t = \text{var}(\mathbf{v}_t)$. The prediction equations of the Kalman filter compute $\mathbf{a}_{t+1}$ and $\mathbf{P}_{t+1}$.

The function KalmanFil implements the Kalman filter recursions in a computationally efficient way. The output of KalmanFil is primarily used by other functions, but it can also be used directly to evaluate the appropriateness of a given state space model through the analysis of the innovations $\mathbf{v}_t$. The function SsfMomentEst computes the filtered state and response estimates from a given state space model and observed data with the optional argument task="STFIL". Predicted state and response estimates are computed using SsfMomentEst with the optional argument task="STPRED".

13.3.2 Kalman filter initialisation

When the initial conditions $\mathbf{a}_1$ and $\mathbf{P}_1$ can be properly defined, the Kalman filter can be used for prediction and filtering. For example, for a stationary ARMA process in state space, the initial conditions of the state vector are obtained from the unconditional mean and variance equations of the ARMA process. For nonstationary processes, however, the initial conditions cannot be properly defined for the obvious reasons. The same applies to fixed unknown state elements. In such cases the applicable initial state element is taken as diffuse which implies that its mean is arbitrary, usually zero, and its variance is infinity.

13.3.3 Kalman smoother

The Kalman filtering algorithm is a forward recursion which computes one-step ahead estimates $\mathbf{a}_{t+1}$ and $\mathbf{P}_{t+1}$ based on $\mathbf{Y}_t$ for $t = 1, \ldots, n$. The Kalman smoothing algorithm is a backward recursion which computes the mean and variance of specific conditional distributions based on the full data set $\mathbf{Y}_n = (y_1, \ldots, y_n)$. The state smoothing residuals are denoted $\mathbf{r}_t$ and the response smoothing residuals are denoted by $\mathbf{e}_t$.

The function KalmanSmo implements the Kalman smoother recursions in a computationally efficient way and it computes the smoothing residuals together with the diagonal elements of their variances. The output of KalmanSmo is primarily used by other functions for computing smoothed estimates of the state and disturbance vectors but it can also be used to compute score information for parameter estimation and to evaluate diagnostics for the detection of outliers and structural breaks.

13.3.4 Smoothed state, response and disturbance estimates

The smoothed estimates of the state vector α_t and its variance matrix are denoted $\hat{\alpha}_t = E[\alpha_t|\mathbf{Y}_n]$ (or $\mathbf{a}_{t|n}$) and $\mathrm{var}(\alpha_t|\mathbf{Y}_n)$, respectively. The smoothed estimate $\hat{\alpha}_t$ is the optimal estimate of α_t using all available information $\mathbf{Y}_n$, whereas the filtered estimate $\mathbf{a}_{t|t}$ is the optimal estimate only using information available at time t, $\mathbf{Y}_t$. The computation of $\hat{\alpha}_t$ and its variance from the Kalman smoother algorithm is described in Durbin and Koopman (2001). The smoothed estimate of the response $\mathbf{y}_t$ and its variance are computed using

$$\hat{\mathbf{y}}_t = \hat{\boldsymbol{\theta}}_t = E[\boldsymbol{\theta}_t|\mathbf{Y}_n] = \mathbf{c}_t + \mathbf{Z}_t\hat{\alpha}_t, \quad \mathrm{var}(\mathbf{y}_t|\mathbf{Y}_n) = \mathbf{Z}_t\mathrm{var}(\alpha_t|\mathbf{Y}_n)\mathbf{Z}_t'. \quad (13.16)$$

Smoothed estimates of states and responses may be computed using the functions SsfCondDens and SsfMomentEst with the optional argument task="STSMO". The function SsfCondDens only computes the smoothed states and responses whereas SsfMomentEst also computes the associated variances.

The smoothed disturbance estimates are the estimates of the measurement equations innovations ε_t and transition equation innovations η_t based on all available information $\mathbf{Y}_n$, and are denoted $\hat{\varepsilon}_t = E[\varepsilon_t|\mathbf{Y}_n]$ (or $\varepsilon_{t|n}$) and $\hat{\eta}_t = E[\eta_t|\mathbf{Y}_n]$ (or $\eta_{t|n}$), respectively. The computation of $\hat{\varepsilon}_t$ and $\hat{\eta}_t$ from the Kalman smoother algorithm is described in Durbin and Koopman (2001). These smoothed disturbance estimates can be useful for parameter estimation by maximum likelihood and for diagnostic checking. The functions SsfCondDens and SsfMomentEst, with the optional argument task="DSSMO", may be used to compute smoothed estimates of the measurement and transition equation disturbances. The function SsfCondDens only computes the smoothed states and responses whereas SsfMomentEst also computes the associated variances.

13.3.5 Missing values and forecasting

The Kalman filter prediction equation produces one-step ahead predictions of the state vector, $\mathbf{a}_{t+1} = E[\boldsymbol{\alpha}_{t+1}|\mathbf{Y}_t]$, along with prediction variance matrices $\mathbf{P}_{t+1}$. In the Kalman filter recursions, if there are missing values in $\mathbf{y}_t$ then $\mathbf{v}_t = \mathbf{0}$, $\mathbf{F}_t^{-1} = \mathbf{0}$ and $\mathbf{K}_t = \mathbf{0}$. This allows out-of-sample forecasts of α_t and $\mathbf{y}_t$ to be computed from the updating and prediction equations. Out-of-sample predictions, together with associated mean square errors, can be computed from the Kalman filter prediction equations by extending the data set $\mathbf{y}_1, \ldots, \mathbf{y}_n$ with a set of missing values. When $\mathbf{y}_\tau$ is missing, the Kalman filter reduces to the prediction step described above. As a result, a sequence of m missing values at the end of the sample will produce a set of h-step ahead forecasts for $h = 1, \ldots, m$. Forecasts with their variances based on a given state space model may be computed using the function `SsfMomentEst` with the optional argument `task="STPRED"`.

13.3.6 Simulation smoothing

The joint simulation of state and response vectors $\boldsymbol{\alpha}_t$ and $\mathbf{y}_t$, $t = 1, \ldots, n$, or disturbance vectors $\boldsymbol{\eta}_t$ and $\boldsymbol{\varepsilon}_t$, $t = 1, \ldots, n$, conditional on the observations $\mathbf{Y}_n$ is called simulation smoothing. Simulation smoothing is useful for evaluating the appropriateness of a proposed state space model, for the Bayesian analysis of state space models using Markov chain Monte Carlo (MCMC) techniques and for the evaluation of the likelihood function using importance sampling techniques. Initial work on simulation smoothing has been developed by Frühwirth-Schnatter (1994c), Carter and Kohn (1994) and de Jong and Shephard (1995). The simulation smoothing method of Durbin and Koopman (2002) is used for the `SsfPack` implementation. The function `SimSmoDraw` generates random draws from the distributions of the state and response variables (argument `task="STSIM"`) or from the distributions of the state and response disturbances (argument `task="DSSIM"`).

13.3.7 Prediction error decomposition and log-likelihood

The prediction error decomposition of the log-likelihood function for the unknown parameters φ of a state space model may be conveniently computed using the output of the Kalman filter

$$\ln L(\varphi|\mathbf{Y}_n) = \sum_{t=1}^{n} \ln f(\mathbf{y}_t|\mathbf{Y}_{t-1};\varphi)$$

$$= -\frac{nN}{2}\ln(2\pi) - \frac{1}{2}\sum_{t=1}^{n}\left(\ln|\mathbf{F}_t| + \mathbf{v}_t'\mathbf{F}_t^{-1}\mathbf{v}_t\right), \quad (13.17)$$

where $f(\mathbf{y}_t|\mathbf{Y}_{t-1};\varphi)$ is a conditional Gaussian density implied by the state space model (13.1)–(13.2). The vector of prediction errors $\mathbf{v}_t$ and prediction error variance matrices $\mathbf{F}_t$ are computed from the Kalman filter recursions.

The functions KalmanFil and SsfLoglike may be used to evaluate the prediction error decomposition of the log-likelihood function for a given set of parameters φ. The S+FinMetrics function SsfFit, which evaluates (13.17) using SsfLoglike, may be used to find the maximum likelihood estimators of the unknown parameters φ using the S-PLUS optimisation function nlminb[5].

In some models, e.g., linear regression models and ARMA models, it is possible to solve explicitly for one scale factor and concentrate it out of the log-likelihood function (13.17). The resulting log-likelihood function is called the concentrated log-likelihood or profile log-likelihood and is denoted $\ln L^c(\varphi|\mathbf{Y}_n)$. Following Koopman, Shephard and Doornik (1999), let σ denote such a scale factor, and let

$$\mathbf{y}_t = \boldsymbol{\theta}_t + \mathbf{G}_t^c \varepsilon_t^c$$

with $\varepsilon_t^c \sim$ iid $N(\mathbf{0}, \sigma^2\mathbf{I})$ denote the scaled version of the measurement equation (13.3). The state space form (13.1)–(13.3) applies but with $\mathbf{G}_t = \sigma\mathbf{G}_t^c$ and $\mathbf{H}_t = \sigma\mathbf{H}_t^c$. This formulation implies that one nonzero element of $\sigma\mathbf{G}_t^c$ or $\sigma\mathbf{H}_t^c$ is kept fixed, usually at unity, which reduces the dimension of the parameter vector φ by one. The solution for σ^2 from (13.17) is given by

$$\tilde{\sigma}^2(\varphi) = \frac{1}{Nn}\sum_{t=1}^{n} \mathbf{v}_t' \left(\mathbf{F}_t^c\right)^{-1} \mathbf{v}_t$$

and the resulting concentrated log-likelihood function is

$$\ln L^c(\varphi|\mathbf{Y}_n) = -\frac{nN}{2}\ln(2\pi) - \frac{nN}{2}\ln\left(\tilde{\sigma}^2(\varphi) + 1\right) - \frac{1}{2}\sum_{t=1}^{n}\ln|\mathbf{F}_t^c|. \quad (13.18)$$

For a given set of parameters φ, the concentrated log-likelihood may be evaluated using the functions KalmanFil and SsfLoglike. Maximisation of

---

[5] There are several optimisation algorithms available in S-PLUS besides nlminb. Most notable are the functions nlmin, ms and optim (in the MASS library of S-PLUS).

Table 13.4. *General* `SsfPack` *state space functions*

function	description	tasks
KalmanIni	initialise Kalman filter	all
KalmanFil	Kalman filtering and likelihood eval	all
KalmanSmo	Kalman smoothing	none
SsfCondDens	conditional density/mean calculation	STSMO,DSSMO
SsfMomentEst	moment estimation and smoothing	STFIL,STPRED, STSMO,DSSMO
SimSmoDraw	simulation smoother draws	STSIM,DSSIM
SsfLoglike	log-likelihood of state space model	none
SsfFit	estimate state space model parameters	none

Table 13.5. *Task argument to* `SsfPack` *functions*

task	description
KFLIK	Kalman filtering and loglikelihood evaluation
STFIL	state filtering
STPRED	state prediction
STSMO	state smoothing
DSSMO	disturbance smoothing
STSIM	state simulation
DSSIM	disturbance simulation

the concentrated log-likelihood function may be specified in the function `SsfFit` by setting the optional argument `conc=T`.

13.4 The capital asset pricing model

This section illustrates the use of the `SsfPack` state space modelling and analysis functions for an empirical analysis of the CAPM.

13.4.1 Recursive least squares

Consider the typical CAPM regression model

$$y_t = \alpha + \beta_M x_{M,t} + \xi_t, \quad \xi_t \sim \text{iid } N(0, \sigma_\xi^2),$$

where y_t denotes the return on an asset in excess of the risk free rate, and $x_{M,t}$ is the excess return on a market index. This is a linear regression model of the form

$$y_t = \mathbf{x}_t'\boldsymbol{\beta} + \xi_t, \quad \xi_t \sim \text{iid } N(0, \sigma_\xi^2),$$

where $\mathbf{x}_t = (1, x_{M,t})'$ is a 2×1 vector of data, and $\boldsymbol{\beta} = (\alpha, \beta_M)'$ is a 2×1 fixed parameter vector. The state space representation is given by

$$\begin{pmatrix} \boldsymbol{\alpha}_{t+1} \\ y_t \end{pmatrix} = \begin{pmatrix} \mathbf{I}_k \\ \mathbf{x}_t' \end{pmatrix} \boldsymbol{\alpha}_t + \begin{pmatrix} \mathbf{0} \\ \sigma_\xi \xi_t \end{pmatrix}. \quad (13.19)$$

The state vector satisfies

$$\boldsymbol{\alpha}_{t+1} = \boldsymbol{\alpha}_t = \boldsymbol{\beta} = (\alpha, \beta_M)'.$$

The state space system matrices are $\mathbf{T}_t = \mathbf{I}_2$, $\mathbf{Z}_t = \mathbf{x}_t'$, $\mathbf{G}_t = \sigma_\xi$ and $\mathbf{H}_t = 0$ and may be compared with the ones in Example 13.2. The coefficient vector β is fixed and unknown so that the initial conditions are $\boldsymbol{\alpha}_1 \sim N(\mathbf{0}, \kappa \mathbf{I}_2)$, where κ is infinity. The monthly excess return data on Microsoft and the S&P 500 index over the period February 1990–December 2000 are used as in Listing 13.2. The state space form for the CAPM with fixed regressors may be created using the function `GetSsfReg` as in Listing 13.4.

An advantage of analysing the linear regression model in state space form is that recursive least squares (RLS) estimates of the regression coefficient vector β are readily computed from the Kalman filter. The RLS estimates are based on estimating the model

$$y_t = \boldsymbol{\beta}_t'\mathbf{x}_t + \xi_t, \quad t = 1, \ldots, n \quad (13.20)$$

by least squares recursively for $t = 3, \ldots, n$ giving $n - 2$ least squares (RLS) estimates $(\hat{\boldsymbol{\beta}}_3, \ldots, \hat{\boldsymbol{\beta}}_T)$. If $\boldsymbol{\beta}$ is constant over time then the recursive estimates $\hat{\boldsymbol{\beta}}_t$ should quickly settle down near a common value. If some of the elements in $\boldsymbol{\beta}$ are not constant then the corresponding RLS estimates should show instability. Hence, a simple graphical technique for uncovering parameter instability is to plot the RLS estimates $\hat{\beta}_{it}$ ($i = 1, 2$) and look for instability in the plots.

The RLS estimates are simply the filtered state estimates from the model (13.19), and may be computed using the function `SsfMomentEst` with the optional argument `task="STFIL"` as in Listing 13.4.

The component `state.moment` contains the filtered state estimates $\mathbf{a}_{t|t}$ for $t = 1, \ldots, n$, which are equal to the RLS estimates of the linear regression coefficients, and the component `response.moment` contains the filtered response estimates $y_{t|t}$. The first column of the component `state.moment` contains the RLS estimates of α, and the second column contains the RLS

```
> ssf.reg = GetSsfReg(X.mat)
> ssf.reg
$mPhi:      $mOmega:              $mSigma:        $mJPhi:
    [,1] [,2]     [,1] [,2] [,3]      [,1] [,2]       [,1] [,2]
[1,]  1   0  [1,]  0   0   0   [1,] -1    0    [1,]  -1   -1
[2,]  0   1  [2,]  0   0   0   [2,]  0   -1    [2,]  -1   -1
[3,]  0   0  [3,]  0   0   1   [3,]  0    0    [3,]   1    2
> filteredEst.reg = SsfMomentEst(msft.ret,ssf.reg,task="STFIL")
> class(filteredEst.reg)
[1] "SsfMomentEst"
> names(filteredEst.reg)
[1] "state.moment"   "state.variance"   "response.moment"
[4] "response.variance"  "task"   "positions"
> filteredEst.reg$state.moment
numeric matrix: 131 rows, 2 columns.
state.1 state.2
[1,] 0.06186 0.0001756
[2,] 0.05179 3.5482887
[3,] 0.07811 1.2844189
...
[131,] 0.01751 1.568
> ols.fit = OLS(MSFT~SP500,data=excessReturns.ts)
> coef(ols.fit)
(Intercept) SP500
0.01751 1.568
> colIds(filteredEst.reg$state.moment) = c("alpha","beta")
> plot(filteredEst.reg,main="Filtered estimates: RLS")
```

Listing 13.4. A RLS analysis of the CAPM.

estimates of β_M. The last row contains the full sample least squares estimates of α and β_M. The RLS estimates can be visualised using the generic `plot` method for objects of the class `SsfMomentEst`. The resulting plot is illustrated in Figure 13.2. Notice that the RLS estimates of β_M seem fairly constant whereas the RLS estimates of α do not.

13.4.2 Tests for constant parameters

Formal tests for structural stability of the regression coefficients, such as the CUSUM test of Brown, Durbin and Evans (1975), may be computed from the standardised one-step ahead recursive residuals

$$w_t = \frac{v_t}{\sqrt{f_t}} = \frac{y_t - \hat{\boldsymbol{\beta}}'_{t-1}\mathbf{x}_t}{\sqrt{f_t}},$$

where f_t is an estimate of the recursive error variance

$$\sigma^2 \left[1 + \mathbf{x}'_t(\mathbf{X}'_{t-1}\mathbf{X}_{t-1})^{-1}\mathbf{x}_t\right]$$

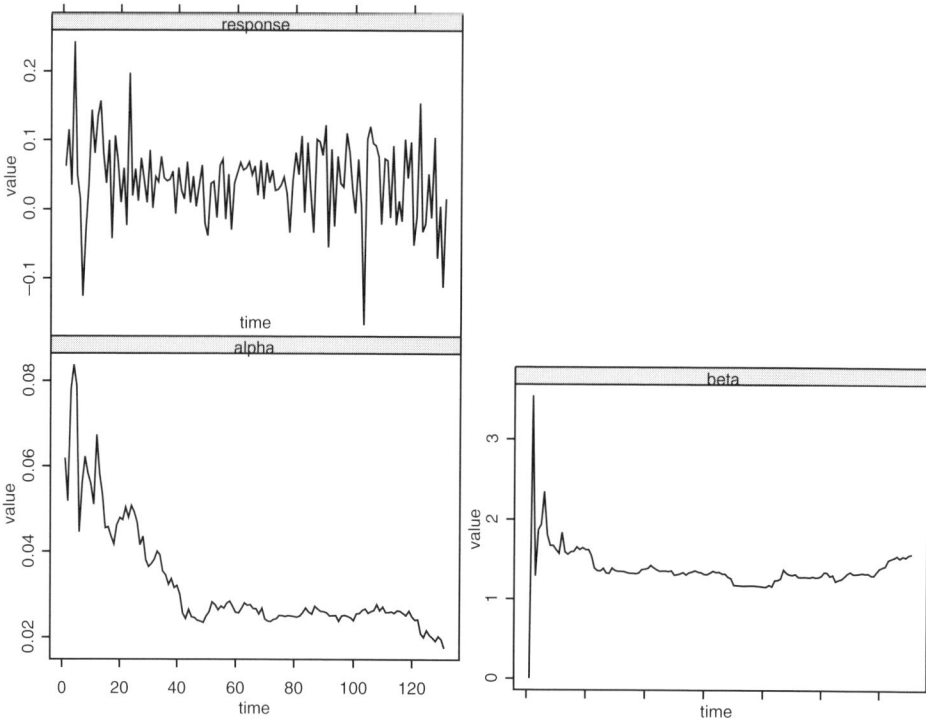

Fig. 13.2. RLS estimates of CAPM for Microsoft using the Kalman filter.

and $\mathbf{X}_t$ is the $(t \times k)$ matrix of observations on $\mathbf{x}_s$ using data from $s = 1, \ldots, t$. These standardised recursive residuals result as a by-product of the Kalman filter recursions and may be extracted using the `SsfPack` function `KalmanFil` as in Listing 13.5. Diagnostic plots of the standardised innovations may be created using the `plot` method for objects of class 'KalmanFil'. Selection 3 produces the graph shown in Figure 13.3.

The CUSUM test is based on the cumulated sum of the standardised recursive residuals

$$CUSUM_t = \sum_{j=k+1}^{t} \frac{\hat{w}_j}{\hat{\sigma}_w},$$

where $\hat{\sigma}_w$ is the sample standard deviation of $\hat{w}_j$ and k denotes the number of estimated coefficients. Under the null hypothesis that β in (13.20) is constant, $CUSUM_t$ has mean zero and variance that is proportional to $t - k - 1$. Brown, Durbin and Evans (1975) show that approximate 95%

```
> kf.reg = KalmanFil(msft.ret,ssf.reg)
> class(kf.reg)
[1] "KalmanFil"
> names(kf.reg)
[1] "mOut"    "innov"   "std.innov" "mGain"
[5] "loglike" "loglike.conc" "dVar"   "mEst"
[9] "mOffP"   "task"    "err"      "call"
[13] "positions"
> plot(kf.reg)
```

Listing 13.5. Kalman filter for CAPM.

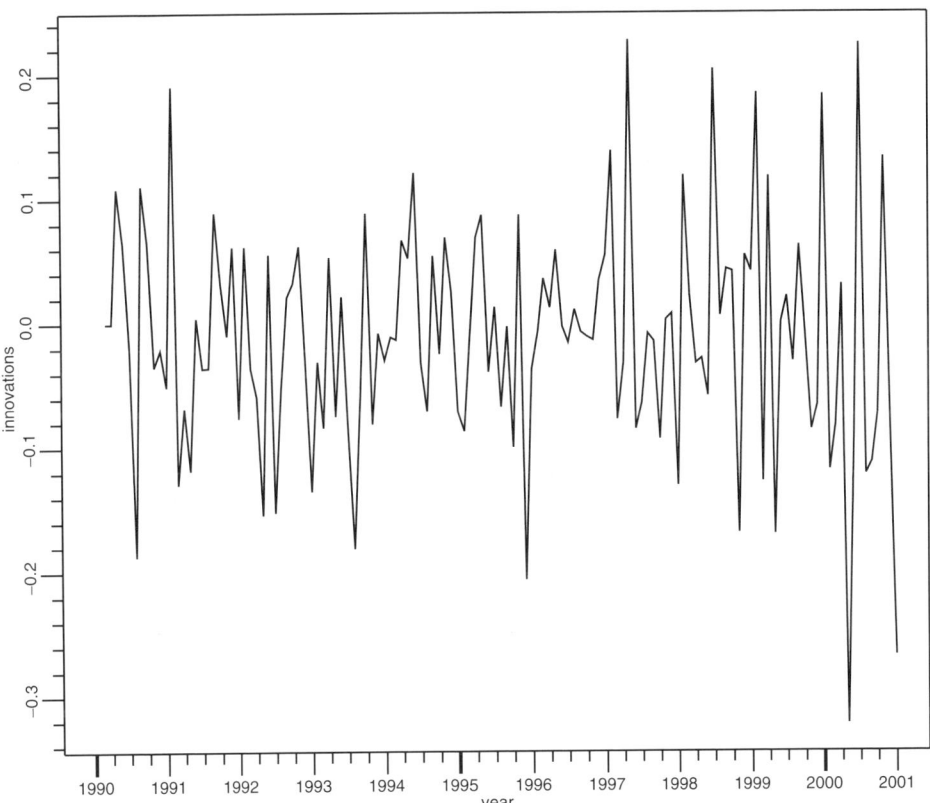

Fig. 13.3. Scaled innovations, $w_t = v_t/\sqrt{f_t}$, computed from the RLS estimates of the CAPM for Microsoft.

confidence bands for $CUSUM_t$ are given by the two lines which connect the points $(k, \pm 0.948\sqrt{n-k-1})$ and $(n, \pm 0.948 \times 3\sqrt{n-k-1})$. If $CUSUM_t$ wanders outside of these bands, then the null of parameter stability may be rejected. The S-PLUS commands to compute $CUSUM_t$ are in Listing 13.6.

The CUSUM plot is given in Figure 13.4 and it indicates that the CAPM for Microsoft has stable parameters.

```
> w.t = kf.reg\$std.innov[-c(1,2)] \# first two innovations are
equal to zero
> cusum.t = cumsum(w.t)/stdev(w.t)
> nobs = length(cusum.t)
> tmp = 0.948*sqrt(nobs)
> upper = seq(tmp,3*tmp,length=nobs)
> lower = seq(-tmp,-3*tmp,length=nobs)
> tmp.ts = timeSeries(pos=kf.reg\$positions[-c(1,2)],
+ data=cbind(cusum.t,upper,lower))
> plot(tmp.ts,reference.grid=F,
+ plot.args=list(lty=c(1,2,2),col=c(1,2,2)))
```

Listing 13.6. Computing the CUSUM test.

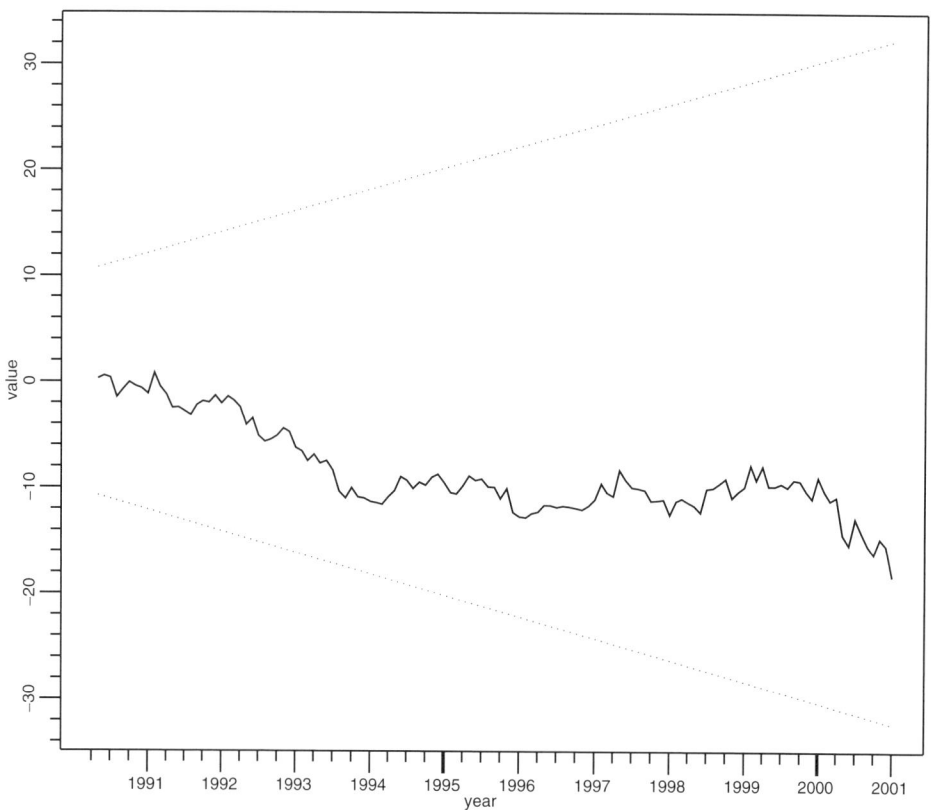

Fig. 13.4. CUSUM test for parameter constancy in CAPM regression for Microsoft. The area between the dotted lines represents a 95% confidence interval.

13.4.3 CAPM with time varying parameters

Consider estimating the CAPM with time varying coefficients (13.14) subject to random walk evolution, using monthly data on Microsoft and the S&P 500 index over the period February 1990–December 2000 contained in the S+FinMetrics 'timeSeries' object excessReturns.ts. Neumann (2002) surveyed several estimation strategies for time varying parameter models and concluded that the state space model with random walk specifications for the evolution of the time varying parameters generally performs very well. The parameters of the model are the variances of the innovations to the transition and measurement equations: $\boldsymbol{\sigma} = (\sigma_\xi^2, \sigma_\varsigma^2, \sigma_\nu^2)'$. Since these variances must be positive the log-likelihood is parameterised using $\boldsymbol{\varphi} = (\ln(\sigma_\xi^2), \ln(\sigma_\varsigma^2), \ln(\sigma_\nu^2))'$, so that $\boldsymbol{\sigma} = (\exp(\varphi_1), \exp(\varphi_2), \exp(\varphi_3))'$. The state space form for the CAPM with time varying coefficients requires a data matrix $\mathbf{X}$ containing the excess returns on the S&P 500 index and therefore the function SsfFit has $\boldsymbol{\varphi}$ and $\mathbf{X}$ as input and it returns the appropriate state space form. Listing 13.7 provides an example of an implementation for estimating the time varying CAPM.

Starting values for $\boldsymbol{\varphi}$ are specified by tvp.start. The maximum likelihood estimates for $\boldsymbol{\varphi}$ are computed using tvp.mle. The print method gives estimates of $\boldsymbol{\varphi} = (\ln(\sigma_\xi^2), \ln(\sigma_\varsigma^2), \ln(\sigma_\nu^2))'$ while the summary method provides estimates for the standard deviations σ_ξ, σ_ς and σ_ν as well as estimated standard errors, from the delta method;[6] see Listing 13.7. It appears that the estimated standard deviations for the time varying parameter CAPM are close to zero, suggesting a constant parameter model.

Given the estimated parameters, the filtered estimates of the time varying parameters α_t and $\beta_{M,t}$ may be computed using SsfMomentEst. The filtered moments, without standard error bands, may be visualised using the plot method for objects of class 'SsfMomentEst' as illustrated in Figure 13.5. The filtered estimates of the parameters from the CAPM with time varying parameters look remarkably like the RLS estimates computed earlier.

The smoothed estimates of the time varying parameters α_t and $\beta_{M,t}$ as well as the expected returns may be extracted using SsfCondDens as in Listing 13.7. The plot method is then used to visualise the smoothed estimates as illustrated in Figure 13.6. The smoothed state estimates appear quite different from the filtered state estimates shown in Figure 13.5, but this difference is primarily due to the erratic behavior of the first few filtered estimates. The function SsfCondDens does not compute estimated variances for the smoothed state and response variables. If standard error

---

[6] If $\sqrt{n}(\hat{\theta} - \theta) \xrightarrow{d} N(0, V)$ and g is a continuous function then $\sqrt{n}(g(\hat{\theta}) - g(\theta)) \xrightarrow{d} N(0, \partial g/\partial\theta \, V \partial g/\partial\theta')$.

```
> tvp.mod = function(parm,mX=NULL) {
parm = exp(parm) # 3 x 1 vector containing log variances
ssf.tvp = GetSsfReg(mX=mX)
diag(ssf.tvp$mOmega) = parm
CheckSsf(ssf.tvp)
}
> tvp.start = c(0,0,0)
> names(tvp.start) = c("ln(s2.alpha)","ln(s2.beta)","ln(s2.y)")
> tvp.mle = SsfFit(tvp.start,msft.ret,"tvp.mod",mX=X.mat)
Iteration 0 : objective = 183.2
...
Iteration 22 : objective = -123
RELATIVE FUNCTION CONVERGENCE
> class(tvp.mle)
[1] "SsfFit"
> names(tvp.mle)
[1] "parameters" "objective" "message" "grad.norm" "iterations"
[6] "f.evals" "g.evals" "hessian" "scale" "aux"
[11] "call" "vcov"
> summary(tvp.mle)
Log-likelihood: -122.979
131 observations
Parameters:
Value Std. Error t value
ln(s2.alpha) -12.480 2.8020 -4.453
ln(s2.beta) -5.900 3.0900 -1.909
ln(s2.y) -4.817 0.1285 -37.480
> tvp2.mle = tvp.mle
> tvp2.mle$parameters = exp(tvp.mle$parameters/2)
> names(tvp2.mle$parameters) = c("s.alpha","s.beta","s.y")
> dg = diag(tvp2.mle$parameters/2)
> tvp2.mle$vcov = dg %*% tvp.mle\$vcov %*% dg
> summary(tvp2.mle)
Log-likelihood: -122.979
131 observations
Parameters:
Value Std. Error t value
s.alpha 0.001951 0.002733 0.7137
s.beta 0.052350 0.080890 0.6472
s.y 0.089970 0.005781 15.5600
> smoothedEst.tvp = SsfCondDens(msft.ret,
+ tvp.mod(tvp.mle$parameters,mX=X.mat),
+ task="STSMO")
```

Listing 13.7. An analysis of the time varying CAPM.

bands for the smoothed estimates are desired, then SsfMomentEst with task="STSMO" must be used and the state variances are available in the component state.variance.

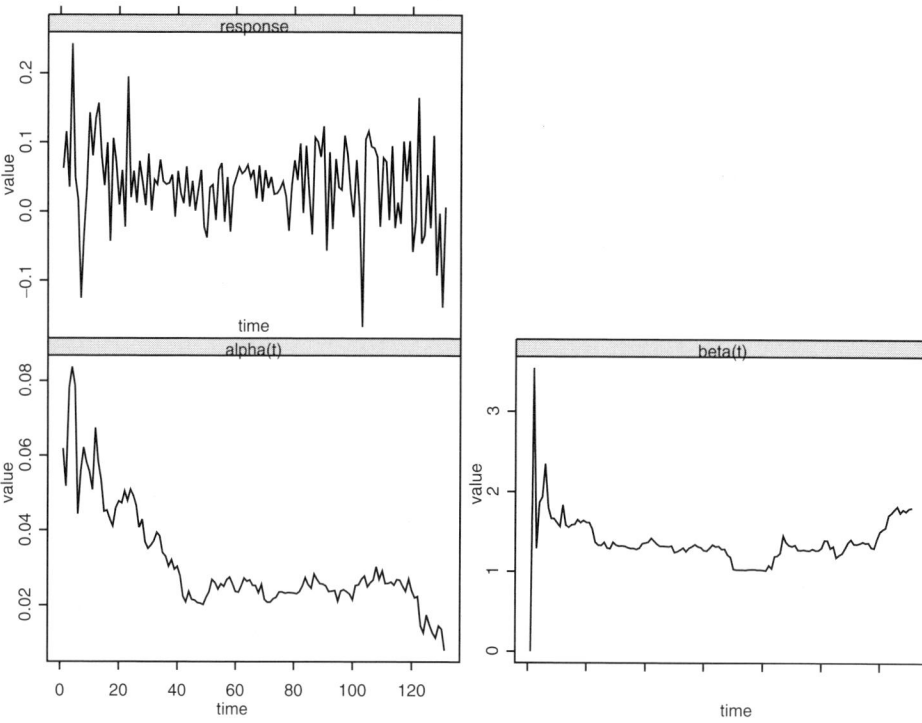

Fig. 13.5. Filtered estimates of CAPM for Microsoft with time varying parameters.

13.5 Time series decompositions

In this section we use state space methods to compute some common trend-cycle decompositions of US postwar quarterly real GDP. These decompositions are used to estimate the long-run trend in output as well as business cycles. We illustrate the well-known Beveridge–Nelson decomposition as well as several unobserved components decompositions.

13.5.1 ARMA modelling and Beveridge–Nelson decompositions

Consider the problem of decomposing the movements in the natural logarithm of US postwar quarterly real GDP into permanent and transitory (cyclical) components. Beveridge and Nelson (1981) proposed a definition for the permanent component of an $I(1)$ time series y_t with drift μ as the limiting forecast as the horizon goes to infinity, adjusted for the mean rate

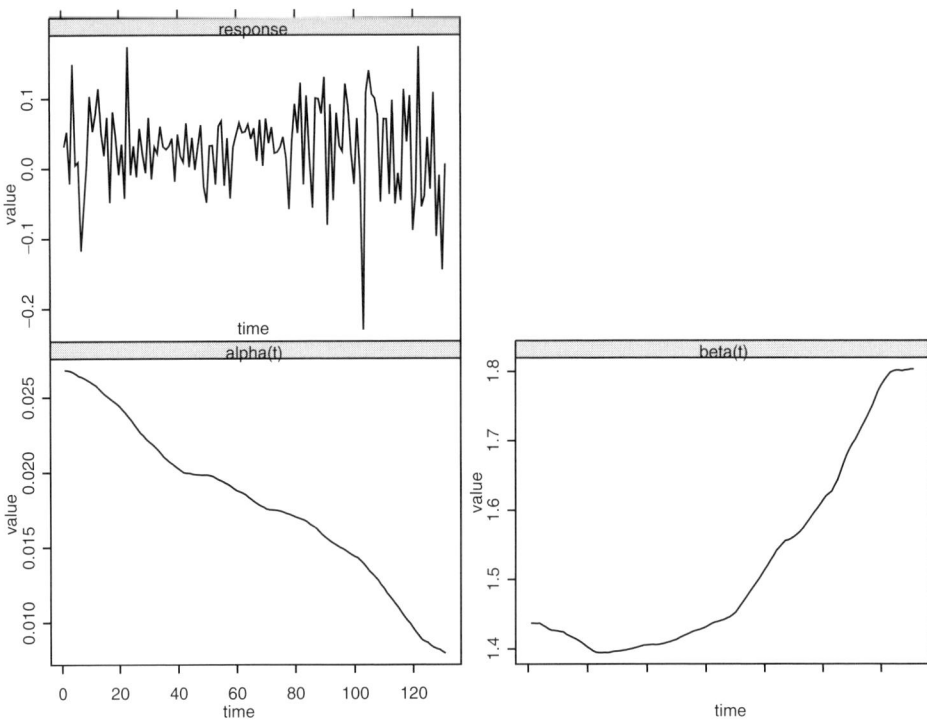

Fig. 13.6. Smoothed estimates of CAPM for Microsoft with time varying parameters.

of growth:

$$BN_t = \lim_{h \to \infty} E_t[y_{t+h} - \mu h],$$

where $E_t[\cdot]$ denotes expectation conditional on information available at time t. The transitory or cycle component is then defined as the gap between the present level of the series and its long-run forecast:

$$c_t = y_t - BN_t.$$

This permanent–transitory decomposition is often referred to as the 'BN decomposition'. In practice, the BN decomposition is obtained by fitting an ARMA(p, q) model to Δy_t, where $\Delta = 1 - L$ is the difference operator, and then computing BN_t and c_t from the fitted model.

As shown by Morley (2002), the BN decomposition may be easily computed using the Kalman filter by putting the forecasting model for $\Delta y_t - \mu$ in state space form. In particular, suppose $\Delta y_t - \mu$ is a linear combination of the elements of the $m \times 1$ state vector $\boldsymbol{\alpha}_t$:

$$\Delta y_t - \mu = \begin{bmatrix} z_1 & z_2 & \cdots & z_m \end{bmatrix} \boldsymbol{\alpha}_t,$$

where z_i $(i = 1, \ldots, m)$ is the weight of the ith element of $\boldsymbol{\alpha}_t$ in determining $\Delta y_t - \mu$. Suppose further that

$$\boldsymbol{\alpha}_{t+1} = \mathbf{T}\boldsymbol{\alpha}_t + \boldsymbol{\eta}_t^*, \qquad \boldsymbol{\eta}_t^* \sim \text{iid } N(\mathbf{0}, \mathbf{V}),$$

such that all of the eigenvalues of $\mathbf{T}$ have modulus less than unity, and $\mathbf{T}$ is invertible. Then, Morley shows that

$$\begin{aligned} BN_t &= y_t + \begin{bmatrix} z_1 & z_2 & \cdots & z_m \end{bmatrix} \mathbf{T}(\mathbf{I}_m - \mathbf{T})^{-1} \mathbf{a}_{t|t}, \\ c_t &= y_t - BN_t = -\begin{bmatrix} z_1 & z_2 & \cdots & z_m \end{bmatrix} \mathbf{T}(\mathbf{I}_m - \mathbf{T})^{-1} \mathbf{a}_{t|t}, \end{aligned} \qquad (13.21)$$

where $\mathbf{a}_{t|t}$ denotes the filtered estimate of α_t.

To illustrate the process of constructing the BN decomposition for US postwar quarterly real GDP over the period 1947:I–1998:II, we follow Morley, Nelson and Zivot (2003) (hereafter MNZ) and consider fitting the ARMA(2,2) model

$$\begin{aligned} \Delta y_t - \mu &= \phi_1(\Delta y_{t-1} - \mu) + \phi_2(\Delta y_{t-2} - \mu) + \varepsilon_t + \theta_1 \varepsilon_{t-1} + \theta_2 \varepsilon_{t-2}, \\ \varepsilon_t &\sim \text{iid } N(0, \sigma^2), \end{aligned}$$

where y_t denotes the natural log of real GDP multiplied by 100. In `SsfPack`, the ARMA(p, q) model for a demeaned stationary variable y_t^* has a state space representation with transition and measurement equations

$$\begin{aligned} \boldsymbol{\alpha}_{t+1} &= \mathbf{T}\boldsymbol{\alpha}_t + \mathbf{H}\xi_t, \ \xi_t \sim N(0, \sigma_\varepsilon^2), \\ y_t^* &= \mathbf{Z}\boldsymbol{\alpha}_t \end{aligned}$$

and time invariant system matrices

$$\mathbf{T} = \begin{pmatrix} \phi_1 & 1 & 0 & \cdots & 0 \\ \phi_2 & 0 & 1 & & 0 \\ \vdots & & & \ddots & \vdots \\ \phi_{m-1} & 0 & 0 & & 1 \\ \phi_m & 0 & 0 & \cdots & 0 \end{pmatrix}, \ \mathbf{H} = \begin{pmatrix} 1 \\ \theta_1 \\ \vdots \\ \theta_{m-1} \\ \theta_m \end{pmatrix}, \qquad (13.22)$$

$$\mathbf{Z} = \begin{pmatrix} 1 & 0 & \cdots & 0 & 0 \end{pmatrix},$$

where $\mathbf{d}, \mathbf{c}$ and $\mathbf{G}$ of the state space form (13.1)–(13.3) are all zero and

$m = \max(p, q+1)$. The state vector $\boldsymbol{\alpha}_t$ has the form

$$\boldsymbol{\alpha}_t = \begin{pmatrix} y_t^* \\ \phi_2 y_{t-1}^* + \cdots + \phi_p y_{t-m+1}^* + \theta_1 \xi_t + \cdots + \theta_{m-1} \xi_{t-m+2} \\ \phi_3 y_{t-1}^* + \cdots + \phi_p y_{t-m+2}^* + \theta_2 \xi_t + \cdots + \theta_{m-1} \xi_{t-m+3} \\ \vdots \\ \phi_m y_{t-1}^* + \theta_{m-1} \xi_t \end{pmatrix}. \quad (13.23)$$

The exact maximum likelihood estimates of the ARMA(2,2) parameters may be computed using the SsfPack functions GetSsfArma and SsfFit. The function SsfFit requires as input a function which takes the unknown parameters φ and produces the state space form for the ARMA(2,2) as is illustrated in Listing 13.8. Notice that the function arma22.mod parameterises the error variance as $\sigma^2 = \exp(\gamma)$, $-\infty < \gamma < \infty$, to ensure that the estimated value of σ^2 is positive, and utilises the SsfPack function GetSsfArma to create the state space form for the ARMA(2,2) function. Starting values for the estimation are given by (conditional) MLE using S-PLUS function arima.mle. The data used for the estimation is in the 'timeSeries' lny.ts and the demeaned first difference data is in the 'timeSeries' dlny.ts.dm. The exact maximum likelihood estimates for $\varphi = (\phi_1, \phi_2, \theta_1, \theta_2, \gamma)'$ are computed using SsfFit.[7]

Given the maximum likelihood estimates φ, the filtered estimate of the state vector may be computed using the function SsfMomentEst with optional argument task="STFIL". The BN decomposition (13.22) may then be computed as in Listing 13.9. Figure 13.7 illustrates the results of the BN decomposition for US real GDP. The BN trend follows the data very closely and, as a result, the BN cycle behaves much like the first difference of the data.

13.5.2 Unobserved components decompositions: Clark model

Harvey (1985) and Clark (1987) provided an alternative to the BN decomposition of an $I(1)$ time series with drift into permanent and transitory components based on unobserved components structural time series models. For example, Clark's model for the natural logarithm of postwar real GDP specifies the trend as a pure random walk, and the cycle as a stationary

---

[7] One may also estimate the ARMA(2,2) model with SsfFit by maximising the log-likelihood function concentrated with respect to the error variance σ^2.

```
> arma22.mod = function(parm) {
phi.1 = parm[1]
phi.2 = parm[2]
theta.1 = parm[3]
theta.2 = parm[4]
sigma2 = exp(parm[5]) # require positive variance
ssf.mod = GetSsfArma(ar=c(phi.1,phi.2),ma=c(theta.1,theta.2),
sigma=sqrt(sigma2))
CheckSsf(ssf.mod)
}

> arma22.start = c(1.34,-0.70,-1.05,0.51,-0.08)
> names(arma22.start) =
c("phi.1","phi.2","theta.1","theta.2","ln.sigma2")
> arma22.mle = SsfFit(arma22.start,dlny.ts.dm,"arma22.mod")
Iteration 0 : objective = 284.6686
...
Iteration 27 : objective = 284.651
RELATIVE FUNCTION CONVERGENCE

> summary(arma22.mle)
Log-likelihood: 284.651
205 observations
Parameters:
           Value Std. Error  t value
    phi.1  1.34200   0.14480   9.2680
    phi.2 -0.70580   0.14930  -4.7290
  theta.1 -1.05400   0.18030  -5.8490
  theta.2  0.51870   0.19330   2.6830
ln.sigma2 -0.06217   0.09878  -0.6294
```

Listing 13.8. Estimation of an ARMA model.

```
> ssf.arma22 = arma22.mod(arma22.mle$parameters)
> filteredEst.arma22 = SsfMomentEst(dlny.ts.dm,
+ ssf.arma22,task="STFIL")
> at.t = filteredEst.arma22$state.moment
> T.mat = ssf.arma22$mPhi[1:3,1:3]
> tmp = t(T.mat %*% solve((diag(3)-T.mat)) %*% t(at.t))
> BN.t = lny.ts[2:nobs,] + tmp[,1]
> c.t = lny.ts[2:nobs,] - BN.t
```

Listing 13.9. Implementation of BN decomposing using Morley (2002).

AR(2) process:

$$\begin{aligned} y_t &= \tau_t + c_t, \\ \tau_{t+1} &= \mu + \tau_t + v_t, \\ c_{t+1} &= \phi_1 c_t + \phi_2 c_t + w_t, \end{aligned} \quad (13.24)$$

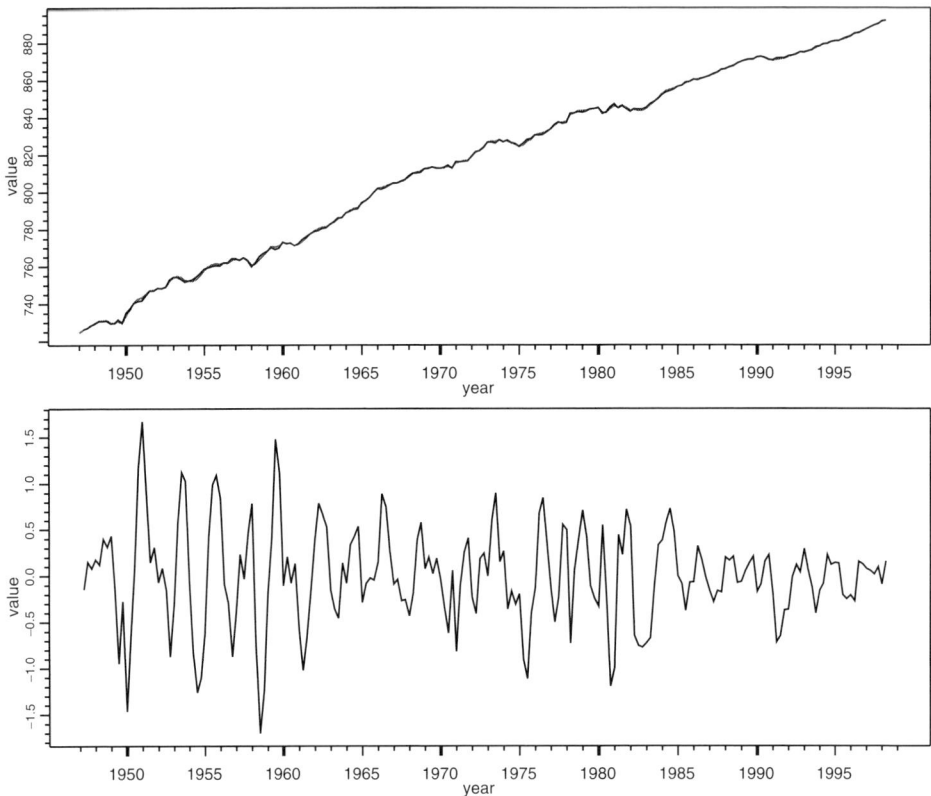

Fig. 13.7. BN decomposition for US postwar quarterly real GDP.

where the roots of $\phi(z) = 1 - \phi_1 z - \phi_2 z^2 = 0$ lie outside the complex unit circle.[8] For identification purposes, Clark assumed that the trend innovations and cycle innovations are uncorrelated and normally distributed:

$$\begin{pmatrix} v_t \\ w_t \end{pmatrix} \sim \text{iid } N\left(\begin{pmatrix} 0 \\ 0 \end{pmatrix}, \begin{pmatrix} \sigma_v^2 & 0 \\ 0 & \sigma_w^2 \end{pmatrix} \right).$$

---

[8] Harvey's model differs from Clark's model in that the stationary AR(2) cycle is restricted to have complex roots. The function `GetSsfStsm` in Table 13.3 may be used to easily construct the state space form for Harvey's model.

The Clark model may be put in state space form (13.5) with

$$\alpha_{t+1} = \begin{pmatrix} \tau_{t+1} \\ c_{t+1} \\ c_t \end{pmatrix}, \delta = \begin{pmatrix} \mu \\ 0 \\ 0 \\ 0 \end{pmatrix}, \Phi = \begin{pmatrix} 1 & 0 & 0 \\ 0 & \phi_1 & \phi_2 \\ 0 & 1 & 0 \\ 1 & 1 & 0 \end{pmatrix}$$

$$\mathbf{u}_t = \begin{pmatrix} \eta_t^* \\ 0 \end{pmatrix}, \eta_t^* = \begin{pmatrix} v_{t+1} \\ w_{t+1} \\ 0 \end{pmatrix}, \Omega = \begin{pmatrix} \sigma_v^2 & 0 & 0 & 0 \\ 0 & \sigma_w^2 & 0 & 0 \\ 0 & 0 & 0 & 0 \\ 0 & 0 & 0 & 0 \end{pmatrix}.$$

Since the trend component is nonstationary, it is given a diffuse initialisation. The initial covariance matrix $\mathbf{P}_*$ of the stationary cycle is determined from

$$\text{vec}(\mathbf{P}_*) = (\mathbf{I}_4 - (\mathbf{F} \otimes \mathbf{F})^{-1})\text{vec}(\mathbf{V}_w),$$

where

$$\mathbf{F} = \begin{pmatrix} \phi_1 & \phi_2 \\ 1 & 0 \end{pmatrix}, \mathbf{V}_w = \begin{pmatrix} \sigma_w^2 & 0 \\ 0 & 0 \end{pmatrix}.$$

The initial value parameter matrix (13.7) is then

$$\Sigma = \begin{pmatrix} -1 & 0 & 0 \\ 0 & p_{11} & p_{12} \\ 0 & p_{21} & p_{22} \\ 0 & 0 & 0 \end{pmatrix},$$

where p_{ij} denotes the (i,j) element of $\mathbf{P}_*$.

The exact maximum likelihood estimates of the Clark model parameters, based on the prediction error decomposition of the log-likelihood function, may be computed using the SsfPack function SsfFit. The function SsfFit requires as input a function which takes the unknown parameters φ and produces the state space form for the Clark model; an example is given in Listing 13.10. Notice that the state variances are parameterised as $\sigma_v^2 = \exp(\gamma_v)$ and $\sigma_w^2 = \exp(\gamma_w)$, $-\infty < \gamma_v, \gamma_w < \infty$, to ensure positive estimates. Starting values for the parameters are based on values near the estimates of the Clark model from MNZ.

The data used for the estimation[9] is in the 'timeSeries' lny.ts, and is the same data used to compute the BN decomposition earlier. The maximum likelihood estimates and asymptotic standard errors of the parameters $\varphi = (\mu, \gamma_v, \gamma_w, \phi_1, \phi_2)'$ using SsfFit are reproduced as output in Listing 13.10.

---

[9] In the estimation, no restrictions were imposed on the AR(2) parameters ϕ_1 and ϕ_2 to ensure that the cycle is stationary. The function SsfFit uses the S-PLUS optimisation algorithm

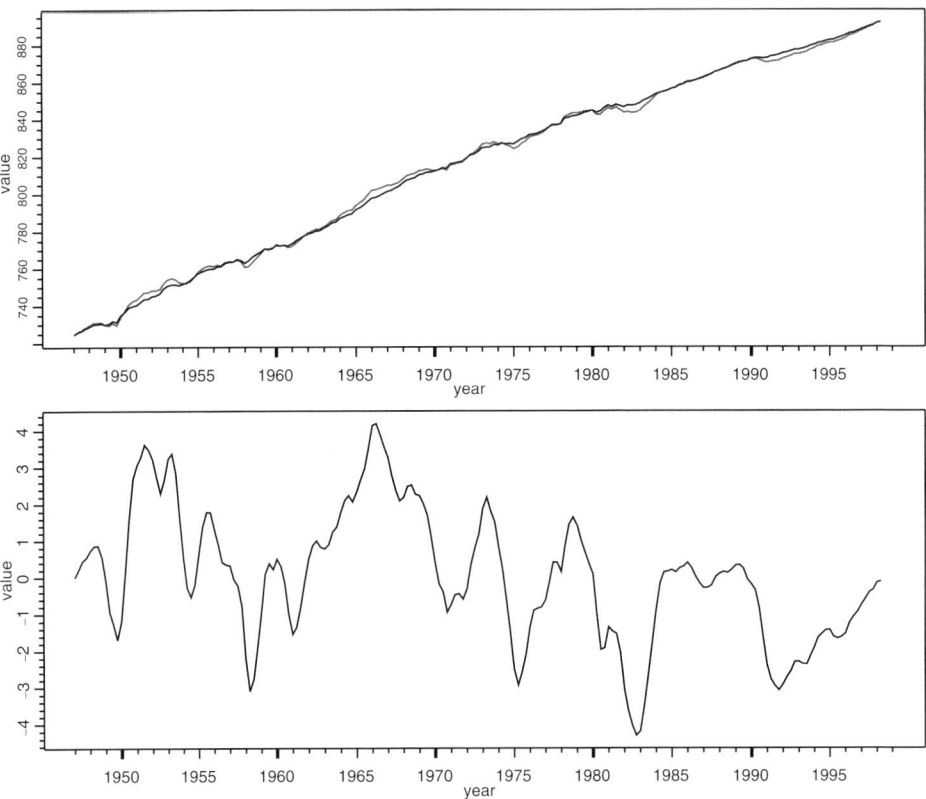

Fig. 13.8. Filtered estimates of the trend (upper graph) and cycle (lower graph) from the Clark model estimated to US real GDP.

The maximum likelihood estimates for the Clark model parameters are almost identical to those found by MNZ [10]. Estimates of the error component standard deviations are also produced.

The filtered estimates of the trend, $\tau_{t|t}$, and cycle, $c_{t|t}$, given the estimated parameters may be computed using the function SsfMomentEst with the optional argument task="STFIL" as in Listing 13.10. The filtered estimates are in the state.moment component and the variances of the filtered estimates are in the state.variance component. The filtered trend estimate is in the first column of the state.moment component and the filtered cycle is in the second column. These filtered estimates are illustrated in Figure 13.8.

nlminb, which performs minimisation of a function subject to box constraints. Box constraints on ϕ_1 and ϕ_2 may be used to constrain their estimated values to be near the appropriate stationary region.

[10] MNZ estimated the Clark model in GAUSS using the prediction error decomposition with the variance of the initial state for the nonstationary component set to a large positive number. The state space representation of the Clark model here utilises an exact initialisation of the Kalman filter.

```
Clark.mod = function(parm) {
  mu = parm[1]
  phi1 = parm[2]
  phi2 = parm[3]
  sigma2.v = exp(parm[4])
  sigma2.w = exp(parm[5])
  bigV = diag(c(sigma2.v,sigma2.w))
  Omega = matrix(0,4,4)
  Omega[1:2,1:2] = bigV
  a1 = matrix(0,3,1)
# solve for initial variance of stationary part
  bigF = matrix(c(phi1,1,phi2,0),2,2)
  vecV = c(sigma2.w,0,0,0)
  vecP = solve(diag(4)-kronecker(bigF,bigF))%*%vecV
  P.ar2 = matrix(vecP,2,2)
  Sigma = matrix(0,4,3)
  Sigma[1,1] = -1
  Sigma[2:3,2:3] = P.ar2
# create state space list
  ssf.mod = list(mDelta=c(mu,0,0,0),
  mPhi=rbind(c(1,0,0),c(0,phi1,phi2),c(0,1,0),c(1,1,0)),
  mOmega=Omega,
  mSigma = Sigma)
  CheckSsf(ssf.mod)
}

> Clark.start=c(0.81,1.53,-0.61,-0.74,-0.96)
> names(Clark.start) = c("mu","phi.1","phi.2",
+ "ln.sigma2.v","ln.sigma2.w")
> Clark.mle = SsfFit(Clark.start,lny.ts,"Clark.mod")
> summary(Clark.mle)
Log-likelihood: 287.524
206 observations
Parameters:
              Value  Std. Error  t value
         mu   0.8119    0.05005   16.220
      phi.1   1.5300    0.10180   15.030
      phi.2  -0.6097    0.11450   -5.326
ln.sigma2.v  -0.7441    0.30100   -2.472
ln.sigma2.w  -0.9565    0.42490   -2.251

> Clark.sd = sqrt(exp(coef(Clark.mle)[4:5]))
> names(Clark.sd) = c("sigma.v","sigma.w")
> Clark.sd
 sigma.v sigma.w
  0.6893  0.6199

> ssf.Clark = Clark.mod(Clark.mle$parameters)
> filteredEst.Clark =
SsfMomentEst(lny.ts,ssf.Clark,task="STFIL")
```

Listing 13.10. Estimating Clark's model.

The filtered trend estimate is fairly smooth and is quite similar to a linear trend. The filtered cycle estimate is large in amplitude and has a period of about eight years. In comparison to the BN decomposition, the trend-cycle decomposition based on the Clark model gives a much smoother trend and longer cycle, and attributes a greater amount of the variability of log output to the transitory cycle.

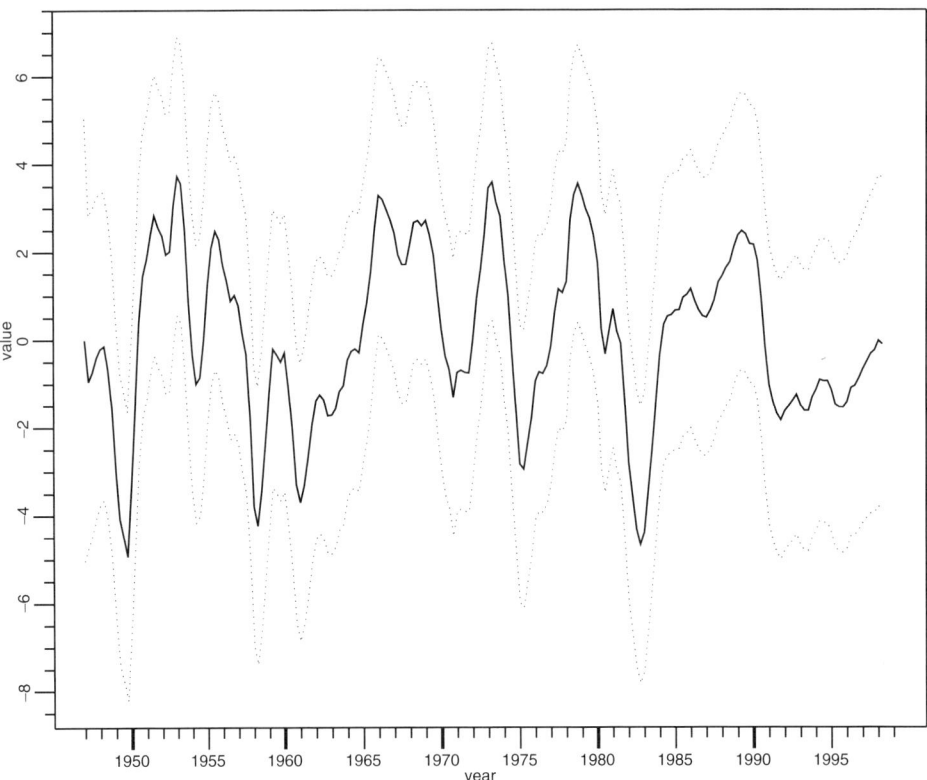

Fig. 13.9. Smoothed cycle estimate, $c_{t|n}$, with 95% error bands from Clark model for US real GDP.

The smoothed estimates of the trend, $\tau_{t|n}$, and cycle, $c_{t|n}$, along with estimated standard errors, given the estimated parameters, may be computed using the function `SsfMomentEst` with the optional argument `task="STSMO"`. The smoothed cycle estimates with 95% standard error bands are illustrated in Figure 13.9.

The Clark model assumes that the unobserved trend evolves as a random walk with drift . That is, the unobserved trend component is nonstationary. If the variance of the drift innovation, σ_v^2, is zero then the trend becomes deterministic. A number of statistics have been proposed to test the null hypothesis that $\sigma_v^2 = 0$, see Harvey and Streibel (1998a) and Harvey (2001) for reviews. For the Clark model, a Lagrange multiplier (LM) test of the null hypothesis $\sigma_v^2 = 0$, against the alternative that $\sigma_v^2 > 0$, can be formulated as

$$\eta = T^{-1} \sum_{i=1}^{T} \left[\sum_{t=1}^{i} e_t \right]^2 > c,$$

where e_t is the standardised innovation at time t from the model assuming that τ_0 is fixed, and c is the desired critical value. Harvey and Streibel (1998a) show that the statistic η has a second-level Cramér–von Mises distribution. Table I(b) of Harvey (2001) gives $c = 0.149$ as the 5% critical value.

To compute the standardised innovations assuming that τ_0 is fixed, Harvey and Streibel (1998a) suggested the following procedures. Start by estimating the unrestricted model by maximum likelihood and compute the smoothed estimates of τ_t. Then the standardised innovations e_t, assuming that τ_0 is fixed, are computed from the Kalman filter algorithm by setting $\sigma_v^2 = 0$ and initialising τ_0 at the smoothed estimate of $\tau_1 - \mu$. The commands to compute the LM test statistic are given in Listing 13.11. Since the test statistic is greater than the 5% critical value of 0.149, the null hypothesis of a deterministic trend is rejected.

```
> n = nrow(lny.ts)
> ssf.Clark0 = ssf.Clark
> ssf.Clark0$mOmega[1,1] = 0
> ssf.Clark0$mSigma[1,1] = 0
> ssf.Clark0$mSigma[4,1] =
smoothedEst.Clark$state.moment[1,1] -
> Clark.mle$parameters["mu"]
> kf.Clark0 = KalmanFil(lny.ts,ssf.Clark0)
> test.stat = sum(cumsum(kf.Clark0$std.innov)^2)/n^2
> test.stat
[1] 56.71
```

Listing 13.11. Computing the LM test for the null of $\sigma_v^2 = 0$.

13.5.3 Unobserved components decompositions: MNZ model

Morley, Nelson and Zivot (2003) have shown that the apparent difference between BN decomposition and the Clark model trend-cycle decomposition is due to the assumption of independence between trend and cycle innovations in the Clark model. In particular, they showed that the independence assumption is actually an overidentifying restriction in the Clark model, and once this assumption is relaxed to allow correlated components the difference between the decompositions disappears.

The MNZ model is simply Clark's model (13.24) where the trend and cycle innovations are allowed to be correlated with correlation coefficient ρ_{vw}:

$$\begin{pmatrix} v_t \\ w_t \end{pmatrix} \sim \text{iid } N\left(\begin{pmatrix} 0 \\ 0 \end{pmatrix}, \begin{pmatrix} \sigma_v^2 & \rho_{vw}\sigma_v\sigma_w \\ \rho_{vw}\sigma_v\sigma_w & \sigma_w^2 \end{pmatrix}\right).$$

The new state space system matrix Ω becomes

$$\Omega = \begin{pmatrix} \sigma_v^2 & \rho_{vw}\sigma_v\sigma_w & 0 & 0 \\ \rho_{vw}\sigma_v\sigma_w & \sigma_w^2 & 0 & 0 \\ 0 & 0 & 0 & 0 \\ 0 & 0 & 0 & 0 \end{pmatrix}.$$

An S-PLUS function, to be passed to SsfFit, to compute the new state space form is given in Listing 13.12. No restrictions are placed on the correlation coefficient ρ_{vw} in the function MNZ.mod. A box constraint $-0.999 < \rho_{vw} < 0.999$ will be placed on ρ_{vw} during the estimation. Starting values for the parameters are based on values near the estimates of the Clark model from MNZ. Box constraints are enforced on the AR parameters ϕ_1 and ϕ_2, to encourage stationarity, and on the correlation coefficient ρ_{vw}, to keep validity. The ML estimates are almost identical to those reported by MNZ. Notice that the estimated value of ρ_{vw} is -0.91 and that the estimated standard deviation of the trend innovation is much larger than the estimated standard deviation of the cycle innovation.

The filtered estimates of the trend, $\tau_{t|t}$, and cycle, $c_{t|t}$, given the estimated parameters are computed using the SsfMomentEst function and are illustrated in Figure 13.10. Notice that the filtered estimates of trend and cycle, when the correlation between the error components is estimated, are identical to the estimated trend and cycle from the BN decomposition. The smoothed estimates of trend and cycle are much more variable than the filtered estimates. Figure 13.11 shows the smoothed estimate of the cycle. For more discussion see Proietti (2002).

13.6 The stochastic volatility model

Financial returns data can be characterised by noise processes with volatility clustering and non-Gaussian features. In the financial and econometrics

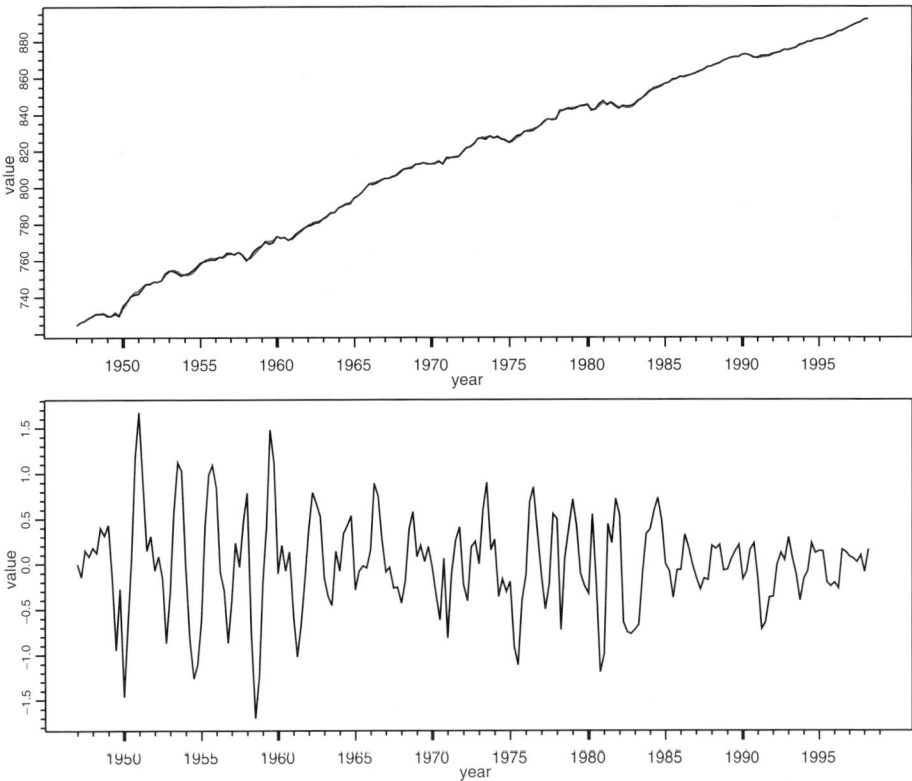

Fig. 13.10. Filtered estimates of trend (upper graph) and cycle (lower graph) from the MNZ model with correlated components for US real GDP.

literature much attention is devoted to the empirical modelling and analysing of volatility since volatility is important for the pricing of financial securities and their associated derivatives. Much applied work on this topic is focused on generalised autoregressive conditional heteroskedasticity (GARCH) models and on a long list of GARCH variants. Although most of these models are relatively straightforward to estimate, they are not necessarily convincing in empirical analyses and do not necessarily produce satisfactory forecasts. In this section we focus on the stochastic volatility model that describes volatility as a stochastic process with its own independent source of randomness. Such descriptions are in nature expressed in continuous time but they can also be formulated in discrete time. The resulting model can then be regarded as the discrete time analogue of the continuous time model used in papers on option pricing, see Hull and White (1987) and Ghysels, Harvey and Renault (1996). The discrete SV model is intrinsically a nonlinear model. The parameters can be estimated by using approximating methods or by

```
MNZ.mod = function(parm) [
  delta = parm[1]
  phi1 = parm[2]
  phi2 = parm[3]
  sigma.v = exp(parm[4])
  sigma.w = exp(parm[5])
  rho.vw = parm[6]
  sigma.vw = sigma.v*sigma.w*rho.vw
  bigV = matrix(c(sigma.v^2,sigma.vw,sigma.vw,sigma.w^2),2,2)
  Omega = matrix(0,4,4)
  Omega[1:2,1:2] = bigV
  a1 = matrix(0,3,1)
# solve for initial variance of stationary part
  bigF = matrix(c(phi1,1,phi2,0),2,2)
  vecV = c(sigma.w^2,0,0,0)
  vecP = solve(diag(4)-kronecker(bigF,bigF))%*%vecV
  P.ar2 = matrix(vecP,2,2)
  Sigma = matrix(0,4,3)
  Sigma[1,1] = -1
  Sigma[2:3,2:3] = P.ar2
  ssf.mod= list(mDelta=c(delta,0,0,0),
    mPhi=rbind(c(1,0,0),c(0,phi1,phi2),c(0,1,0),c(1,1,0)),
    mOmega=Omega,
    mSigma = Sigma)
  CheckSsf(ssf.mod)
}
> MNZ.start=c(0.81,1.34,-0.70,0.21,-0.30,-0.9)
> names(MNZ.start) = c("mu","phi.1","phi.2",
+ "ln.sigma.v","ln.sigma.w","rho")
> MNZ.mle = SsfFit(MNZ.start,lny.ts,"MNZ.mod",
+ lower=low.vals,upper=up.vals)
> summary(MNZ.mle)
Log-likelihood: 285.57
206 observations
Parameters:
            Value  Std. Error  t value
     delta  0.8156    0.08651   9.4280
     phi.1  1.3420    0.14550   9.2250
     phi.2 -0.7059    0.15050  -4.6890
ln.sigma.v  0.2125    0.13100   1.6230
ln.sigma.w -0.2895    0.38570  -0.7505
       rho -0.9062    0.12720  -7.1260
> MNZ.sd = exp(coef(MNZ.mle)[4:5])
> names(MNZ.sd) = c("sigma.v","sigma.w")
> MNZ.sd
sigma.v sigma.w
  1.237  0.7487
```

Listing 13.12. Estimating the MNZ model.

using exact methods based on simulation which are subject to Monte Carlo error. Both estimation approaches will be illustrated in the next sections.

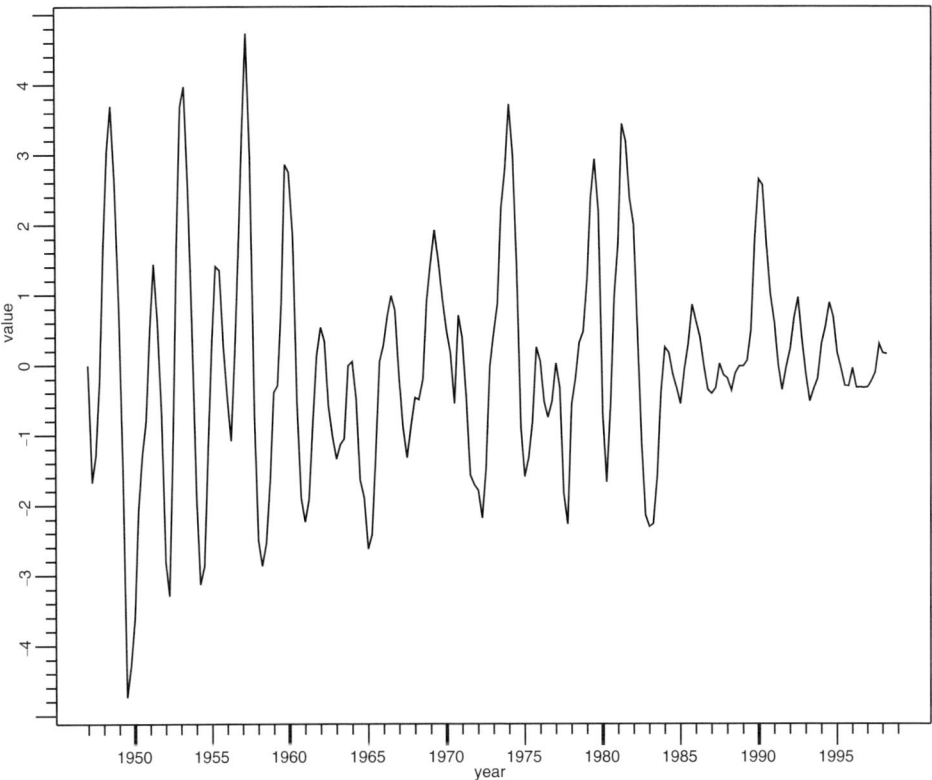

Fig. 13.11. Smooth cycle estimate, $c_{t|n}$, from the MNZ model for US real GDP.

13.6.1 Quasi-maximum likelihood estimation

Let r_t denote the continuously compounded return on an asset between times $t-1$ and t. Following Harvey, Ruiz and Shephard 1994, hereafter HRS, a simple SV model has the form

$$\begin{aligned} r_t &= \sigma_t \varepsilon_t, \quad \varepsilon_t \sim \text{iid } N(0,1), \\ h_{t+1} &= \ln \sigma_t^2 = \gamma + \phi h_t + \eta_t, \quad \eta_t \sim \text{iid } N(0, \sigma_\eta^2), \quad (13.25) \\ E[\varepsilon_t \eta_t] &= 0. \end{aligned}$$

Defining $y_t = \ln r_t^2$, and noting that $E[\ln \varepsilon_t^2] = -1.27$ and $\text{var}(\ln \varepsilon_t^2) = \pi^2/2$ an unobserved components state space representation for y_t has the form

$$\begin{aligned} y_t &= -1.27 + h_t + \xi_t, \quad \xi_t \sim \text{iid } (0, \pi^2/2), \\ h_{t+1} &= \gamma + \phi h_t + \eta_t, \quad \eta_t \sim \text{iid } N(0, \sigma_\eta^2), \\ E[\xi_t \eta_t] &= 0. \end{aligned}$$

If ξ_t were iid Gaussian then the parameters $\varphi = (\gamma, \sigma_\eta^2, \phi,)'$ of the SV model could be efficiently estimated by maximising the prediction error decomposition of the log-likelihood function constructed from the Kalman filter recursions. However, since $\xi_t = \ln \varepsilon_t^2$ is not normally distributed the Kalman filter only provides minimum mean squared error linear estimators of the state and future observations. Nonetheless, HRS pointed out that even though the exact log-likelihood cannot be computed from the prediction error decomposition based on the Kalman filter, consistent estimates of φ can still be obtained by treating ξ_t as though it were iid $N(0, \pi^2/2)$ and maximising the quasi-log-likelihood function constructed from the prediction error decomposition.

The state space representation of the SV model has system matrices

$$\boldsymbol{\delta} = \begin{pmatrix} \gamma \\ -1.27 \end{pmatrix}, \quad \boldsymbol{\Phi} = \begin{pmatrix} \phi \\ 1 \end{pmatrix}, \quad \boldsymbol{\Omega} = \begin{pmatrix} \sigma_\eta^2 & 0 \\ 0 & \pi^2/2 \end{pmatrix}.$$

Assuming that $|\phi| < 1$, the initial value matrix has the form

$$\boldsymbol{\Sigma} = \begin{pmatrix} \sigma_\eta^2/(1-\phi^2) \\ \gamma/(1-\phi) \end{pmatrix}.$$

If $\phi = 1$ then use

$$\boldsymbol{\Sigma} = \begin{pmatrix} -1 \\ 0 \end{pmatrix}.$$

The function to obtain the state space form of the SV model given a vector of parameters, assuming $|\phi| < 1$, is given in Listing 13.13. The logit transformation is used to impose the restriction $0 < \phi < 1$.

The analysis reported in Listing 13.13 starts with simulating $T = 1000$ observations from the SV model using the parameters $\gamma = -0.3556$, $\sigma_\eta^2 = 0.0312$ and $\phi = 0.9646$. The simulated squared returns, r_t^2, and latent squared volatility, σ_t^2, are shown in Figure 13.12. To estimate the underlying parameters of the simulated realisations, the starting values of $\varphi = (\gamma, \sigma_\eta^2, \phi,)'$ are chosen close to the true values. The quasi-maximum likelihood (QML) estimates are obtained using `SsfFit`. The QML estimates of σ_η^2 and ϕ are 0.02839 and 0.95088, respectively.[11] These values are fairly close to the true values.

The filtered and smoothed estimates of log-volatility and volatility may be computed using `SsfMomentEst` and `SsfCondDens`. One disadvantage of the QML approach is that the variances for log-volatility computed from the Kalman filter and smoother are not valid.

---

[11] Currently, `SsfFit` does not compute the 'sandwich' covariance matrix estimator required for the quasi-MLE.

```
sv.mod = function(parm) {
    g = parm[1]
    sigma2.n = exp(parm[2])
    phi = exp(parm[3])/(1+exp(parm[3]))
    ssf.mod = list(mDelta=c(g,-1.27),
        mPhi=as.matrix(c(phi,1)),
        mOmega=matrix(c(sigma2.n,0,0,0.5*pi^2),2,2),
        mSigma=as.matrix(c((sigma2.n/(1-phi^2)),g/(1-phi))))
    CheckSsf(ssf.mod)
}

> parm.hrs = c(-0.3556,log(0.0312),log(0.9646/0.0354))
> nobs = 1000
> set.seed(179)
> e = rnorm(nobs)
> xi = log(e^2)+1.27
> eta = rnorm(nobs,sd=sqrt(0.0312))
> sv.sim = SsfSim(sv.mod(parm.hrs),
+ mRan=cbind(eta,xi),a1=(-0.3556/(1-0.9646)))

> sv.start = c(-0.3,log(0.03),0.9)
> names(sv.start) = c("g","ln.sigma2","exp(phi)/(1+exp(phi))")

> sv.mle = SsfFit(sv.start,sv.sim[,2],"sv.mod")
Iteration 0 : objective = 5147.579
...
Iteration 32 : objective = 2218.26
RELATIVE FUNCTION CONVERGENCE
> sv.mle
Log-likelihood: 2218
1000 observations
Parameter Estimates:
        g ln.sigma2 exp(phi)/(1+exp(phi))
-0.4815574 -3.561574         2.963182
```

Listing 13.13. Estimating the SV model.

13.6.2 Simulated maximum likelihood estimation

We will consider here the estimation of the parameters of the SV model by exact maximum likelihood methods using Monte Carlo importance sampling techniques. For this purpose, we will use the following reparameterisation of the SV model

$$r_t = \sigma \exp\left(\frac{1}{2}\theta_t\right)\varepsilon_t, \quad \varepsilon_t \sim N(0,1), \qquad (13.26)$$
$$\theta_{t+1} = \phi\theta_t + \eta_t, \quad \eta_t \sim N(0,\sigma_\eta^2),$$

where σ represents average volatility. The likelihood function of this SV model can be constructed using simulation methods developed by Shephard and Pitt (1997) and Durbin and Koopman (1997) (see also Kim, Shephard

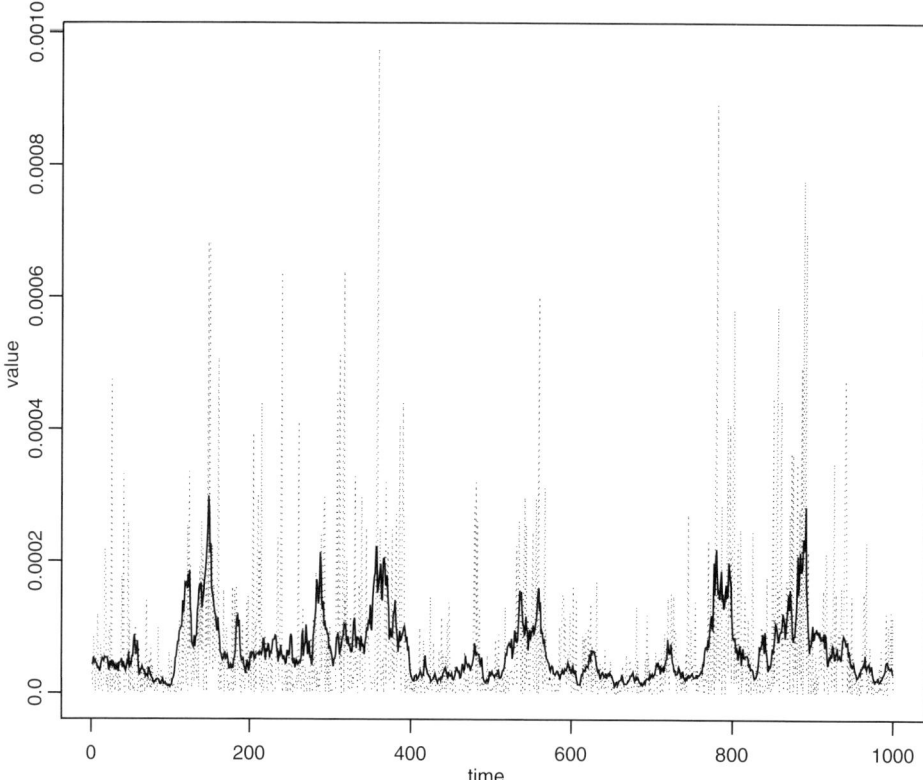

Fig. 13.12. Simulated data from SV model with volatility (solid line) and squared returns (dotted line).

and Chib (1998) for some alternative methods). The nonlinear relation between log-volatility θ_t and the observation equation of r_t does not allow the computation of the exact likelihood function by linear methods such as the Kalman filter. However, for the SV model (13.26) we can express the likelihood function as

$$L(\boldsymbol{\psi}) = p(\mathbf{y}|\boldsymbol{\psi}) = \int p(\mathbf{y}, \boldsymbol{\theta}|\boldsymbol{\psi})\mathrm{d}\boldsymbol{\theta} = \int p(\mathbf{y}|\boldsymbol{\theta}, \boldsymbol{\psi})p(\boldsymbol{\theta}|\boldsymbol{\psi})\mathrm{d}\boldsymbol{\theta}, \qquad (13.27)$$

where

$$\mathbf{y} = (r_1, \ldots, r_n)', \qquad \boldsymbol{\psi} = (\sigma, \phi, \sigma_\eta^2)', \qquad \boldsymbol{\theta} = (\theta_1, \ldots, \theta_n)'.$$

An efficient way of evaluating such expressions is by using importance sampling; see Ripley (1987, Chapter 5). A simulation device is required to sample from an importance density $\tilde{p}(\boldsymbol{\theta}|\mathbf{y}, \boldsymbol{\psi})$ which we prefer to be as close as possible to the true density $p(\boldsymbol{\theta}|\mathbf{y}, \boldsymbol{\psi})$. An obvious choice for the importance density is the conditional Gaussian density since in this case it is relatively straightforward to sample from $\tilde{p}(\boldsymbol{\theta}|\mathbf{y},\boldsymbol{\psi}) = g(\boldsymbol{\theta}|\mathbf{y},\boldsymbol{\psi})$ using simulation smoothers such as the ones developed by de Jong and Shephard (1995) and Durbin and Koopman (2002). For constructing the likelihood function using this approach, the following three steps are important:

(i) The likelihood function (13.27) is obtained by writing

$$L(\boldsymbol{\psi}) = \int p(\mathbf{y}|\boldsymbol{\theta},\boldsymbol{\psi})\frac{p(\boldsymbol{\theta}|\boldsymbol{\psi})}{g(\boldsymbol{\theta}|\mathbf{y},\boldsymbol{\psi})}g(\boldsymbol{\theta}|\mathbf{y},\boldsymbol{\psi})\mathrm{d}\boldsymbol{\theta} = \tilde{E}\left\{p(\mathbf{y}|\boldsymbol{\theta},\boldsymbol{\psi})\frac{p(\boldsymbol{\theta}|\boldsymbol{\psi})}{g(\boldsymbol{\theta}|\mathbf{y},\boldsymbol{\psi})}\right\}, \tag{13.28}$$

where $\tilde{E}$ denotes expectation with respect to the importance density $g(\boldsymbol{\theta}|\mathbf{y},\boldsymbol{\psi})$. Expression (13.28) can be simplified using a suggestion of Durbin and Koopman (1997). This leads to

$$L(\boldsymbol{\psi}) = L_g(\boldsymbol{\psi})\tilde{E}\left\{\frac{p(\mathbf{y}|\boldsymbol{\theta},\boldsymbol{\psi})}{g(\mathbf{y}|\boldsymbol{\theta},\boldsymbol{\psi})}\right\}, \tag{13.29}$$

which is a convenient expression that we will use in the calculations. The likelihood function of the approximating Gaussian model $L_g(\boldsymbol{\psi})$ can be calculated via the Kalman filter. The conditional density functions $p(\mathbf{y}|\boldsymbol{\theta},\boldsymbol{\psi})$ and $g(\mathbf{y}|\boldsymbol{\theta},\boldsymbol{\psi})$ are obviously easy to compute for given values of $\boldsymbol{\theta}$ and $\boldsymbol{\psi}$. It follows that the likelihood function of the SV model is equivalent to the likelihood function of an approximating Gaussian model, multiplied by a correction term. This correction term only needs to be evaluated via simulation.

(ii) An obvious estimator for the likelihood of the SV model is

$$\hat{L}(\boldsymbol{\psi}) = L_g(\boldsymbol{\psi})\bar{w}, \tag{13.30}$$

where

$$\bar{w} = \frac{1}{M}\sum_{i=1}^{M} w_i, \quad w_i = \frac{p(\mathbf{y}|\boldsymbol{\theta}^i,\boldsymbol{\psi})}{g(\mathbf{y}|\boldsymbol{\theta}^i,\boldsymbol{\psi})}, \tag{13.31}$$

and $\boldsymbol{\theta}^i$ denotes a draw from the importance density $g(\boldsymbol{\theta}|\mathbf{y},\boldsymbol{\psi})$. The accuracy of this estimator depends on M, the number of simulation samples. In practice, we usually work with the log of the likelihood function to manage the magnitude of density values. The log transformation of $\hat{L}(\boldsymbol{\psi})$ introduces bias for which we can correct up to order $O(M^{-3/2})$. We obtain

$$\ln \hat{L}(\boldsymbol{\psi}) = \ln L_g(\boldsymbol{\psi}) + \ln \bar{w} + \frac{s_w^2}{2M\bar{w}^2}, \tag{13.32}$$

with $s_w^2 = (M-1)^{-1}\sum_{i=1}^{M}(w_i - \bar{w})^2$.

(iii) Denote

$$p(\mathbf{y}|\theta) = \prod_{t=1}^{T} p_t, \quad g(\mathbf{y}|\theta) = \prod_{t=1}^{T} g_t,$$

where $p_t = p(y_t|\theta_t)$ and $g_t = g(y_t|\theta_t)$. The importance density is based on the linear Gaussian model

$$y_t = h_t + u_t, \quad u_t \sim N(c_t, d_t), \quad (13.33)$$

where c_t and d_t are chosen such that the first and second derivatives of p_t and g_t are equal for $t = 1, \ldots, n$. These conditions lead to n nonlinear equations which we solve by a Newton–Raphson scheme of optimisation. This involves a sequence of Kalman filter smoothers. Convergence is usually fast.

Once the maximum likelihood estimates are obtained, smoothed estimates of volatility may be computed using Monte Carlo integration with importance sampling based on the weights w_i in (13.31).

The S-PLUS code to estimate the SV model (13.26) by maximising the simulated log-likelihood function (13.32) is somewhat involved so we do not list all of the details here.[12] The key computations involve evaluating (13.33) and (13.32). For (13.33), the linear Gaussian model is an AR(1) model with autoregressive coefficient ϕ and heteroskedastic errors. This may be constructed using the `SsfPack` function `GetSsfArma` and then modifying the state space representation to allow for time varying variances. The solution for c_t and d_t involves a loop in which `SsfCondDens` is called repeatedly until convergence is achieved. To evaluate (13.32), the term $\ln L_g(\psi)$ is computed using `SsfLoglike` and the weights w_i are computed by drawing θ^i using `SimSmoDraw`. As noted in Durbin and Koopman (2002), antithetic variates may be used to improve the efficiency of $\bar{w}$.

To illustrate, consider estimating the SV model (13.26) for daily log returns on the UK/US spot exchange rate over the period 1 October 1981 through 28 June 1985. The MLEs for the elements of $\psi = (\sigma, \phi, \sigma_\eta^2)'$ are found to be $\hat{\sigma} = 0.6352$, $\hat{\phi} = 0.9744$ and $\hat{\sigma}_\eta^2 = 0.0278$. Figure 13.13 shows the absolute log returns along with the smoothed volatility estimates and 95% error bands.

---

[12] The code is contained in the file `ssStochasticVolatility.ssc`, available on Eric Zivot's web page, and replicates the Ox code in the files `ssfnong.ox` and `sv_mcl_est.ox` written by Siem Jan Koopman.

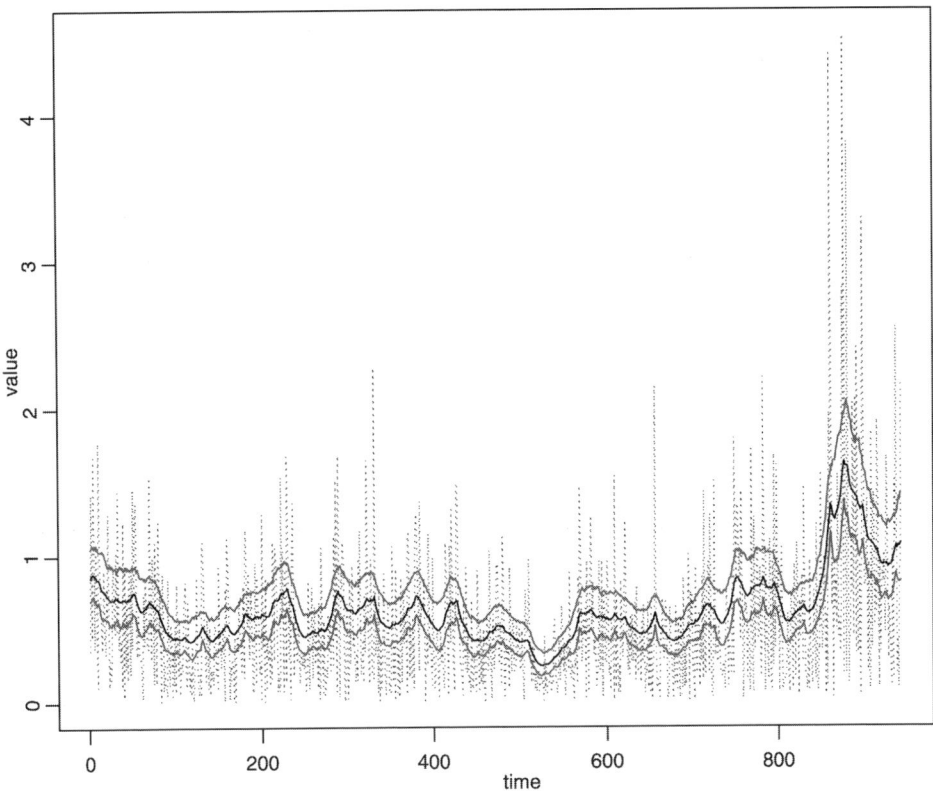

Fig. 13.13. Smoothed volatility (solid line) with 95% confidence interval from maximum simulated likelihood estimates of SV model fit to daily UK/US log returns (dotted line).

13.7 Term structure models

This section illustrates how some common affine term structure models may be expressed in state space form, estimated and evaluated using the Kalman filter and smoothing algorithms in `SsfPack`. The notation and examples are taken from Duan and Simonato (1999).

13.7.1 Affine term structure models

Traditionally the study of the term structure of interest rates focuses on either the cross sectional aspect of the yield curve, or the time series properties of the interest rate. Recently, researchers have utilised state space models and Kalman filtering techniques to estimate affine term structure models by combining both time series and cross sectional data. For simple models, the state space representation is often linear and Gaussian and analysis is straightforward. For more general models, the unobserved state variables generally influence the variance of the transition equation errors making the errors non-Gaussian. In these cases, nonstandard state space methods are necessary.

Duffie and Kan (1996) showed that many of the theoretical term structure models, such as the Vasicek (1977) Ornstein–Uhlenbeck model, the Cox, Ingersoll and Ross (1985) square root diffusion model and its multifactor extensions (for example, see Chen and Scott (1993)), the Longstaff and Schwartz (1992) two-factor model, and the Chen (1996) three-factor model, are special cases of the class of affine term structure models. The class of affine term structure models is one in which the yields to maturity on default free pure discount bonds and the instantaneous interest rate are affine (constant plus linear term) functions of m unobservable state variables $\mathbf{X}_t$, which are assumed to follow an affine diffusion process

$$d\mathbf{X}_t = \mathbf{U}(\mathbf{X}_t; \mathbf{\Psi})dt + \Sigma(\mathbf{X}_t; \mathbf{\Psi})d\mathbf{W}_t, \qquad (13.34)$$

where $\mathbf{W}_t$ is an $m \times 1$ vector of independent Wiener processes; $\mathbf{\Psi}$ is a $p \times 1$ vector of model specific parameters; $\mathbf{U}(\cdot)$ and $\Sigma(\cdot)$ are affine functions in $\mathbf{X}_t$ such that (13.34) has a unique solution. In general, the functions $\mathbf{U}(\cdot)$ and $\Sigma(\cdot)$ can be obtained as the solution to some ordinary differential equations. Only in special cases are closed form solutions available. In this class of models, the price at time t of a default free pure discount bond with time to maturity τ has the form

$$P_t(\mathbf{X}_t; \mathbf{\Psi}, \tau) = A(\mathbf{\Psi}, \tau) \exp\left\{-\mathbf{B}(\mathbf{\Psi}, \tau)'\mathbf{X}_t\right\}, \qquad (13.35)$$

where $A(\tau, \mathbf{\Psi})$ is a scalar function and $\mathbf{B}(\tau, \mathbf{\Psi})$ is an $m \times 1$ vector function. The time-t continuously compounded yield-to-maturity on a pure discount bond with time to maturity τ is defined as

$$Y_t(\mathbf{X}_t; \mathbf{\Psi}, \tau) = -\frac{\ln P_t(\mathbf{X}_t; \mathbf{\Psi}, \tau)}{\tau}, \qquad (13.36)$$

which, using (13.35), has the affine form

$$Y_t(\mathbf{X}_t; \mathbf{\Psi}, \tau) = -\frac{\ln A(\mathbf{\Psi}, \tau)}{\tau} + \frac{\mathbf{B}(\mathbf{\Psi}, \tau)'\mathbf{X}_t}{\tau}. \qquad (13.37)$$

13.7.2 State space representation

Although (13.37) dictates an exact relationship between the yield $Y_t(\tau)$ and the state variables $\mathbf{X}_t$, in econometric estimation it is usually treated as an approximation giving rise to the measurement equation

$$Y_t(\tau) = -\frac{\ln A(\boldsymbol{\Psi}, \tau)}{\tau} + \frac{\mathbf{B}(\boldsymbol{\Psi}, \tau)'\mathbf{X}_t}{\tau} + \epsilon_t(\tau), \qquad (13.38)$$

where ϵ_t is a normally distributed measurement error with zero mean and variance σ_τ^2. For any time to maturity τ, the above equation can be naturally treated as the measurement equation of a state space model, with $\mathbf{X}_t$ being the unobserved state variable. To complete the state space representation, the transition equation for $\mathbf{X}_t$ over a discrete time interval h needs to be specified. Defining $\boldsymbol{\Phi}(\mathbf{X}_t; \boldsymbol{\Psi}, h) = \text{var}(\mathbf{X}_{t+h}|\mathbf{X}_t)$, Duan and Simonato (1999) showed that the transition equation for $\mathbf{X}_t$ has the form

$$\mathbf{X}_{t+h} = \mathbf{a}(\boldsymbol{\Psi}, h) + b(\boldsymbol{\Psi}, h)\mathbf{X}_t + \boldsymbol{\Phi}(\mathbf{X}_t; \boldsymbol{\Psi}, h)^{1/2}\boldsymbol{\eta}_{t+h}, \qquad (13.39)$$

where $\boldsymbol{\eta}_t \sim$ iid $N(\mathbf{0}, \mathbf{I}_m)$, and $\boldsymbol{\Phi}(\mathbf{X}_t; \boldsymbol{\Psi}, h)^{1/2}$ represents the Cholesky factorization of $\boldsymbol{\Phi}(\mathbf{X}_t; \boldsymbol{\Psi}, h)$.

In general, the state space model defined by (13.38) and (13.39) is non-Gaussian because the conditional variance of $\mathbf{X}_{t+h}$ in (13.39) depends on $\mathbf{X}_t$. Only for the special case in which $\boldsymbol{\Sigma}(\cdot)$ in (13.34) is not a function of $\mathbf{X}_t$ is the conditional variance term $\boldsymbol{\Phi}(\mathbf{X}_t; \boldsymbol{\Psi}, h)$ also not a function of $\mathbf{X}_t$ and the state space model is Gaussian.[13] See Lund (1997) for a detailed discussion of the econometric issues associated with estimating affine term structure models using the Kalman filter. Although the quasi-maximum likelihood estimator of the model parameters based on the modified Kalman filter is inconsistent, the Monte Carlo results in Duan and Simonato (1999) and de Jong (2000) show that the bias is very small even for the moderately small samples likely to be encountered in practice.

13.7.3 Estimation of Vasicek's model

The data used for this example are in the S+FinMetrics 'timeSeries' fama.bliss, and consist of four monthly yield series over the period April 1964–December 1997 for the US treasury debt securities with maturities of 3, 6, 12 and 60 months, respectively. These data were also used by Duan and

---

[13] To estimate the non-Gaussian state space model, Duan and Simonato (1999) modified the Kalman filter recursions to incorporate the presence of $\boldsymbol{\Phi}(\mathbf{X}_t; \boldsymbol{\Psi}, h)$ in the conditional variance of $\boldsymbol{\eta}_{t+h}$. The SsfPack functions KalmanFil and SsfLoglike can be modified to accommodate this modification.

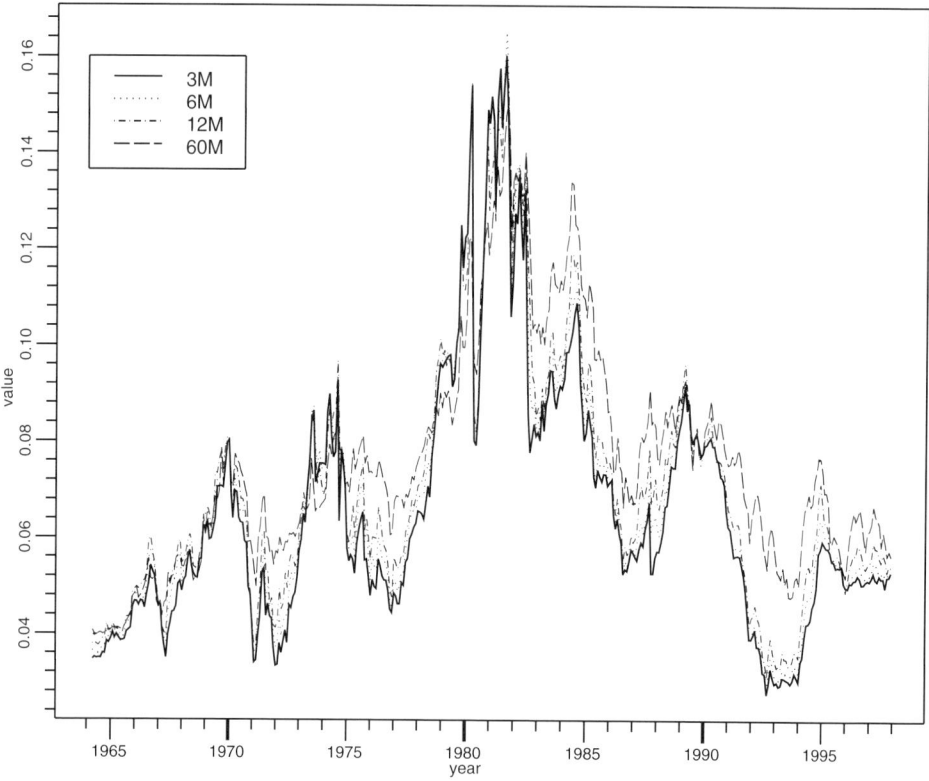

Fig. 13.14. Monthly yields on US treasury debt securities.

Simonato (1999). All rates are continuously compounded rates expressed on an annual basis. These rates are displayed in Figure 13.14.

In the Vasicek (1977) model, the state variable driving the term structure is the instantaneous (short) interest rate, r_t, and is assumed to follow the mean-reverting diffusion process

$$\mathrm{d}r_t = \kappa(\theta - r_t)\mathrm{d}t + \sigma \mathrm{d}W_t, \quad \kappa \geq 0, \sigma > 0, \qquad (13.40)$$

where W_t is a scalar Wiener process, θ is the long-run average of the short rate, κ is a speed of adjustment parameter and σ is the volatility of r_t. Duan and Simonato (1999) showed that the functions $A(\cdot), B(\cdot), a(\cdot), b(\cdot)$ and

$\Phi(\cdot)$ have the forms

$$\ln A(\Psi, \tau) = \gamma(B(\Psi, \tau) - \tau) - \frac{\sigma^2 B^2(\Psi, \tau)}{4\kappa},$$

$$B(\Psi, \tau) = \frac{1}{\kappa}(1 - \exp(-\kappa\tau)),$$

$$\gamma = \theta + \frac{\sigma\lambda}{\kappa} - \frac{\sigma^2}{2\kappa^2},$$

$$a(\Psi, h) = \theta(1 - \exp(-\kappa h)), \; b(\Psi, h) = \exp(-\kappa h),$$

$$\Phi(X_t; \Psi, h) = \Phi(\Psi, h) = \frac{\sigma^2}{2\kappa}(1 - \exp(-2\kappa h)),$$

where λ is the risk premium parameter. The model parameters are $\Psi = (\kappa, \theta, \sigma, \lambda)'$. Notice that for the Vasicek model, $\Phi(X_t; \Psi, h) = \Phi(\Psi, h)$ so that the state variable r_t does not influence the conditional variance of transition equation errors, the state space model is Gaussian.

The state space representation of the Vasicek model has system matrices

$$\delta = \begin{pmatrix} a(\Psi, h) \\ -\ln A(\Psi, \tau_1)/\tau_1 \\ \vdots \\ -\ln A(\Psi, \tau_4)/\tau_4 \end{pmatrix}, \; \Phi = \begin{pmatrix} b(\Psi, h) \\ B(\Psi, \tau_1)/\tau_1 \\ \vdots \\ B(\Psi, \tau_4)/\tau_4 \end{pmatrix}, \quad (13.41)$$

$$\Omega = \text{diag}(\Phi(\Psi, h), \sigma^2_{\tau_1}, \ldots, \sigma^2_{\tau_4})$$

and initial value matrix

$$\Sigma = \begin{pmatrix} \theta \\ \sigma^2/2\kappa \end{pmatrix}$$

based on the stationary distribution of the short rate in (13.40). Notice that this a multivariate state space model.

A function to compute the state space form of the Vasicek model for a given set of parameters Ψ, number of yields $\tau_1, \ldots, \tau_N$, and sampling frequency h is given in Listing 13.14. Notice that the exponential transformation is used for those parameters that should be positive, and, since the data in `fama.bliss` are monthly, the default length of the discrete sampling interval, h, is set to $1/12$.

An implementation of the Vasicek model is provided in Listing 13.15. Specific starting values for the parameters

$$\varphi = (\ln \kappa, \ln \theta, \ln \sigma, \lambda, \ln \sigma_{\tau_1}, \ln \sigma_{\tau_2}, \ln \sigma_{\tau_3}, \ln \sigma_{\tau_4})'$$

and the maturity specification for the yields are given and maximum

```
vasicek.ssf = function(param, tau=NULL, freq=1/52)
{
## 1. Check for valid input.
  if (length(param) < 5)
    stop("param must have length greater than 4.")
  N = length(param) - 4
  if (length(tau) != N)
    stop("Length of tau is inconsistent with param.")
## 2. Extract parameters and impose constraints.
  Kappa = exp(param[1]) ## Kappa > 0
  Theta = exp(param[2]) ## Theta > 0
  Sigma = exp(param[3]) ## Sigma > 0
  Lamda = param[4]
  Var = exp(param[1:N+4]) ## meas eqn stdevs
## 3. Compute Gamma, A, and B.
  Gamma = Theta + Sigma * Lamda / Kappa - Sigma^2 / (2 * Kappa^2)
  B = (1 - exp(-Kappa * tau)) / Kappa
  lnA = Gamma * (B - tau) - Sigma^2 * B^2 / (4 * Kappa)
## 4. Compute a, b, and Phi.
  a = Theta * (1 - exp(-Kappa * freq))
  b = exp(-Kappa * freq)
  Phi = (Sigma^2 / (2 * Kappa)) * (1 - exp(-2 * Kappa * freq))
## 5. Compute the state space form.
  mDelta = matrix(c(a, -lnA/tau), ncol=1)
  mPhi = matrix(c(b, B/tau), ncol=1)
  mOmega = diag(c(Phi, Var^2))
## 6. Duan and Simonato used this initial setting.
  A0 = Theta
  P0 = Sigma * Sigma / (2*Kappa)
  mSigma = matrix(c(P0, A0), ncol=1)
## 7. Return state space form.
  ssf.mod =
  list(mDelta=mDelta, mPhi=mPhi, mOmega=mOmega, mSigma=mSigma)
  CheckSsf(ssf.mod)
}
```

Listing 13.14. Computing the state space form for Vasicek model.

likelihood estimates for the parameters are obtained using SsfFit. The maximum likelihood estimates and asymptotic standard errors for the model parameters

$$\boldsymbol{\theta} = (\kappa, \theta, \sigma, \lambda, \sigma_{\tau_1}, \sigma_{\tau_2}, \sigma_{\tau_3}, \sigma_{\tau_4})'$$

computed using the delta method are reported as output in Listing 13.15.

These results are almost identical to those reported by Duan and Simonato (1999). All parameters are significant at the 5% level except the measurement equation standard deviation for the six-month maturity yield. The largest measurement equation error standard deviation is for the sixty-month yield, indicating that the model has the poorest fit for this yield. The short rate is mean reverting since $\hat{\kappa} > 0$, and the long-run average short rate is $\hat{\theta} = 5.74\%$ per year. The estimated risk premium parameter, $\hat{\lambda} = 0.3477$, is positive indicating a positive risk premium for bond prices.

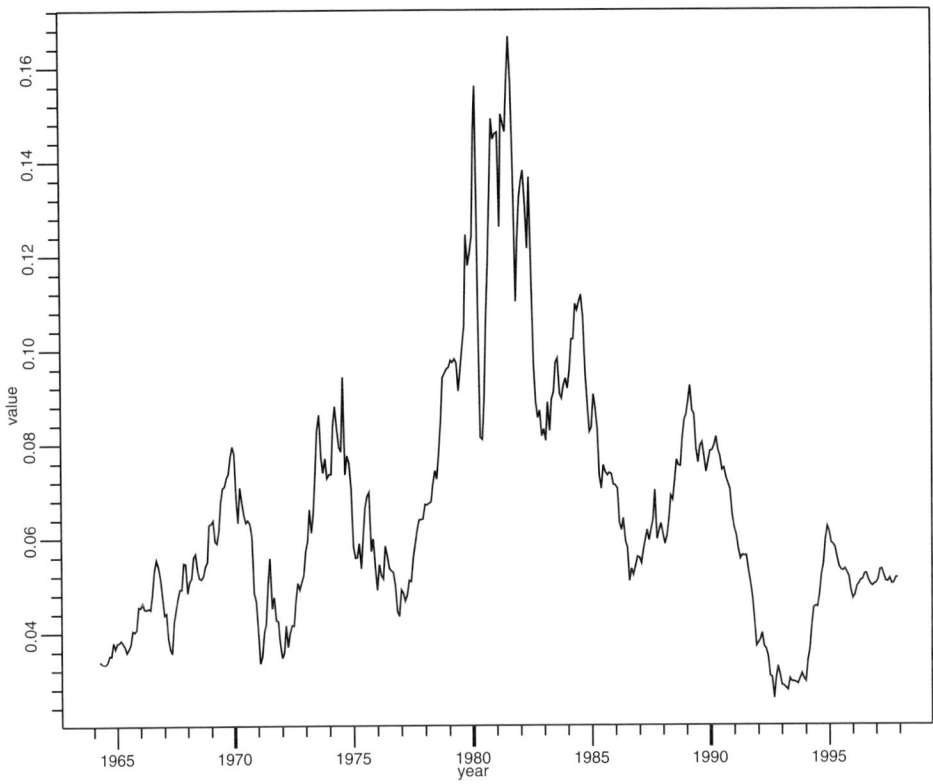

Fig. 13.15. Smoothed estimate of short rate r_t from (13.40).

The smoothed estimates of the short rate and the yields are computed using SsfCondDens with task="STSMO". Figure 13.15 gives the smoothed estimate of the instantaneous short rate r_t from (13.40). The differences between the actual and smoothed yield estimates are displayed in Figure 13.16. The model fits well on the short end of the yield curve but poorly on the long end.

As another check on the fit of the model, the presence of serial correlation in the standardised innovations is tested using the Box–Ljung modified Q-statistic (computed using the S+FinMetrics function autocorTest). The null of no serial correlation is easily rejected for the standardised innovations of all yields.

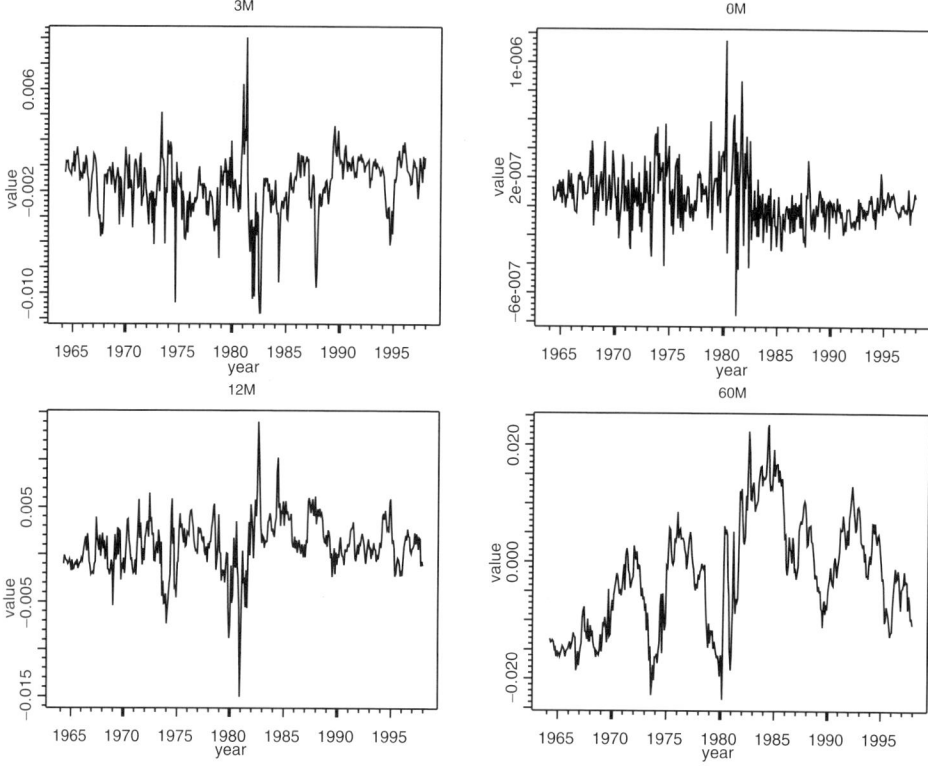

Fig. 13.16. Smoothed estimated of yields from Vasicek term structure model.

13.8 Conclusion

This paper provides an overview of the SsfPack state space functions available in S+FinMetrics. The functions can be used to implement the state space methods and algorithms presented in, for example, Durbin and Koopman (2001). We also illustrate the use of the SsfPack functions through several examples from macroeconomics and finance. With powerful easy-to-use software, a wide variety of state space models are straightforward to estimate and analyse.

State space modelling in macroeconomics and finance 335

```
> start.vasicek = c(log(0.1), log(0.06), log(0.02), 0.3,
log(0.003),
+ log(0.001), log(0.003), log(0.01))
> names(start.vasicek) =
c("ln.kappa","ln.theta","ln.sigma","lamda",
+ "ln.sig.3M","ln.sig.6M","ln.sig.12M","ln.sig.60M")
> start.tau = c(0.25, 0.5, 1, 5)

> ans.vasicek = SsfFit(start.vasicek, fama.bliss, vasicek.ssf,
+ tau=start.tau, freq=1/12, trace=T,
+ control=nlminb.control(abs.tol=1e-6, rel.tol=1e-6,
+ x.tol=1e-6, eval.max=1000, iter.max=500))
Iteration 0 : objective = -6347.453
...
Iteration 37 : objective = -6378.45

> ssf.fit =
  vasicek.ssf(ans.vasicek$parameters,tau=start.tau,freq=1/12)

Log-likelihood: -6378.45
1620 observations
Parameters:
            Value    Std. Error  t value
   kappa  0.11880000 0.0106300  11.1700
   theta  0.05729000 0.0269000   2.1300
   sigma  0.02139000 0.0007900  27.0800
   lamda  0.34800000 0.1500000   2.3200
  sig.3M  0.00283500 0.0001011  28.0500
  sig.6M  0.00001773 0.0001148   0.1544
 sig.12M  0.00301700 0.0001083  27.8600
 sig.60M  0.00989800 0.0003703  26.7300

> autocorTest(KalmanFil(fama.bliss,ssf.fit)$std.innov)

Test for Autocorrelation: Ljung-Box
Null Hypothesis: no autocorrelation
Test Statistics:
                3M        6M        12M       60M
Test Stat   80.9471  282.4316   756.3304  3911.7736
p.value      0.0000    0.0000     0.0000     0.0000

Dist. under Null: chi-square with 26 degrees of freedom
    Total Observ.: 405
```

Listing 13.15. An analysis of the Vasicek model.

14

Finding genes in the human genome with hidden Markov models

Richard Durbin
Wellcome Trust Sanger Institute

Abstract

Large scale genome sequencing generates very long sequences of DNA bases, which we can view as sequences over a four letter alphabet. An important analysis goal is to identify the genes or subsequences that code for proteins. We can view this as a statistical inference problem, where we have a model for gene structure and the sequence properties of gene-encoding regions, and many examples of known genes. Here I review the approach taken in most recent gene finding methods, based on hidden Markov models, which are a type of discrete state space model. I also discuss ways that further information can be incorporated, including a comparative sequence from a related organism, such as mouse for human sequence analysis.

State Space and Unobserved Component Models: Theory and Applications, eds. Andrew C. Harvey, Siem Jan Koopman and Neil Shephard. Published by Cambridge University Press. © Cambridge University Press 2004

14.1 Introduction and background

I would like to preface this short review by saying what a pleasure it is to be able to contribute to this collection of papers associated with the work of my father, Jim Durbin. Although on the surface I have followed a very different career path than my father, working on computational and mathematical issues connected to molecular biology and genetics, there are some surprising parallels between the methods we have been using in the last few years. Just as the formalism of state space models has proved powerful and flexible for encompassing and generalising a variety of older approaches to time series analysis, a discrete space, discrete observation version of the state space model, the hidden Markov model (HMM), has come to play a significant role in unifying earlier approaches to biological sequence analysis and stimulating new developments, see Durbin, Eddy, Krogh and Mitchison (1998). In this paper I will illustrate this by describing the use of HMMs and related methods for the statistical prediction of gene structures in genomic DNA sequence, which is a key problem in making use of all the new data from the human genome project.

We are now in the era of sequencing genomes. Within a period of ten years or so starting in 1995, we will have obtained reference genome sequences of most of the organisms studied intensively in biological research, including of course man (see Table 14.1). This outpouring of scientific data is providing many millions of base pairs of DNA sequence, which for each species contain the genetic information typically inherited by an individual from its parents. There are of course genetic variations within a species, meaning that individuals have distinct sequences, but for man these are of the order of 0.1% variation, and here we are concerned with what is in common for the species, as represented by the reference sequence.

Amongst the millions of bases in the sequence are the genes, which are local regions of the genome sequence that have specific, separable functions. For the current purposes we will only consider protein coding genes, and furthermore only consider the sections of them that code for protein. Figure 14.1 shows an example, and introduces the terms *exon* and *intron*, which describe respectively the segments of the genomic sequence that contribute to the protein product, and the intervening sequences. (For simplicity, and in common with much of the gene finding literature, I am restricting the term 'exon' to mean only the part that is protein coding, rather than the fuller segments of the mature RNA transcript as standardly used by molecular biologists.)

If we know the structure of a protein coding gene as indicated in Figure

Table 14.1. *Sizes and times of completion for various genomes that have been, or are being, sequenced*

	size (Mb)	genes			completion date
H. influenzae	2	1 700	1/1kb	bacterium	1995
yeast	13	6 000	1/2kb	eukaryotic cell	1996
nematode	100	18 000	1/6kb	metazoon	1998
human	3000	30 000	1/100kb	mammal	2000/3
mouse, fish (3), rat, another worm, fly				drafts available in 2002/3	
20 others, e.g. apes, agricultural species				drafts available in 2003/5	

14.1, that is, the position of the exon boundaries including the start at the beginning of the first exon and stop at the end of the last exon, then we can deterministically predict the protein sequence encoded by the gene, by translating the concatenated exon sequence using the standard triplet code (each set of three bases, called a *codon*, specifies a particular amino acid). Therefore, if we can find all the gene structures, we can predict the sequences of all the proteins, a very major step in providing a 'parts list' of cellular components used to build an organism. Typically it is these proteins that are, for example, the targets of drugs.

There are experimental ways to determine gene structures, by obtaining the sequence of the spliced product, so we know the structures of many genes. However, these experimental approaches don't find all genes, and so purely computational approaches to identifying genes are important in finding new genes.

14.2 Hidden Markov models (HMMs)

The approach that is taken by essentially all current computational methods for finding genes is to build a probabilistic model for gene structures, from which a probability of a genomic sequence can be calculated given the gene structure. Then either the likelihood of the sequence is maximised over all structures, or a Bayesian approach can be taken to find posterior probabilities for specific components of gene structures such as individual exons. The natural form of model is a discrete state space, discrete observation space state space model, which is called an HMM. We can think of an HMM as generating the sequence from left to right. The simplest form of gene-finding

Fig. 14.1. An example of a protein coding gene. Following transcription into RNA, the shaded regions, called exons, are concatenated by the splicing process, then translated into protein, each triplet determining one amino acid. There are partially conserved signals at the start and end, and at the boundaries between the exons and the intervening sequences, called introns. These boundaries are called splice sites. In addition, the coding regions have different statistical properties from the surrounding DNA, typically modelled by a fifth order Markov process. This gene is from the worm *C. elegans*. Typical human genes have more exons, on average nearer ten, and longer introns (see Figures 14.2 and 14.3), and frequently are spread across tens of thousands of bases.

HMM might have three states, intergenic, exon and intron, corresponding to the three types of region in a gene structure. These states would be connected by a simple Markov chain, with a single base being generated at each time point with the distribution conditional on the current state, and then a transition to the state for the next time point (which is often the same as the current state). Thinking of this as a generative model for sequence, we say that a base is 'emitted' from a state at each time point.

This simple model is very weak, sensitive only to nucleotide frequency differences in the three different types of region, and implying a geometric length distribution for each type of region. To capture important properties of known gene structures we should model higher order sequence properties (e.g., frequencies of triplets or longer short strings), nongeometric length distributions, and some long range information between states (e.g., the triplet phase must be maintained between neighbouring exons, bridging across the intervening intron). These more complex properties can be achieved by a combination of increasing the number of states, allowing multiple bases to be emitted in a single time step, and making emissions and/or state transitions conditional on the past history of the emitted sequence as well as the current state. Practical methods use a combination of these techniques, varying from method to method and I will discuss some of them further below. However, I will first give the standard equations for the simplest form of HMM, since these are easiest to state and consider, and for the most part the versions required for the more complex models are natural extensions. A classic exposition of HMM theory is given in Rabiner (1989b), whilst a version in the context of biological sequence analysis is available in Durbin, Eddy, Krogh and Mitchison (1998). A version that may be more accessible to statisticians and econometricians can be found in MacDonald and Zucchini (1997).

Let us assume we have an observed sequence

$$x = x_1, x_2, x_3, \ldots, x_L,$$

where

$$x_i \in \{A, C, G, T\}.$$

There is a corresponding state path

$$\pi = \pi_1, \pi_2, \pi_3, \ldots, \pi_i, \ldots, \pi_{L+1}, \quad \pi_i \in 0, \ldots, K,$$

where I will use a convention that state 0 is a special nonemitting state used only at the start and end of the sequence (this simplifies many of the equations). The state path π evolves according to the stationary Markov chain:

$$P(\pi_{i+1} = k \mid \pi_i = j) = a_{jk}.$$

Base b is observed (emitted) from state k with probability $e_k(b)$, so

$$P(x_i = b) = e_{\pi_i}(b).$$

The joint probability of a path π and sequence x is therefore the product of the path probability independent of sequence and the probability of the sequence given the path. In gene prediction terms this is the probability of the gene structure *per se* times the probability of the sequence given that structure. We can calculate the full probability $P(x)$ of the sequence integrating over all possible paths using the following recursion:

Algorithm Forward algorithm

$$\text{Initialisation } (i=0): \quad f_0(0)=1, \quad f_k(0)=0 \text{ for } k>0,$$
$$\text{Recursion } (i=1,\ldots,L): \quad f_l(i)=e_l(x_i)\sum_k f_k(i-1)a_{kl}, \quad (14.1)$$
$$\text{Termination:} \quad P(x)=\sum_k f_k(L)a_{k0},$$

where

$$f_k(i) = P(x_1 x_2 x_3 \ldots x_i, \pi_i = k).$$

Frequently we are interested in the maximum likelihood path π^*, which can be obtained using the Viterbi algorithm:

Algorithm Viterbi

$$\text{Initialisation } (i=0): \quad v_0(0)=1, v_k(0)=0 \text{ for } k>0,$$
$$\text{Recursion } (i=1,\ldots,L): \quad \begin{aligned} v_l(i) &= e_l(x_i)\max_k(v_k(i-1)a_{kl}), \\ \text{ptr}_i(l) &= \arg\max_k(v_k(i-1)a_{kl}), \end{aligned}$$
$$\text{Termination:} \quad \begin{aligned} P(x,\pi^*) &= \max_k(v_k(L)a_{k0}), \\ \pi^*_L &= \arg\max_k(v_k(L)a_{k0}), \end{aligned} \quad (14.2)$$
$$\text{Traceback } (i=L,\ldots,1): \quad \pi^*_{i-1} = \text{ptr}_i(\pi^*_i),$$

where $\arg\max_k(F(k))$ is the value of k that maximises $F(k)$. The state space valued variable $\text{ptr}_i(l)$ therefore holds the state that would need to have been used at the preceding point $i-1$ to achieve the most likely subpath ending in state l at point i. These 'pointer' variables are used in the traceback step to retrieve the optimal path. (If more than one choice has the same maximum probability at any point an arbitrary choice is normally made; it is possible to keep track of all maximal paths with some extra bookkeeping.)

The Viterbi algorithm is an example of what is known as a 'dynamic programming' recursion in computer science. It corresponds to the smoothing process for continuous state space models. It is also possible to calculate posterior probabilities for state assignments and transitions using the forward variables $f_k(i)$ in conjunction with corresponding backward variables

$$b_k(i) = P(\pi_i = k, x_{i+1}, x_{i+2}, x_{i+3}, \ldots, x_L),$$

which are obtained by an analogous process to the forward algorithm but starting at the end of the sequence. For example,

$$P(\pi_i = k \mid x) = f_k(i)b_k(i)/P(x).$$

The parameters a_{jk} and $e_k()$ are normally estimated from labelled data for which the state assignments are known, but there is an expectation-maximisation (EM) algorithm known as Baum–Welch that can be applied to estimate them from unlabelled data, treating the state assignments π_i as missing data. Longer expositions of the Viterbi and Baum–Welch algorithms are presented in the books mentioned above.

These algorithms are all linear in the length of the sequence x, and quadratic in the number of states (in fact linear in the number of non-zero transitions). Although simple implementations require storage space linear in the product of the sequence length and the number of states, there are so called 'linear space' implementations that use space linear in the number of states independent of sequence length, at the cost of a factor of 2 increase in time. These considerations are important when a typical sequence analysed might be millions of bases long.

14.3 Enhancements to the basic model

As mentioned above, a straightforward implementation of an HMM for gene finding is very weak. One reason is that consecutive bases are not independent, even conditioned on state. All practical gene-finding models therefore support a higher order probabilistic model for the sequence, where the distribution of emission probabilities for a base depends on recent history. At first sight this appears to break the Markov property, or equivalently require a combinatorial explosion of the state space by maintaining a state vector of the last k state values. This would be true if the probability were dependent on the state history. However, most implementations make the emission distribution dependent on the recent history of the sequence x, and only the current value of the state. This keeps the size of the state space small, although it increases the number of parameters, and hence the requirement for training data. The amount of history used varies. Typically five bases are used in the exons, with a cyclical three state model to represent the three positions in a codon because these have different properties, whereas often only one base of history might be used in introns and intergenic sequences.

Another enhancement can be achieved by the notion that there is a lot of information in the imperfect consensus sequence at the splice sites between exons and introns. One way to capture this is to have special states just before and for a little time after the splice sites. These states can then be higher order in the sense of the previous paragraph, and hence can capture an extended pattern, perhaps 10 or 20 bases long. This approach is taken in some models, but an alternative is to adjust the transition probabilities between exon and intron states, by precalculating sequence-dependent values using some other pattern recognition process such as a neural network, decision tree or discriminant based approach. This allows the problem of modelling splice sites to be separated from the gene finding model. To correctly maintain the validity of the algorithms, care should be taken to preserve conditional independence when using these externally derived transition probabilities. In practice there is frequently some sloppiness here, although it is not believed that this causes significant problems. Similar approaches can be taken at the start and stop of the gene, which are the boundaries between intergenic regions and exons.

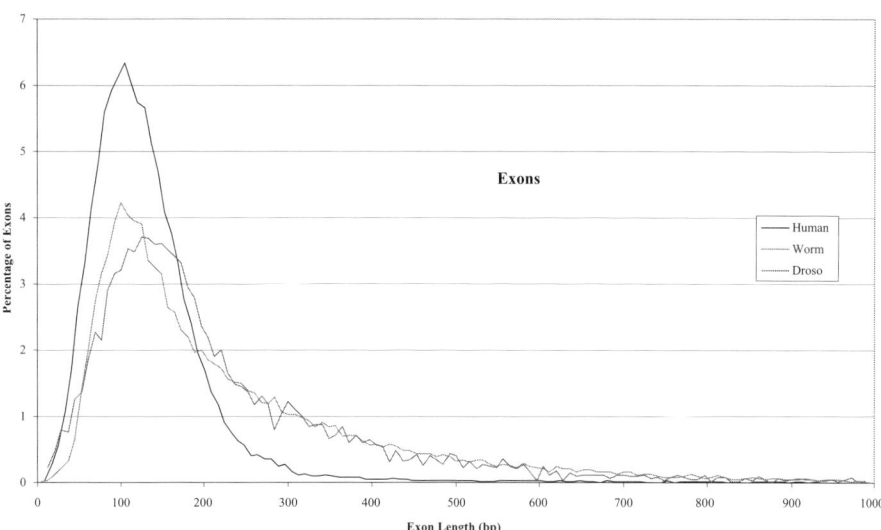

Fig. 14.2. The length distributions of *exons* in human, worm (*Caenorhabditis elegans*) and fruit fly (*Drosophila melanogaster*). Exons have the same mode in all species, but longer tails in worm and fly. Data reproduced with modification from International Human Genome Sequencing Consortium (2001).

A third area of complexity concerns the length distributions of some of the labelled regions. By default, the length distribution for staying in a simple HMM state with a self-recurring loop is geometrically distributed. Figures 14.2 and 14.3 show the length distributions of exons and introns from three species including human; they clearly have modes well away from zero and are not geometrically distributed. One possibility would be to model a

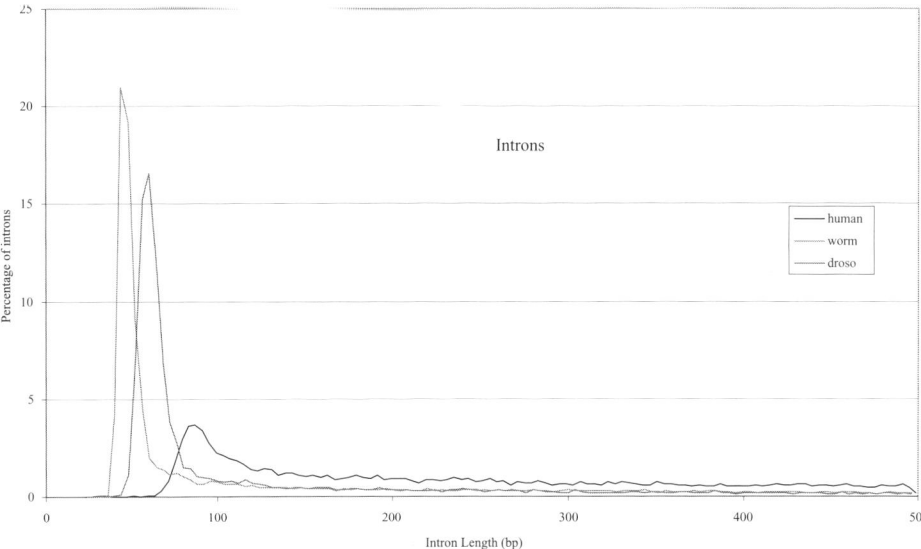

Fig. 14.3. The length distributions of *introns* in human, worm (*Caenorhabditis elegans*) and fruit fly (*Drosophila melanogaster*). Intron lengths all have a sharp minimum well away from zero, an initial peak that contains about half the density for worms and flies but less than 10% for humans, then a long tail. For humans the tail extends to tens of thousands of bases, with some introns known to be longer than 100 000 bases. Data reproduced with modification from International Human Genome Sequencing Consortium (2001).

length distribution explicitly by having a sequence of chained states with the same label (such as 'exon') and the same emission characteristics, but different 'exit' transition probabilities to the next state. This would allow an arbitrary distribution. However, it is costly in the size of the state space. In practice two other approaches are frequently used. The first is to have two or three chained states with the same label, then to search for the most likely assignment of labels to the sequence, integrating over all paths with the same labelling. This amounts to a hybrid between the forward and Viterbi algorithms, and is the approach taken by HMMgene, see Krogh (2001). It corresponds to using a beta type distribution, which is quite a good fit, particularly for exon distributions. The more widely used approach to modelling length distributions is to allow a single state to generate an entire region of sequence, such as a complete exon. This is then generated with kth order Markov properties and a fully defined length distribution. The down side is that the algorithms become potentially quadratic in the sequence length, since the subsequence ending at some base that is generated by a state can start at any earlier position. In practice this unacceptable increase in time complexity is resolved first by restricting the possible set of start positions to a set of precomputed potential splice sites, and second by bounding the search backwards, which is easy with exons since they cannot cross a stop codon (three of the 64 possible codons encode the stop, or end of the protein). It is also possible to bound the search for nonexon states for the Viterbi algorithm (and in practice for almost all the density for the forward algorithm) if certain conditions are imposed on the emission distributions, see Howe, Chothia and Durbin (2002). Models in which a state emits a region rather than a single base or a defined number of bases are sometimes called *generalised HMMs* or *semi-HMMs*. The best known implementation of this approach is Genscan, see Burge and Karlin (1997), which has been the *de facto* standard for mammalian gene finding for the past few years.

A few other tricks are used in practical implementations. For example Genscan conditions some of the parameters on the overall GC density of the sequence (the proportion of human sequence that is G or C as opposed to A or T varies between 35% and 60% in different regions of the genome for reasons we do not fully understand).

14.4 Practical performance of HMM gene-finding methods

When tested on known gene sequences that do not overlap their training sets, human-gene-finding methods typically find about 70 – 80% of the exons correctly: both the sensitivity, the fraction of true exons found correctly, and the specificity, the fraction of predicted exons that are correct, are around this level. However the frequency with which whole genes are predicted correctly is only around 20%. These figures vary depending on how the test set is obtained, with whole gene performance sometimes reported as up to or above 50%, e.g., Rogic, Mackworth and Ouellette (2001); however, in general these higher figures are obtained when using genes with short introns, which for ascertainment reasons were over-represented in sequence collections before the systematic genome sequencing projects.

Performance is better on simpler genomes, such as those of the invertebrate genetic model organisms *Caenorhabditis elegans* (a roundworm, the first animal to have its genome sequenced) and *Drosophila melanogaster* (a fruit fly). For these the exon accuracy can be around 90%, with whole gene accuracy over 50%.

This level of performance is useful in practice for some purposes, but far from ideal. In particular, it frequently happens that the ends of genes are mispredicted, so that exons from neighbouring genes are joined together or a single gene is split into two or more pieces. Unfortunately this is quite a confusing sort of error. It is partly for this reason that our current estimates of gene number in the human genome are so uncertain, which is also unfortunate because this type of simple headline figure attracts a lot of attention.

14.5 Comparative methods

How can we improve on the performance of stand alone gene finders? One approach is to make use of further information provided by related sequences. There are a number of potential sources of information and methods for using it. Methods that use another sequence to provide extra information are known as comparative methods because they involve comparison of two sequences. Here I give descriptions of two comparative gene-finding approaches to give an indication of possible approaches.

One type of related sequence is the corresponding DNA sequence from a related organism. When two species separate in evolution they have (nearly) identical DNA sequences. As time passes, the corresponding sequences diverge, but in general two things remain true: first the sequences can be aligned to each other, showing which particular bases correspond to each other in being derived from a common ancestor, and second regions that are functionally important such as coding regions and splice sites tend to be more conserved than nonfunctional regions, in a way that reflects their function. For example, in protein coding regions the first two bases of each codon triplet tend to be more conserved than the third base, because the genetic code is redundant, with the amino acid that is encoded by a triplet frequently being completely determined by the first two bases. If we therefore look for a region of sequence that has an alignment pattern in which, averaging over consecutive triplets, there is a tendency for two bases to be conserved, then one less conserved, then two more conserved then one less conserved etc., it is likely to be part of a coding exon. We can formalise this observation by extending the HMM approach.

One way to do this is to expand the set of possible observables at each position to be the values taken not by one base from one sequence, but by a pair of aligned bases from two aligned sequences. Extra cases have to be considered when a gap has been introduced into the alignment where part of a sequence has been deleted or inserted with respect to the common ancestor. This is known as a pair HMM, see Durbin, Eddy, Krogh and Mitchison (1998), and is relatively straightforward in the case where the alignment has been determined previously, as in the ROSETTA program of Batzoglu, Pachter, Mesirov, Berge and Lander (2000).

However, in many cases the correct alignment is not completely clear because there are multiple choices about which bases share common ancestors and where gaps should be placed, and if an incorrect alignment is used the prediction can be seriously misled. A solution to this problem is to search for the alignment at the same time as a gene structure, that is, to look for an alignment that is likely in terms of descent from a common ancestral sequence, and shows a conserved gene structure. This can be achieved with a pair HMM that allows arbitrary placement of gap states, which increases the number of states and, more importantly, increases the time and space complexity of the search algorithms by a factor proportional to the length of the second sequence. Two implementations of this type of strategy have been published, Doublescan by Meyer and Durbin (2002) and SLAM by Pachter, Alexandersson and Cawley (2002), each providing different heuristics to limit the search complexity, as well as being different in other details. To give some idea of the state configurations used, Figure 14.4 shows the state diagram from Doublescan. In reality twice as many states would be needed, to allow for genes on the reverse strand as well as the forward strand.

A second type of related sequence information that can be used is protein sequence from a protein that is evolutionarily related to the one derived from the gene being predicted. Protein sequences tend to be more conserved than DNA sequences, in ways that have been very well characterised, because they are the functional molecules in the cell. There are a number of sequence comparison programs that compare protein sequences to DNA sequences looking for conserved matches; the best known of these is BLASTX, one of the BLAST suite of programs, see Altschul, Gish, Miller, Myers and Lipman (1990). BLASTX generates matches that indicate regions of the DNA sequence that match protein, with a score related to the log-likelihood ratio of the match being true compared to a null model of random alignment.

BLAST scores can be used simply as evidence for a segment of sequence being part of a coding exon, in the same sort of way as fifth order Markov log-likelihood ratios for coding potential. Methods such as Genomescan of Yeh, Lim and Burge (2001) work this way. However, BLASTX does not know anything about gene structure, for example, about splice sites. Just as with SLAM and Doublescan, it is possible to search jointly for a gene structure and for an alignment of the product of the gene to a protein sequence. This is the approach taken by GeneWise of Birney and Durbin (2000), which is effectively another form of pair HMM. GeneWise tends to be conservative,

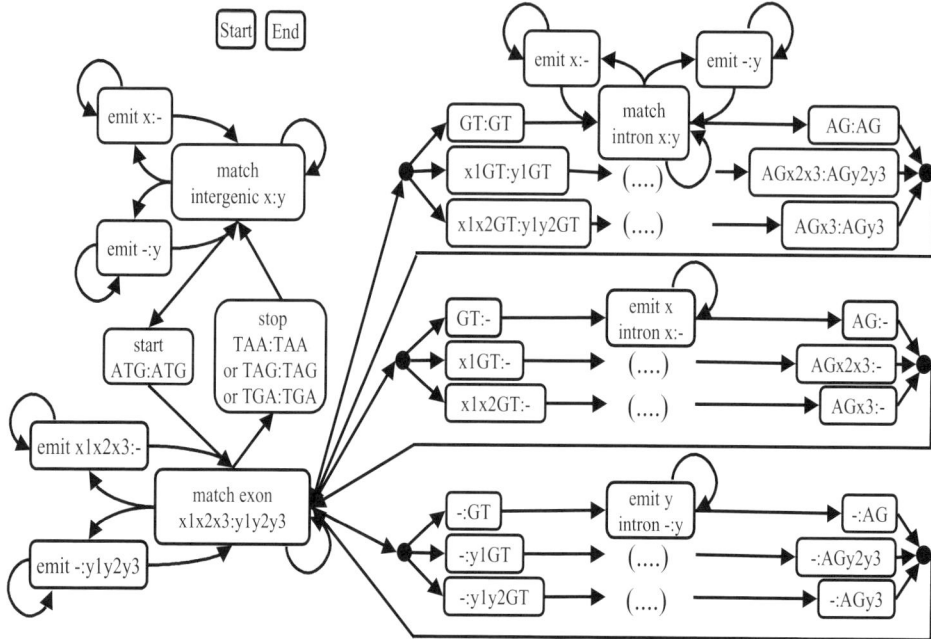

Fig. 14.4. A schematic state diagram for the Doublescan pair HMM for comparative gene prediction. This covers genes on the forward strand only. States are represented by boxes, with a label showing how many bases are matched by the state in each sequence, and legal transitions indicated by arrows. Upper left are a group of three states that handle intergenic regions, with two extra 'emit' states responsible for gaps in the alignment. The lower left group of states handle matching coding exons. Note that exon gaps must come in groups of three for coding regions. The upper right set of states handles matching introns in each of three phases. The splice site states to the left and right sides of this set are treated in a special way to use extra information about the splice site consensus. Lower right are two corresponding sets of states handling introns present in only the x sequence or only the y sequence respectively. The Start and End states at top must be linked to all other states to allow for the possibility of starting and ending sequences at any point in a gene structure (transition arrows omitted for clarity). See Meyer and Durbin (2002) for further details.

having a very high specificity (that is, its predictions are very likely to be correct) but a lower sensitivity (it doesn't make predictions where the evidence is comparatively weak). This behaviour is desirable for some applications, but not others. Also, GeneWise is computationally expensive, and in practice an initial search is made with BLASTX, or a similar very fast program, against a database of all known proteins (perhaps 500 000 sequences of average length 350), and then the top scoring one or few matching proteins are used with GeneWise to predict a gene structure.

All of the methods described in this section have used a single comparative sequence. Frequently there will be several related sequences available, and one of the current challenges in the field is to use these jointly to improve performance. There are two major issues. First, the sequences will not be related in an independent fashion, so it is not legitimate to sum information from all the sources. Rather, all the sequences, including the one of interest, will be related by an evolutionary tree, having an original common ancestor. Statistical methods exist for working with sequences related by a tree in this way, known as phylogenetic methods, see Durbin, Eddy, Krogh and Mitchison (1998) for further discussion and references, and their combination with gene-finding methods seems a promising area for future investigation. Second, many of the methods, including in particular pair HMMs, scale badly with increasing numbers of sequences, effectively as the length of one of the sequences to the power of the number of sequences. This means that heuristic simplifications will need to be developed, or possibly as an alternative Markov chain Monte Carlo methods that also limit the effective search space, albeit stochastically.

14.6 Discussion

This survey is necessarily very brief. Other reviews, written primarily for biologists but explaining the issues and methods, include Burge and Karlin (1998) and Zhang (2002). Further details are available in many of the papers on specific methods.

In particular, a number of important topics have been skipped entirely. One of the most important of these is that for many real genes there is not a single unambiguous sequence of exons. Rather, there are two or more alternative sequences of exons that make up what are known as alternative transcripts. These normally share many of the exons, but vary at one or more places. It used to be believed that producing many transcripts was relatively rare, but it seems now that perhaps the majority of genes have alternative transcripts, with some having many variants. Predicting alternative transcripts is hard, since looking for the most likely single gene structure is no longer appropriate. In practice current approaches can only suggest possible alternative exons and splicing patterns, perhaps based on posterior probabilities above some threshold, and leave the final decisions on alternative structures to be made using empirical evidence.

Although there has been progressive improvement in performance during the last decade, it is somewhat disappointing that we cannot more confidently predict gene structures. Unlike for many statistical inference problems, where there is implicit uncertainty and one does not expect error-free prediction, in this case there is an existence proof that the sequence does contain enough information in itself to determine the gene structure. This is provided by biology itself: a living cell will correctly transcribe, splice and translate a DNA sequence that is inserted into it by genetic engineering. Furthermore, it does this without using much of the information that we have described using above, for example the comparative sequence data from other organisms. In a sense, if we fully understood the biological processes gene prediction could be deterministic rather than statistical. However, that goal still seems a long way off, and the prospects for the time being are that we will continue to combine our increasing understanding of the biological processes with additional information from wherever we can get it, using statistical techniques, so as to improve progressively the accuracy of gene prediction methods.

Finally, as well as developing methodology as described here, data resources are of equal importance to the biologist. The application of methods to generate candidate genes for genomes of interest such as human genomes requires high-quality data. Frequently in practice heuristic approaches combining gene-finding predictions with other information are the most useful to an end user. An example of such a resource is Ensembl, see Hubbard *et al.* (2002),

$$\texttt{http://www.ensembl.org/},$$

which provides gene predictions and other genomic information for a number of the most important genomes.

Acknowledgments

This work was supported by the Wellcome Trust. I thank Ian Korf and the editors for comments.

References

Ahtola, J. and G. C. Tiao (1987). Distributions of least squares estimators of autoregressive parameters for a process with complex roots on the unit circle. *Journal of Time Series Analysis* 8, 1–14.

Akaike, H. (1978). Covariance matrix computation of the state variable of a stationary Gaussian process. *Annals of the Institute of Statistical Mathematics, Part B* 30, 499–504.

(1980). Seasonal adjustment by a Bayesian modeling. *Journal of Time Series Analysis* 1, 1–13.

Altschul, S., W. Gish, W. Miller, E. Myers and D. Lipman (1990). Basic local alignment search tool. *Journal of Molecular Biology* 215, 403–410.

Andersen, T. G., T. Bollerslev and F. X. Diebold (2004). Parametric and nonparametric measurement of volatility. In Y. Ait-Sahalia and L. P. Hansen (Eds.), *Handbook of Financial Econometrics*. Amsterdam: North Holland. Forthcoming.

Andersen, T. G., T. Bollerslev, F. X. Diebold and H. Ebens (2001). The distribution of realized stock return volatility. *Journal of Financial Economics* 61, 43–76.

Andersen, T. G., T. Bollerslev, F. X. Diebold and P. Labys (2001). The distribution of exchange rate volatility. *Journal of the American Statistical Association* 96, 42–55.

(2003). Modeling and forecasting realized volatility. *Econometrica* 71, 579–625.

Andersen, T. G., T. Bollerslev, and N. Meddahi (2004). Analytic evaluation of volatility forecasts. *International Economic Review* 45, forthcoming.

Anderson, B. D. O. and J. B. Moore (1979). *Optimal Filtering*. Englewood Cliffs, NJ: Prentice-Hall.

Anderson, E., Z. Bai, C. Bischof, J. Demmel, J. J. Dongarra, J. D. Croz, A. Greenbaum, S. Hammarling, A. McKenney, S. Ostrouchov and D. Sorensen (1992). *Lapack User's Guide*. Philadelphia, PA: Society for Industrial and Applied Mathematics.

Anderson, T. W.(1993). Goodness of fit tests for spectral distributions. *Annals of Statistics* 21, 830–847.

— (1997). Goodness-of-fit tests for autoregressive processes. *Journal of Time Series Analysis* 18, 321–339.

Anderson, T. W. and D. A. Darling (1952). Asymptotic theory of certain goodness-of-fit criteria based on stochastic processes. *Annals of Mathematical Statistics* 23, 193–212.

Anderson, T. W. and M. A. Stephens (1993). The modified Cramér–von Mises goodness-of-fit criterion for time series. *Sankhya* 55 A, 357–369.

Anderson, T. W., R. A. Lockhart and M. A. Stephens (1995). Goodness-of-fit tests for the time series models AR(1) and MA(1). Research report 95-09. Department of Mathematics and Statistics, Simon Fraser University.

— (1997). Goodness-of-fit tests for the time series models AR(1) and MA(1):2. Research report 97-05. Department of Mathematics and Statistics, Simon Fraser University.

— (2003). A test for the AR(1) time series model, and power against state-space alternatives. Research report 03-01. Department of Statistics and Actuarial Science, Simon Fraser University.

— (2004). An omnibus test for the time series model AR(1). *Journal of Econometrics*, 118, 111–127.

Andrews, D. W. K. (1991). Heteroskedasticity and autocorrelation consistent covariance matrix estimation. *Econometrica* 59, 817–858.

Andrieu, C., M. Davy, and A. Doucet (2002). Efficient particle filtering for jump Markov systems. In *Proceedings of IEEE ICASSP 2002, Orlando, USA*. IEEE.

Ansley, C. F. (1979). An algorithm for the exact likelihood of a mixed autoregressive-moving average process. *Biometrika* 64, 59–65.

Ansley, C. F. and R. Kohn (1985). Estimation, filtering and smoothing in state space models with incompletely specified initial conditions. *Annals of Statistics* 13, 1286–1316.

Ansley, C. F. and P. Newbold (1980). Finite sample properties of estimators for autoregressive moving average processes. *Journal of Econometrics* 13, 159–183.

Aoki, M. (1989). *Optimization of Stochastic Systems: Topics in Discrete-time Dynamics* (2 edn). New York, NY: Academic Press.

Arulampalam, S. and B. Ristic (2000). Comparison of the particle filter with range parameterised and modified polar EKF's for angle-only tracking. *Proceedings of SPIE Conference on Signal Processing of Small Targets* 4048, 288–299.

Arulampalam, S., S. Maskell, N. J. Gordon and T. C. Clapp (2002). A tutorial on particle filters for on-line non-linear/non-Gaussian Bayesian tracking. *IEEE Transactions on Signal Processing* 50, 174–188.

Avitzour, D. (1995). Stochastic simulation Bayesian approach to multi-target tracking. *IEE Proceedings-Radar Sonar and Navigation* 142, 41–44.

Back, K. (1991). Asset pricing for general processes. *Journal of Mathematical Economics* 20, 371–395.

Bar-Shalom, Y. and X. Li (1995). *Multitarget-Multisensor Tracking: Principles and Techniques*. Storrs, CT: YBS Publishing.

Barndorff-Nielsen, O. E. and D. R. Cox (1994). *Inference and Asymptotics*. London: Chapman & Hall.

Barndorff-Nielsen, O. E. and N. Shephard (2001). Non-Gaussian Ornstein–Uhlenbeck-based models and some of their uses in financial economics (with discussion). *Journal of the Royal Statistical Society, Series B* 63, 167–241.

(2002a). Econometric analysis of realised volatility and its use in estimating stochastic volatility models. *Journal of the Royal Statistical Society, Series B* 64, 253–280.

(2002b). Estimating quadratic variation using realised variance. *Journal of Applied Econometrics* 17, 457–477.

(2003). Realised power variation and stochastic volatility. *Bernoulli* 9, 243–265. Correction is available at www.levyprocess.org.

(2004). Econometric analysis of realised covariation: high frequency covariance, regression and correlation in financial economics. *Econometrica* 72, forthcoming.

Bartlett, M. (1949). Some evolutionary stochastic processes. *Journal of the Royal Statistical Society, Series B* 11, 211–229.

Batzoglu, S., L. Pachter, J. Mesirov, B. Berge and E. Lander (2000). Human and mouse gene structure: comparative analysis and application to exon prediction. *Genome Research* 10, 950–958.

Bayarri, M., M. DeGroot and J. Kadane (1988). What is the likelihood function? In G. S and J. Berger (Eds.), *Statistical Decision Theory and Related Topics IV*, Volume 1, pp. 1–27. New York, NY: Springer-Verlag.

Bell, W. R. (1984). Signal extraction for nonstationary time series. *Annals of Statistics* 12, 646–664.

———(1987). A note on overdifferencing and the equivalence of seasonal time series models with monthly means and models with (0,1,1)12 seasonal parts when $\Theta = 1$. *Journal of Business and Economic Statistics* 5, 383–387.

———(1991). Correction: Signal extraction for nonstationary time series. *Annals of Statistics* 19, 2280.

———(1993). Empirical comparisons of seasonal ARIMA and ARIMA component (structural) time series models. *American Statistical Association, Proceedings of the Business and Economic Statistics Section*, pp. 226–231. Alexandria, VA: American Statistical Association.

Bell, W. R. and S. C. Hillmer (1983). Modeling time series with calendar variation. *Journal of the American Statistical Association* 78, 526–534.

———(1988). A matrix approach to signal extraction and likelihood evaluation for ARIMA component time series models. Research Report Number 88/22, Statistical Research Division, US Bureau of the Census.

———(1989). Modeling time series subject to sampling error. Research Report Number 89/01, Statistical Research Division, US Bureau of the Census.

———(1990). The time series approach to estimation for repeated surveys. *Survey Methodology* 16, 195–216.

———(1991). Initializing the Kalman filter for nonstationary time series models. *Journal of Time Series Analysis* 12, 283–300.

Bell, W. R. and M. Kramer (1999). Towards variances for X-11 seasonal adjustments. *Survey Methodology* 25, 13–29.

Bell, W. R. and M. C. Otto (1992). Bayesian assessment of uncertainty in seasonal adjustment with sampling error present. Research Report Number 92/12, Statistical Research Division, US Bureau of the Census.

Bell, W. R. and M. G. Pugh (1990). Alternative approaches to the analysis of time series components. In A. C. Singh and P. Whitridge (Eds.), *Analysis of Data in Time: Proceedings of the 1989 International Symposium*, pp. 105–116. Ottawa: Statistics Canada.

Bell, W. R. and D. W. Wilcox (1993). The effect of sampling error on the time series behavior of consumption data. *Journal of Econometrics* 55, 235–265.

Bernardo, J. M. and A. F. M. Smith (1994). *Bayesian Theory*. Chichester: John Wiley.

Bertsekas, D. (1987). *Dynamic Programming: Deterministic and Stochastic Models*. Englewood Cliffs, NJ: Prentice-Hall.

Beveridge, S. and C. R. Nelson (1981). A new approach to decomposition of economic time series into permanent and transitory components with particular attention to measurement of the business cycle. *Journal of Monetary Economics* 7, 151–174.

Bierens, H. J. (2001). Complex unit roots and business cycles: Are they real? *Econometric Theory* 17, 962–983.

Binder, D. A. and J. P. Dick (1989). Modelling and estimation for repeated surveys. *Survey Methodology* 15, 29–45.

(1990). A method for the analysis of seasonal ARIMA models. *Survey Methodology* 16, 239–253.

Birney, E. and R. Durbin (2000). Using GeneWise in the Drosophila annotation experiment. *Genome Research* 10, 547–548.

Blake, A. and M. Isard (1998). *Active Contours*. Berlin: Springer-Verlag.

Box, G. E. P. and G. M. Jenkins (1976). *Time Series Analysis: Forecasting and Control* (2 edn). San Francisco, CA: Holden-Day.

Breidt, J., R. A. Davis and W. Dunsmuir (1992). On backcasting in linear time series. In D. Brillinger, P. Caines, J. Geweke, E. Parzen, M. Rosenblatt and M. S. Taqqu (Eds.), *New Directions in Time Series Analysis*, Volume 1, pp. 25–40. New York, NY: Springer-Verlag.

(1995). Time-reversibility, identifiability, and independence of innovations in stationary time series. *Journal of Time Series Analysis* 13, 377–390.

Brockett, R. W. (1970). *Finite-dimensional Linear Systems*. New York: Wiley.

Brockwell, P. J. and R. A. Davis (1987). *Time Series: Theory and Methods*. New York, NY: Springer-Verlag.

Brown, R. L., J. Durbin and J. M. Evans (1975). Techniques of testing the constancy of regression relationships over time. *Journal of the Royal Statistical Society, Series B* 37, 141–192.

Burge, C. B. and S. Karlin (1997). Prediction of complete gene structures in human genomic DNA. *Journal of Molecular Biology* 268, 78–94.

(1998). Finding the genes in genomic data. *Current Opinion in Structural Biology* 8, 346–354.

Burman, J. P. (1980). Seasonal adjustment by signal extraction. *Journal of the Royal Statistical Society, Series A* 143, 321–337.

Busetti, F. and A. C. Harvey (2002). Testing for trend. DAE working paper 0237, Cambridge University.

(2003). Seasonality testing. *Journal of Business and Economic Statistics* 21, 420–436.

Caines, P. E. (1988). *Linear Stochastic Systems*. New York, NY: John Wiley & Sons.

Canova, F. and B. E. Hansen (1995). Are seasonal patterns constant over time? A test for seasonal stability. *Journal of Business and Economic Statistics* 13, 237–252.

Cantwell, P. J. and J. R. Black (1998). Redesigning the monthly surveys of retail and wholesale trade for the year 2000. *American Statistical Association, Proceedings of the Section on Survey Research Methods*, 481–486.

Carlin, B. P., N. G. Polson and D. S. Stoffer (1992a). A Monte Carlo approach to nonnormal and nonlinear state-space modeling. *Journal of the American Statistical Association* 87, 493–500.

(1992b). A Monte Carlo approach to nonnormal and nonlinear state-space modelling. *Journal of the American Statistical Association* 87, 493–500.

Carpenter, J. R., P. Clifford and P. Fearnhead (1999). An improved particle filter for non-linear problems. *IEE Proceedings on Radar, Sonar and Navigation* 146, 2–7.

Carter, C. K. and R. Kohn (1994). On Gibbs sampling for state space models. *Biometrika* 81, 541–553.

Chan, N. H. (2002). *Time Series: Application to Finance*. New York, NY: John Wiley & Sons.

Chang, I., G. C. Tiao and C. Chen (1988). Estimation of time series parameters in the presence of outliers. *Technometrics* 30, 193–204.

Chen, L. (1996). Stochastic mean and stochastic volatility – a three-factor model of the term structure of interest rates and its application to the pricing of interest rate derivatives. *Financial Markets, Institution and Instruments* 5, 1–88.

Chen, R. and L. Scott (1993). Maximum likelihood estimation for a multifactor equilibrium model of the term structure of interest rates. *Journal of Fixed Income* 3, 14–31.

Chib, S. and E. Greenberg (1994). Bayes inference in regression models with ARMA(p,q) errors. *Journal of Econometrics* 64, 183–206.

Chopin, N. (2002). A sequential particle filter method for static models. *Biometrika* 89, 539–552.

Christensen, B. J. and N. R. Prabhala (1998). The relation between implied and realized volatility. *Journal of Financial Economics* 37, 125–150.

Clapp, T. C. and S. J. Godsill (1999). Fixed-lag smoothing using sequential importance sampling. In J. M. Bernardo, J. O. Berger, A. P. Dawid, and A. F. M. Smith (Eds.), *Bayesian Statistics* Volume 6, pp. 743–752. Oxford: Oxford University Press.

Clark, P. K. (1987). The cyclical component of U.S. economic activity. *Quarterly Journal of Economics* 102, 797–814.

Cleveland, W. S. and S. J. Devlin (1980). Calendar effects in monthly time series: detection by spectrum and graphical methods. *Journal of the American Statistical Association* 75, 487–496.

Cox, D. R. (1980). Local ancillarity. *Biometrika* 67, 279–286.

Cox, D. R. and N. Wermuth (1996). *Multivariate dependencies*. London: Chapman and Hall.

Cox, J. C., J. E. Ingersoll and S. A. Ross (1985). A theory of the term structure of interest rates. *Econometrica* 53, 385–407.

Dacorogna, M. M., R. Gencay, U. A. Müller, R. B. Olsen and O. V. Pictet (2001). *An Introduction to High-Frequency Finance*. San Diego, CA: Academic Press.

Dan, C., P. D. Moral and T. Lyons (1999). Discrete filtering using branching and interacting particle systems. *Markov Processes and Related Fields* 5, 293–318.

Danielsson, J. (1994). Stochastic volatility in asset prices: estimation with simulated maximum likelihood. *Journal of Econometrics* 61, 375–400.

Daubechies, I. (1992). *Ten Lectures on Wavelets*. Philadelphia, PA: SIAM.

Davis, M. H. A. (1977). *Linear Estimation and Stochastic Control*. London: Chapman and Hall.

——— (1993). *Markov Models and Optimization*. London: Chapman and Hall.

de Jong, F. (2000). Time series and cross-section information in affine term – structure models. *Journal of Business and Economic Statistics* 18, 300–314.

de Jong, P. (1991). The diffuse Kalman filter. *Annals of Statistics* 19, 1073–1083.

de Jong, P. and N. Shephard (1995). The simulation smoother for time series models. *Biometrika* 82, 339–350.

Dent, W. and A. S. Min (1978). A Monte Carlo study of autoregressive-integrated-moving average processes. *Journal of Econometrics* 7, 23–55.

Deriche, M. and A. H. Tewfik (1993). Maximum likelihood estimation of the parameters of discrete fractionally differenced Gaussian noise process. *IEEE Transactions on Signal Processing* 41, 2977–2989.

Dey, D., P. Müller and D. Sinha (Eds.) (1998). *Practical Nonparametric and Semiparametric Bayesian Statistics*, New York. Springer-Verlag.

Dongarra, J. J., C. B. Moler, J. R. Bunch and G. W. Stewart (1979). *Linpack User's Guide*. Philadelphia, PA: Society for Industrial and Applied Mathematics.

Doornik, J. A. (1999). *Object-Oriented Matrix Programming using Ox* (3rd edn). London: Timberlake Consultants Ltd. See http://www.nuff.ox.ac.uk/Users/Doornik.

— (2001). *Ox: Object Oriented Matrix Programming, 3.0*. London: Timberlake Consultants Press.

Doornik, J. A. and M. Ooms (2003). Computational aspects of maximum likelihood estimation of autoregressive fractionally integrated moving average models. *Computational Statistics and Data Analysis* 42, 333–348.

Doucet, A., N. de Freitas and N. J. Gordon (2001). *Sequential Monte Carlo Methods in Practice*. New York, NY: Springer-Verlag.

Doucet, A., S. J. Godsill and C. Andrieu (2000). On sequential Monte Carlo sampling methods for Bayesian filtering. *Statistics and Computing* 10, 197–208.

Doucet, A., N. J. Gordon and V. Krishnamurthy (2001). Particle filters for state estimation of jump Markov linear systems. *IEEE Transactions on Signal Processing* 49, 613–624.

Duan, J. C.-. and J. G. Simonato (1999). Estimating exponential-affine term structure models by Kalman filter. *Review of Quantitative Finance and Accounting* 13, 111–135.

Duffie, D. and R. Kan (1996). A yield-factor model of interest rates. *Mathematical Finance* 6, 379–406.

Durbin, J. (1969). Tests for serial correlation in regression analysis based on the periodogram of least-squares residuals. *Biometrika* 56, 1–15.

— (1970). Testing for serial correlation in least squares regression when some of the regressors are lagged dependent variables. *Econometrica* 38, 410–421.

— (1973). *Distribution Theory for Tests Based on the Sample Distribution Function*, Volume 9 of Regional Conference Series in Applied Mathematics. Philadelphia, PA: SIAM.

— (1975). Kolmogorov-Smirnov tests when parameters are estimated with applications to tests of exponentiality and tests on spacings. *Biometrika* 62, 5–22.

— (1980). Approximations for densities of sufficient estimators. *Biometrika* 67, 311–333.

Durbin, J. and A. C. Harvey (1985). The effects of seat belt legislation on road casualties in Great Britain: report on assessment of statistical evidence. Annexe to Compulsary Seat Belt Wearing Report, Department of Transport, London, HMSO.

Durbin, J. and S. J. Koopman (1997). Monte Carlo maximum likelihood estimation of non-Gaussian state space model. *Biometrika* 84, 669–684.

(2001). *Time Series Analysis by State Space Methods*. Oxford: Oxford University Press.

(2002). A simple and efficient simulation smoother for state space time series analysis. *Biometrika* 89, 603–616.

Durbin, R., S. Eddy, A. Krogh and G. Mitchison (1998). *Biological Sequence Analysis: Probabilistic Models of Proteins and Nucleic Acids*. Cambridge: Cambridge University Press.

Efron, B. (1987). Better bootstrap confidence intervals. *Journal of the American Statistical Association* 82, 171–185.

Escobar, M. and M. West (1995). Bayesian density estimation and inference using mixtures. *Journal of the American Statistical Association* 90, 577–588.

Fearnhead, P. (2002). MCMC, sufficient statistics and particle filter. *Journal of Computational and Graphical Statistics* 11, 848–862.

Fernández, C., J. Osiewalski and M. F. J. Steel (1997). On the use of panel data in stochastic frontier models with improper priors. *Journal of Econometrics* 79, 169–193.

Findley, D. F. (1986). On bootstrap estimates of forecast mean square errors for autoregressive processes. In D. Allen (Ed.), *Computer Science and Statistics: The Interface*, pp. 11–17. Amsterdam: North-Holland.

Findley, D. F., B. C. Monsell, W. R. Bell, M. C. Otto and B.-C. Chen (1998). New capabilities of the X-12-ARIMA seasonal adjustment program (with discussion). *Journal of Business and Economic Statistics* 16, 127–177.

Fisher, R. A. (1929). Tests of significance in harmonic analysis. *Proceedings of the Royal Society of London* 125, 54–59.

Fong, W., S. J. Godsill, A. Doucet and M. West (2002). Monte Carlo smoothing with application to audio signal enhancement. *IEEE Transactions on Signal Processing* 50, 438–449.

Frühwirth-Schnatter, S. (1994a). Applied state space modelling of non-Gaussian time series using integration-based Kalman filtering. *Statistics and Computing* 4, 259–269.

(1994b). Bayesian model discrimination and Bayes factors for state space models. *Journal of the Royal Statistical Society, Series B* 56, 237–246.

(1994c). Data augmentation and dynamic linear models. *Journal of Time Series Analysis* 15, 183–202.

Frühwirth-Schnatter, S. and A. Geyer (1996). Bayesian estimation of econometric multi-factor Cox–Ingersoll–Ross-models of the term structure of interest rates via MCMC-methods. Working paper.

Frühwirth-Schnatter, S. and L. Sögner (2002). Bayesian estimation of the Heston stochastic volatility model. Working paper.

Gallant, A. R., D. Hsieh and G. Tauchen (1997). Estimation of stochastic volatility models with diagnostics. *Journal of Econometrics* 81, 159–192.

Gastwirth, J. L. and H. Rubin (1975). The asymptotic distribution theory of the empiric c.d.f. for mixing processes. *Annals of Statistics* 3, 809–824.

Gelfand, A. E., S. K. Sahu and B. P. Carlin (1995). Efficient parametrisations for normal linear mixed models. *Biometrika* 82, 479–488.

Gersch, W. and G. Kitagawa (1983). The prediction of time series with trends and seasonalities. *Journal of Business and Economic Statistics* 1, 253–264.

Geweke, J. and S. Porter-Hudak (1983). The estimation and application of long memory time series models. *Journal of Time Series Analysis* 4, 221–238.

Ghysels, E., A. C. Harvey and E. Renault (1996). Stochastic volatility. In C. R. Rao and G. S. Maddala (Eds.), *Statistical Methods in Finance*, pp. 119–191. Amsterdam: North-Holland.

Gilks, W. and C. Berzuini (2001). Following a moving target – Monte Carlo inference for dynamic Bayesian models. *Journal of the Royal Statistical Society, Series B* 63, 127–146.

Godsill, S. J. and T. C. Clapp (2001). Improvement strategies for Monte Carlo particle filters. In A. Doucet, N. de Freitas and N. J. Gordon (Eds.), *Sequential Monte Carlo Methods in Practice*, pp. 139–158. New York, NY: Springer-Verlag.

Godsill, S. J., A. Doucet and M. West (2001). Maximum a posteriori sequence estimation using Monte Carlo particle filters. *Annals of the Institute of Statistical Mathematics* 53, 82–96.

Gordon, N. J., D. J. Salmond and A. F. M. Smith (1993). A novel approach to nonlinear and non-Gaussian Bayesian state estimation. *IEE-Proceedings F* 140, 107–113.

Green, P. and B. W. Silverman (1994). *Nonparameteric Regression and Generalized Linear Models: A Roughness Penalty Approach.* London: Chapman & Hall.

Gustafsson, F., F. Gunnarsson, N. Bergman, U. Forssell, J. Jansson, R. Karlsson and P.-J. Nordlund (2002). Particle filers for positioning, navigation and tracking. *IEEE Transactions on Signal Processing* 50, 425–437.

Hamilton, J. (1994). *Time Series Analysis.* Princeton, NJ: Princeton University Press.

Hannan, E. J. (1970). *Multiple Time Series.* New York, NY: Wiley.

(1973). The asymptotic theory of linear time series models. *Journal of Applied Probability* 10, 130–145.

Hannan, E. J. and M. Deistler (1988). *The Statistical Theory of Linear Systems.* New York, NY: John Wiley.

Harrison, J. and C. F. Stevens (1976). Bayesian forecasting (with discussion). *Journal of the Royal Statistical Society, Series B* 38, 205–247.

Harvey, A. C. (1985). Trends and cycles in macroeconomic time series. *Journal of Business and Economic Statistics* 3, 216–227.

(1989). *Forecasting, Structural Time Series Models and the Kalman Filter.* Cambridge: Cambridge University Press.

(1993). *Time Series Models* (2 edn). Hemel Hempstead: Harvester Wheatsheaf.

(2001). Testing in unobserved components models. *Journal of Forecasting* 20, 1–19.

(2004). A unified approach to testing for stationarity and unit roots. In D. W. K. Andrews, J. Powell, P. Ruud, and J. H. Stock (Eds.), *Identification and Inference for Econometric Models.* Cambridge: Cambridge University Press. Forthcoming.

Harvey, A. C. and C.-H. Chung (2000). Estimating the underlying change in unemployment in the UK (with discussion). *Journal of the Royal Statistical Society, Series A* 163, 303–339.

Harvey, A. C. and J. Durbin (1986). The effects of seat belt legislation on British road casualties: A case study in structural time series modelling. *Journal of the Royal Statistical Society, Series B* 149, 187–227.

Harvey, A. C. and S. J. Koopman (2000). Signal extraction and the formulation of unobserved components models. *Econometrics Journal* 3, 84–107.

Harvey, A. C. and R. G. Pierse (1984). Estimating missing observations in economic time series. *Journal of the American Statistical Association* 79, 125–131.

Harvey, A. C. and M. Streibel (1998a). Testing for nonstationary unobserved components. Unpublished paper: Department of Economics, University of Cambridge.

(1998b). Tests for deterministic versus indeterministic cycles. *Journal of Time Series Analysis* 19, 505–529.

Harvey, A. C. and P. H. J. Todd (1983). Forecasting economic time series with structural and Box-Jenkins models: A case study. *Journal of Business and Economic Statistics* 1, 299–307.

Harvey, A. C. and T. Trimbur (2003). General model-based filters for extracting cycles and trends in economic time series. *Review of Economics and Statistics* 85, 244–255.

Harvey, A. C., E. Ruiz and N. Shephard (1994). Multivariate stochastic variance models. *Review of Economic Studies* 61, 247–264.

Hickman, J. and R. Miller (1981). Bayesian bivariate graduation and forecasting. *Scandinavian Actuarial Journal*, 129–150.

Hijab, O. (1984). Asymptotic Bayesian estimation of a first-order equation with small diffusion. *Annals of Probability* 12, 890–902.

Hillmer, S. C. and G. C. Tiao (1982). An ARIMA-model-based approach to seasonal adjustment. *Journal of the American Statistical Association* 77, 63–70.

Hinkley, D. V. (1980). Likelihood as approximate pivotal distribution. *Biometrika* 67, 287–292.

Ho, Y. and R. Lee (1964). A Bayesian approach to problems in stochastic estimation and control. *IEEE Transactions on Automatic Control* 9, 333–339.

Hotta, L. K. (1989). Identification of unobserved components models. *Journal of Time Series Analysis* 10, 259–270.

Howe, K., T. Chothia and R. Durbin (2002). Gaze: a generic framework for the integration of gene prediction data by dynamic programming. *Genome Research* 12, 1418–1427.

Hubbard, T., D. Barker, E. Birney, G. Cameron, Y. Chen, L. Clark, T. Cox, J. Cuff, V. Curwen, T. Down, R. Durbin, E. Eyras, J. Gilbert, M. Hammond, L. Huminiecki, A. Kasprzyk, H. Lehvaslaiho, P. Lijnzaad, C. Melsopp, E. Mongin, R. Pettett, M. Pocock, S. Potter, A. Rust, E. Schmidt, S. Searle, G. Slater, J. Smith, W. Spooner, A. Stabenau, J. Stalker, E. Stupka, A. Ureta-Vidal, I. Vastrik and M. Clamp (2002). The Ensembl genome database project. *Nucleic Acids Research* 30, 38–41.

Hull, J. and A. White (1987). The pricing of options on assets with stochastic volatilities. *Journal of Finance* 42, 281–300.

Hurvich, C. M., R. Deo and J. Brodsky (1998). The mean squared error of Geweke and Porter–Hudak's estimator of the memory parameter of a long-memory time series. *Journal of Time Series Analysis* 19, 19–46.

Hylleberg, S. (1995). Tests for seasonal unit roots. General to specific or specific to general? *Journal of Econometrics* 69, 5–25.

Iwata, S. (1996). Bounding posterior means by model criticism. *Journal of Econometrics* 75, 239–261.

Jacobson, D. (1973). Optimal stochastic linear systems with exponential performance criteria; their relation to deterministic differential games. *IEEE Transactions on Automatic Control* AC-18, 124–131.

—— (1977). *Extensions of Linear-quadratic Control, Optimization and Matrix Theory*. New York, NY: Academic Press.

Jacod, J. and A. N. Shiryaev (1987). *Limit Theorems for Stochastic Processes*. Springer-Verlag: Berlin.

Jacquier, E., N. G. Polson and P. E. Rossi (1994). Bayesian analysis of stochastic volatility models (with discussion). *Journal of Business and Economic Statistics* 12, 371–417.

—— (2003). Stochastic volatility models: Univariate and multivariate extensions. *Journal of Econometrics*. Forthcoming.

Jazwinski, A. (1970). *Stochastic Processes and Filtering Theory*. New York, NY: Academic Press.

Jensen, M. J. (1999). Using wavelets to obtain a consistent ordinary least squares estimator of the long-memory parameter. *Journal of Forecasting* 18, 17–32.

—— (2000). An alternative maximum likelihood estimator of long-memory processes using compactly supported wavelets. *Journal of Economic Dynamics and Control* 24, 361–387.

Johannes, M., N. G. Polson and J. Stroud (2002a). Nonlinear filtering of stochastic differential equations with jumps. Unpublished paper: Graduate School of Business, Columbia University.

—— (2002b). Sequential optimal portfolio performance: market and volatility timing. Technical report, Graduate School of Business, University of Chicago.

Jones, R. H. (1980). Maximum likelihood fitting of ARIMA models to time series with missing observations. *Technometrics* 22, 389–95.

—— (1984). Fitting multivariate models to unequally spaced data. In E. Parzen (Ed.), *Time Series Analysis of Irregularly Observed Data*, Number 25 in Lecture Notes in Statistics, pp. 158–188. New York, NY: Springer-Verlag.

Julier, S. (1998). A skewed approach to filtering. In Oliver E. Drummond (Ed.), *Proceedings of SPIE Conference on Signal Processing of Small Targets*, Volume 3373, pp. 271–282 Bellingham: SPIE.

Kabaila, P. (1993). On bootstrap predictive inference for autoregressive processes. *Journal of Time Series Analysis* 14, 473–484.

Kalman, R. E. (1960). A new approach to linear filtering and prediction problems. *Journal of Basic Engineering, Transactions ASMA, Series D* 82, 35–45.

Kalman, R. E. and R. S. Bucy (1961). New results in linear filtering and prediction theory. *Transactions ASME Journal of Basic Engineering, Series D* 83, 95–108.

Kim, C.-J. and C. R. Nelson (1999). *State-Space Models with Regime Switching. Classical and Gibbs-Sampling Approaches with Applications.* Cambridge, MA: MIT.

Kim, S., N. Shephard and S. Chib (1998). Stochastic volatility: likelihood inference and comparison with ARCH models. *Review of Economic Studies* 65, 361–393.

King, M. L. and M. Evans (1988). Locally optimal properties of the Durbin-Watson test. *Econometric Theory* 4, 509–516.

Kitagawa, G. (1996). Monte Carlo filter and smoother for non-Gaussian nonlinear state space models. *Journal of Computational and Graphical Statistics* 5, 1–25.

Kitagawa, G. and W. Gersch (1984). A smoothness priors – state space modeling of time series with trend and seasonality. *Journal of the American Statistical Association* 79, 378–89.

(1996). *Smoothness Priors Analysis of Time Series.* New York, NY: Springer Verlag.

Kohn, R. and C. F. Ansley (1985). Efficient estimation and prediction in time series regression models. *Biometrika* 72, 694–697.

(1986). Estimation, prediction, and interpolation for ARIMA models with missing data. *Journal of the American Statistical Association* 81, 751–761.

(1987). Signal extraction for finite nonstationary time series. *Biometrika* 74, 411–421.

Koop, G. and D. J. Poirier (2003). Bayesian variants on some classical semiparametric regression techniques. *Journal of Econometrics*. Forthcoming.

Koop, G. and H. van Dijk (2000). Testing for integration using evolving trend and seasonals models: a Bayesian approach. *Journal of Econometrics* 97, 261–291.

Koopman, S. J. (1997). Exact initial Kalman filtering and smoothing for non-stationary time series models. *Journal of the American Statistical Association* 92, 1630–1638.

Koopman, S. J., A. C. Harvey, J. A. Doornik and N. Shephard (2000). *STAMP 6.0: Structural Time Series Analyser, Modeller and Predictor*. London: Chapman & Hall.

Koopman, S. J., N. Shephard and J. A. Doornik (1999). Statistical algorithms for models in state space using SsfPack 2.2. *Econometrics Journal* 2, 107–166.

— (2001). SsfPack 3.0beta: Statistical algorithms for models in state space. Discussion paper, Free University, Amsterdam.

Krogh, A. (2001). Two methods for improving performance of a HMM and their application for gene finding. In T. Gaasterland (Ed.), *Proceedings of the Fifth International Conference on Intelligent Systems for Molecular Biology*, pp. 179–186. Menlo Park, CA: AAAI Press.

Liu, J. and M. West (2001). Combined parameter and state estimation in simulation-based filtering. In A. Doucet, N. de Freitas, and N. J. Gordon (Eds.), *Sequential Monte Carlo Methods in Practice*, pp. 197–223. New York: Springer-Verlag.

Liu, J. S. and R. Chen (1998). Sequential Monte Carlo methods for dynamic systems. *Journal of the American Statistical Association* 93, 1032–1044.

Longstaff, F. and E. Schwartz (1992). Interest rate volatility and the term structure: A two-factor general equilibrium model. *Journal of Finance* 47, 1259–1282.

Lund, J. (1997). Non-linear Kalman filtering techniques for term – structure models. Discussion paper, The Aarhus School of Business, Denmark.

MacDonald, I. L. and W. Zucchini (1997). *Hidden Markov and Other Models for Discrete-Values Time Series*. London: Chapman and Hall.

Marrs, A. (2001). In-situ ellipsometry solutions using sequential Monte Carlo. In A. Doucet, J. F. G. de Freitas, and N. J. Gordon (Eds.), *Sequential Monte Carlo Methods in Practice*, pp. 465–477. New York, NY: Springer-Verlag.

Marrs, A., S. Maskell and Y. Bar-Shalom (2002). Expected likelihood for tracking in clutter with particle filters. In Oliver E. Drummond (Ed.), *Proceedings of SPIE Conference on Signal Processing of Small Targets*, pp. 230–239. Bellingham: SPIE.

Maskell, S., N. J. Gordon, M. Rollason and D. J. Salmond (2002). Efficient multitarget tracking using particle filters. *Proceedings of SPIE Conference on Signal Processing of Small Targets*, 251–262.

McCoy, E. J. and A. T. Walden (1996). Wavelet analysis and synthesis of stationary long-memory processes. *Journal of Computational and Graphical Statistics* 5, 26–56.

McCullough, B. D. (1994). Bootstrapping forecasting intervals: An application to AR(p) models. *Journal of Forecasting* 13, 51–66.

(1996). Consistent forecast intervals when the forecast-period exogenous variables are stochastic. *Journal of Forecasting* 15, 293–304.

McLeod, I. (1975). Derivation of the theoretical autocovariance function of autoregressive-moving average time series. *Applied Statistics* 24, 255–256.

(1977). Correction to derivation of the theoretical autocovariance function of autoregressive-moving average time series. *Applied Statistics* 26, 194.

Meddahi, N. (2002). A theoretical comparison between integrated and realized volatilities. *Journal of Applied Econometrics* 17, 479–508.

Meng, X. L. and D. van Dyk (1997). The EM algorithm – an old folk-song sung to a fast new tune. *Journal of Royal Statistical Society, Series B* 59, 511–567.

(1998). Fast EM-type implementations for mixed effects models. *Journal of Royal Statistical Society, Series B* 60, 559–578.

(1999). Seeking efficient data augmentation schemes via conditional and marginal data augmentation. *Biometrika* 86, 301–320.

Meyer, I. and R. Durbin (2002). Comparative gene prediction using pair hidden Markov models. *Bioinformatics* 18, 1309–1318.

Montemerlo, M., S. Thrun, D. Koller and B. Wegbreit (2002). FastSLAM: A Factored Solution to the Simultaneous Localization and Mapping Problem. In Erick Cant-Paz (Ed.), *Proceedings of the AAAI National Conference on Artificial Intelligence*. AAAI Press.

More, J. J., B. S. Garbow and K. E. Hillstrom (1980). *User Guide for Minpack-1*. Argonne, IL: Report ANL-80-74, Argonne National Laboratory.

Morley, J. C. (2002). A state-space approach to calculating the Beveridge–Nelson decomposition. *Economic Letters* 75, 123–127.

Morley, J. C., C. R. Nelson and E. Zivot (2003). Why are Beveridge–Nelson and unobserved components decompositions of GDP so different? *Review of Economics and Statistics* 85, 235–243.

Musso, C., N. Oudjane and F. LeGland (2001). Improving regularised particle filters. In A. Doucet, J. F. G. de Freitas, and N. J. Gordon (Eds.), *Sequential Monte Carlo Methods in Practice*, pp. 247–271. New York, NY: Springer-Verlag.

Neumann, T. (2002). Time-varying coefficient models: A comparison of alternative estimation strategies. Discussion paper, DZ Bank, Germany.

Newbold, P. and T. Bos (1985). *Stochastic Parameter Regression Models*. Beverly Hills: Sage Publications.

Newton, J. H., G. R. North and T. J. Crowley (1991). Forecasting global ice volume. *Journal of Time Series Analysis* 12, 255–265.

O'Hagan, A. (1978). Curve fitting and optimal design for prediction. *Journal of the Royal Statistical Society, Series B* 40, 1–42.

Orton, M. and A. Marrs (2001). A Bayesian approach to multi-target tracking data fusion with out-of sequence measurements. Technical report, Unpublished paper: Cambridge University Engineering Department.

Otto, M. C., W. R. Bell and J. P. Burman (1987). An iterative GLS approach to maximum likelihood estimation of regression models with ARIMA errors. *American Statistical Association, Proceedings of the Business and Economic Statistics Section*, 632–637.

Pachter, L., M. Alexandersson and S. Cawley (2002). Applications of generalized pair hidden Markov models to alignment and gene finding problems. *Journal of Computational Biology* 9, 389–400.

Papaspiliopoulos, O., G. O. Roberts and M. Sköld (2003). Non-centered parameterizations for hierarchical models and data augmentation. In J. Bernardo, M. Bayarri, J. Berger, A. Dawid, D. Heckerman, A. Smith, and M. West (Eds.), *Bayesian Statistics* Volume 7, pp. 307–326. Oxford: Oxford University Press.

Peña, D. and I. Guttman (1988). Bayesian approach to robustifying the Kalman filter. In J. C. Spall (Ed.), *Bayesian Analysis of Time Series and Dynamic Models*. Cambridge; MA: Marcel Decker.

Percival, D. B. and A. T. Walden (2000). *Wavelet Methods for Time Series Analysis*. Cambridge: Cambridge University Press.

Pfeffermann, D. (1991). Estimation and seasonal adjustment of population means using data from repeated surveys. *Journal of Business and Economic Statistics* 9, 163–175.

(1994). A general method for estimating the variances of X-11 seasonally adjusted estimators. *Journal of Time Series Analysis* 15, 85–116.

Pitt, M. K. and N. Shephard (1999a). Analytic convergence rates and parameterisation issues for the Gibbs sampler applied to state space models. *Journal of Time Series Analysis* 21, 63–85.

— (1999b). Filtering via simulation: auxiliary particle filter. *Journal of the American Statistical Association* 94, 590–599.

— (2001). Auxiliary variable based particle filters. In N. de Freitas, A. Doucet and N. J. Gordon (Eds.), *Sequential Monte Carlo Methods in Practice*, pp. 273–293. New York, NY: Springer-Verlag.

Poirier, D. J. (1995). *Intermediate Statistics and Econometrics*. Cambridge, MA: MIT.

Polson, N. G., J. Stroud and P. Müller (2002). Practical filtering with sequential parameter learning. Technical report, Graduate School of Business, University of Chicago.

Priestley, M. B. (1981). *Spectral Analysis and Time Series*. London: Academic Press.

Proietti, T. (2000). Comparing seasonal components for structural time series models. *International Journal of Forecasting* 16, 247–260.

— (2002). Some reflections on trend-cycle decompositions with correlated components. Discussion paper, European University Institute, Florence.

Protter, P. (1990). *Stochastic Integration and Differential Equations: A New Approach*. New York, NY: Springer-Verlag.

Rabiner, L. (1989a). A tutorial on hidden Markov models and selected applications in speech recognition. In *Proceedings of the IEEE* 77, 257–285.

— (1989b). A tutorial on hidden Markov models and selected applications in speech recognition. *Proceedings of the IEEE* 77, 257–286.

Ripley, B. D. (1987). *Stochastic Simulation*. New York, NY: Wiley.

Robert, C. (1996). Mixtures of distributions: inference and estimation. In W. Gilks, S. Richardson, and D. Spiegelhalter (Eds.), *Markov Chain Monte Carlo in Practice*, pp. 441–464. New York, NY: Chapman and Hall.

Roberts, G. O. and S. K. Sahu (1997). Updating schemes, correlation structure, blocking and parameterization for the Gibbs sampler. *Journal of the Royal Statistical Society, Series B* 59, 291–317.

Roberts, G. O., O. Papaspiliopoulos and P. Dellaportas (2004). Bayesian inference for non-Gaussian Ornstein-Uhlenbeck stochastic volatility processes. *Journal of the Royal Statistical Society, Series B* 66, forthcoming.

Rodrigues, P. M. M. (2002). On LM type tests for seasonal unit roots in quarterly data. *Econometrics Journal* 5, 176–195.

Rogic, S., A. K. Mackworth and B. F. Ouellette (2001). Evaluation of gene-finding programs on mammalian sequences. *Genome Research* 11, 817–832.

Sandmann, G. and S. J. Koopman (1998). Estimation of stochastic volatility models via Monte Carlo maximum likelihood. *Journal of Econometrics* 87, 271–301.

Schotman, P. C. (1994). Priors for the AR(1) model: parameterisation issues and time series considerations. *Econometric Theory* 10, 579–595.

Schwert, G. W. (1989). Why does stock market volatility change over time? *Journal of Finance* 44, 1115–1153.

Scott, A. J. and T. M. F. Smith (1974). Analysis of repeated surveys using time series methods. *Journal of the American Statistical Association* 69, 674–678.

Scott, A. J., T. M. F. Smith and R. G. Jones (1977). The application of time series methods to the analysis of repeated surveys. *International Statistical Review* 45, 13–28.

Shephard, N. (1994). Partial non-Gaussian state space. *Biometrika* 81, 115–131.

(1996). Statistical aspects of ARCH and stochastic volatility. In D. R. Cox, D. V. Hinkley and O. E. Barndorff-Nielsen (Eds.), *Time Series Models in Econometrics, Finance and Other Fields*, pp. 1–67. London: Chapman & Hall.

Shephard, N. and M. K. Pitt (1997). Likelihood analysis of non-Gaussian measurement time series. *Biometrika* 84, 653–667.

Shively, T. and R. Kohn (1997). A Bayesian approach to model selection in stochastic coefficient regression models and structural time series models. *Journal of Econometrics* 76, 39–52.

Shumway, R. H. and D. S. Stoffer (1982). An approach to time series smoothing and forecasting using the EM algorithm. *Journal of Time Series Analysis* 3, 253–264.

(2000). *Time Series Analysis and Its Applications*. New York, NY: Springer-Verlag.

Silverman, B. W. (1985). Some aspects of the spline smoothing approach to non-parametric regression curve fitting. *Journal of the Royal Statistical Society, Series B* 47, 1–52.

Skovgaard, I. M. (2001). Likelihood asymptotics. *Scandinavian Journal of Statistics* 28, 3–32.

Smith, M. and R. Kohn (1996). Nonparametric regression using Bayesian variable selection. *Journal of Econometrics* 75, 317–343.

Stephens, M. A. (1970). Use of the Kolmogorov–Smirnov, Cramér–von Mises, and related statistics without extensive tables. *Journal of the Royal Statistical Society, Series B* 32, 115–122.

— (1976). Asymptotic results for goodness-of-fit statistics with unknown parameters. *Annals of Statistics* 4, 357–369.

— (1986). Tests based on EDF statistics. In R. B. D'Agostino and M. A. Stephens (Eds.), *Goodness-of-fit Techniques*, Chapter 4, pp. 97–193. New York, NY: Marcel Dekker.

Stine, R. A. (1987). Estimating properties of autoregressive forecasts. *Journal of the American Statistical Association* 82, 1072–1078.

Stoffer, D. S. and K. D. Wall (1991). Bootstrapping state-space models: Gaussian maximum likelihood estimation and the Kalman filter. *Journal of the American Statistical Association* 86, 1024–1033.

Storvik, G. (2002). Particle filters in state space models with the presence of unknown static parameters. *IEEE Transactions on Signal Processing* 50, 281–289.

Stultz, C., J. V. White, and T. F. Smith (1993). Structural analysis based on state space modeling. *Protein Science* 2, 305–331.

Tanaka, K. (2002). A unified approach to the measurement error problem in time series models. *Econometric Theory* 18, 278–296.

Taylor, S. J. (1994). Modelling stochastic volatility. *Mathematical Finance* 4, 183–204.

Taylor, S. J. and X. Xu (1997). The incremental volatility information in one million foreign exchange quotations. *Journal of Empirical Finance* 4, 317–340.

Thombs, L. and W. Schuchany (1990). Bootstrap prediction intervals for autoregression. *Journal of the American Statistical Association* 85, 486–492.

Tiller, R. B. (1992). Time series modeling of sample survey data from the US Current Population Survey. *Journal of Official Statistics* 8, 149–166.

Tunnicliffe-Wilson, G. (1979). Some efficient computational procedures for high order ARMA models. *Journal of Statistical Computation and Simulation* 8, 301–309.

van der Merwe, R., A. Doucet, J. F. G. de Freitas, and E. Wan (2000). The unscented particle filter. In Todd K. Leen, Thomas G. Dietterich and Volker Trap (Eds.), *Advances in Neural Information Processing Systems (NIPS13)*, pp. 584–590. Boston, MA: MIT Press.

van Dyk, D. and X. L. Meng (2001). The art of data augmentation. *Journal of Computational and Graphical Statistics* 10, 1–50.

Vasicek, O. (1977). An equilibrium characterization of the term structure. *Journal of Financial Economics* 5, 177–188.

Velasco, C. and P. M. Robinson (2000). Whittle pseudo-maximum likelihood estimation for nonstationary time series. *Journal of the American Statistical Association* 95, 1229–1243.

Wahba, G. (1983). Bayesian confidence intervals for the cross-validated smoothing spline. *Journal of the Royal Statistical Society, Series B* 45, 133–150.

Wall, K. D. and D. S. Stoffer (2002). A state space approach to bootstrapping conditional forecasts in ARMA models. *Journal of Time Series Analysis* 23, 733–751.

Watanabe, N. (1985). Note on the Kalman filter with estimated parameters. *Journal of Time Series Analysis* 6, 269–278.

Watson, G. S. (1961). Goodness-of-fit tests on a circle. *Biometrika* 48, 109–114.

West, M. and J. Harrison (1997). *Bayesian Forecasting and Dynamic Models* (2 edn). New York, NY: Springer-Verlag.

Whittle, P. (1951). *Hypothesis Testing in Time Series Analysis*. Uppsala: Almquist and Wicksell.

(1953). The analysis of multiple stationary time series. *Journal of the Royal Statistical Society, Series B* 15, 125–139.

(1981). Risk-sensitive linear/quadratic/Gaussian control. *Advances in Applied Probability* 13, 764–777.

(1982). *Optimisation over Time*, Volume 1. Chichester: Wiley.

(1991). A risk-sensitive maximum principle; the case of imperfect state observation. *IEEE Transactions on Automatic Control* AC-36, 793–801.

(1996). *Optimal Control; Basics and Beyond*. Chichester: Wiley.

Wolter, K. M. and N. J. Monsour (1981). On the problem of variance estimation for a deseasonalized series. In D. Krewski, R. Platek, and J. N. K. Rao (Eds.), *Current Topics in Survey Sampling*, pp. 367–403. New York, NY: Academic Press.

Yatchew, A. (1998). Nonparametric regression techniques in economics. *Journal of Economic Literature* 36, 669–721.

Ych, R., L. Lim and C. B. Burge (2001). Computational inference of homologous gene structures in the human genome. *Genome Research* 11, 803–816.

Young, P. (1998). Data-based mechanistic modelling of environmental, ecological, economic and engineering systems. *Environmental Modelling and Software* 13, 105–122.

Zhang, M. Q. (2002). Computational prediction of eukaryotic protein-coding genes. *Nature Reveiws Genetics* 3, 698–709.

Zivot, E. and J. Wang (2003). *Modeling Financial Time Series with S-PLUS*. New York, NY: Springer-Verlag.

Author index

Ahtola, J., 115
Akaike, H., 157, 254, 256
Alexandersson, M., 347
Altschul, S., 347
Andersen, T. G., 210, 223
Anderson, B. D. O., 172, 256
Anderson, E., 282
Anderson, T. W., 93, 96, 98–100
Andrews, D. W. K., 109, 110
Andrieu, C., 53, 61, 62, 69, 237
Ansley, C. F., 157, 162, 174, 254, 256, 258,
    261, 266, 267, 268, 282
Aoki, M., 189
Arulampalam, S., 41, 67
Avitzour, D., 65

Back, K., 210
Bai, Z., 282
Bar-Shalom, Y., 42, 45, 65
Barker, D., 350
Barndorff-Nielsen, O. E., 140, 206, 208,
    210–212, 224–226, 228, 230
Bartlett, M., 38
Batzoglu, S., 347
Bayarri, M., 157
Bell, W. R., 252–254, 256, 261, 265, 267–272,
    276, 280, 281
Berge, B., 347
Bergman, N., 68
Bernardo, J. M., 22
Bertsekas, D., 28
Berzuini, C., 64, 237, 239
Beveridge, S., 307
Bierens, H. J., 104, 113
Binder, D. A., 271,
Birney, E., 347, 351
Bischof, C., 282
Black, J. R., 272
Blake, A., 41, 69
Bollerslev, T., 210, 223
Bos, T., 175
Box, G. E. P., 256, 258
Breidt, J., 188, 189

Brockett, R. W., 28
Brockwell, P. J., 4
Brodsky, J., 78
Brown, R. L., 301, 302
Bucy, R. S., 172
Bunch, J. R., 282
Burge, C. B., 345, 347, 349
Burman, J. P., 261, 272
Busetti, F., 111, 112, 116

Caines, P. E., 174, 189, 200
Cameron, G., 350
Canova, F., 110, 111
Cantwell, P. J., 272
Carlin, B. P., 124, 126–128, 240
Carpenter, J. R., 55, 237, 244
Carter, C. K., 11, 125, 157, 240, 297
Cawley, S., 347
Chan, N. H., 285
Chang, I., 280
Chen, B.-C., 280
Chen, C., 280
Chen, L., 328
Chen, R., 55, 61, 237, 328
Chen, Y., 350
Chib, S., 69, 125, 143, 157, 179, 180, 230, 237,
    241, 242, 324
Chopin, N., 237, 239
Chothia, T., 345
Christensen, B. J., 210
Chung, C.-H., 271
Clamp, M., 350
Clapp, T. C., 41, 239
Clark, L., 350
Clark, P. K., 310
Cleveland, W. S., 112
Clifford, P., 55, 237, 244
Cox, D. R., 206, 217
Cox, J. C., 328
Cox, T., 350
Crowley, T. J., 108, 118
Croz, J. D., 282
Cuff, J., 350

Curwen, V., 350

Dacorogna, M. M., 208, 235
Dan, C., 41
Danielsson, J., 180
Darling, D. A., 93
Daubechies, I., 82
Davis, M. H. A., 28, 32
Davis, R. A., 4, 188, 189
Davy, M., 69
de Freitas, J. F. G., 62
de Freitas, N., 41
de Jong, F., 329
de Jong, P., 11, 125, 145, 157, 256, 297, 325
DeGroot, M., 157
Deistler, M., 200
Dellaportas, P., 125, 126, 140
Demmel, J., 282
Dent, W., 174
Deo, R., 78
Deriche, M., 89
Devlin, S. J., 112
Dick, J. P., 271
Diebold, F. X., 210, 223
Dongarra, J. J., 282
Doornik, J. A., 24, 77, 108, 235, 285, 288, 294, 298
Doucet, A., 41, 53, 59, 61, 62, 69, 70, 237
Down, T., 350
Duan, J. C.-., 327, 329, 330, 332
Duffie, D., 328
Dunsmuir, W., 188, 189
Durbin, J., 4, 10, 20, 27, 33, 34, 43–45, 52, 94–96, 104, 125, 145, 157, 162, 169, 172, 206, 228, 249, 251, 256, 285, 294, 296, 297, 301, 302, 323, 325, 326, 334
Durbin, R., 337, 340, 345–347, 349

Ebens, H., 210
Eddy, S., 337, 340, 346, 349
Efron, B., 193
Escobar, M., 154
Evans, J. M., 301, 302
Evans, M., 114
Eyras, E., 351

Fearnhead, P., 55, 237, 239, 244
Fernández, C., 157
Findley, D. F., 187, 188, 189, 280
Fisher, R. A., 103
Fong, W., 69
Forssell, U., 68
Frühwirth-Schnatter, S., 11, 13, 15, 125, 126, 157, 240, 297

Gallant, A. R., 230
Garbow, B. S., 262
Gastwirth, J. L., 199, 202
Gelfand, A. E., 126–128
Gencay, R., 208, 235

Gersch, W., 4, 172, 252
Geweke, J., 77
Geyer, A., 125
Ghysels, E., 208, 241, 319
Gilbert, J., 350
Gilks, W., 64, 237, 239
Gish, W., 347
Godsill, S. J., 53, 59, 61, 62, 69, 237, 239
Gordon, N. J., 41, 55, 59, 64, 66, 70, 237
Green, P., 154, 168
Greenbaum, A., 282
Greenberg, E., 143
Gunnarsson, F., 68
Gustafsson, F., 68
Guttman, I., 180

Hamilton, J., 4, 285, 288
Hammarling, S., 282
Hammond, M., 350
Hannan, E. J., 78, 200, 254
Hansen, B. E., 110, 111
Harrison, J., 4, 41, 44, 157, 172, 285
Harvey, A. C., 4, 6, 24, 41, 44, 104, 108, 111, 112, 114–117, 157, 162, 172, 178, 206, 208, 228, 241, 249, 251, 252, 256, 257, 271, 294, 310, 317, 319, 321
Hickman, J., 157
Hijab, O., 39
Hillmer, S. C., 256, 267, 270–272, 276
Hillstrom, K. E., 262
Hinkley, D. V., 206
Ho, Y., 44
Hotta, L. K., 263
Howe, K., 345
Hsieh, D., 230
Hubbard, T., 350
Hull, J., 319
Huminiecki, L., 350
Hurvich, C. M., 78
Hylleberg, S., 116

Ingersoll, J. E., 328
Isard, M., 41, 69
Iwata, S., 160

Jacobson, D., 37
Jacod, J., 210
Jacquier, E., 179, 241
Jansson, J., 68
Jazwinski, A., 49
Jenkins, G. M., 256, 258
Jensen, M. J., 85–87
Johannes, M., 237, 238
Jones, R. G., 271
Jones, R. H., 172, 256
Julier, S., 48

Kabaila, P., 188
Kadane, J., 157

Kalman, R. E., 44, 172
Kan, R., 328
Karlin, S., 345, 349
Karlsson, R., 68
Kasprzyk, A., 350
Kim, C.-J., 4, 285
Kim, S., 69, 125, 157, 179, 180, 230, 237, 241, 242, 324
King, M. L., 114
Kitagawa, G., 4, 41, 55, 172, 252
Kohn, R., 11, 125, 154, 157, 162, 240, 254, 256, 258, 261, 266–268, 297
Koller, D., 68
Koop, G., 153, 155, 157, 160, 163
Koopman, S. J., 4, 10, 20, 24, 27, 33, 34, 43–45, 52, 108, 125, 145, 157, 162, 169, 172, 180, 228, 249, 251, 256, 285, 288, 294, 296–298, 323, 325, 326, 334
Kramer, M., 271
Krishnamurthy, V., 70
Krogh, A., 337, 340, 344, 346, 349

Labys, P., 210, 223
Lander, E., 347
Lee, R., 44
LeGland, F., 64
Lehraslaiho, H., 350
Li, X., 42, 45
Lijnzaad, p., 350
Lim, L., 347
Lipman, D., 347
Liu, J., 237
Liu, J. S., 55, 61, 237
Lockhart, R. A., 96, 98–100
Longstaff, F., 328
Lund, J., 329
Lyons, T., 41

MacDonald, I. L., 340
Mackworth, A. K., 345
Marrs, A., 65, 68, 70
Maskell, S., 41, 65, 66
McCoy, E. J., 86
McCullough, B. D., 188
McKenney, A., 282
McLeod, I., 259
Meddahi, N., 210, 227, 228
Melsopp, C., 350
Meng, X. L., 123, 126, 132, 133, 143
Mesirov, J., 347
Meyer, I., 347
Miller, R., 157
Miller, W., 347
Min, A. S., 174
Mitchison, G., 337, 340, 346, 349
Moler, C. B., 282
Mongin, E., 350
Monsell, B. C., 280
Monsour, N. J., 271
Montemerlo, M., 68

Moore, J. B., 172, 256
Moral, P. D., 41
More, J. J., 262
Morley, J. C., 308, 309, 317
Musso, C., 64
Myers, E., 347
Müller, P., 236, 237, 239
Müller, U. A., 208, 235

Nelson, C. R., 4, 285, 307, 309, 317
Neumann, T., 305
Newbold, P., 174, 175
Newton, J. H., 108, 118
Nordlund, P.-J., 68
North, G. R., 108, 118

O'Hagan, A., 169
Olsen, R. B., 208, 235
Ooms, M., 77
Orton, M., 70
Osiewalski, J., 157
Ostrouchov, S., 282
Otto, M. C., 261, 271, 280
Oudjane, N., 64
Ouellette, B. F., 345

Pachter, L., 347
Papaspiliopoulos, O., 125, 126, 129, 131, 140, 151
Percival, D. B., 80, 81, 83
Prttett, R., 350
Peña, D., 180
Pfeffermann, D., 271
Pictet, O. V., 208, 235
Pierse, R. G., 172
Pitt, M. K., 52, 61, 69, 123, 126, 128, 136, 237, 239, 240, 243, 323
Pocock, M., 350
Poirier, D. J., 14, 153, 155, 157–160, 163
Polson, N. G., 124, 179, 236–241
Porter-Hudak, S., 77
Potter, S., 350
Prabhala, N. R., 210
Priestley, M. B., 103, 106, 114, 117
Proietti, T., 254, 318
Protter, P., 210
Pugh, M. G., 271

Rabiner, L., 44
Rabiner, L. R., 340
Renault, E., 208, 241, 319
Ripley, B. D., 19, 55, 324
Ristic, B., 67
Robert, C., 154
Roberts, G. O., 125–129, 131, 140, 151
Robinson, P. M., 78
Rodrigues, P. M. M., 115
Rogic, S., 345
Rollason, M., 66
Ross, S. A., 328
Rossi, P. E., 179, 241

Rubin, H., 199, 202
Ruiz, E., 178, 321
Rust, A., 350

Sahu, S. K., 126, 127, 128
Salmond, D. J., 41, 55, 59, 64, 66, 237
Sandmann, G., 180
Schmidt, E., 350
Schotman, P. C., 141
Schuchany, W., 187
Schwartz, E., 328
Schwert, G. W., 210
Scott, A. J., 271
Scott, L., 328
Searle, S., 350
Shephard, N., 11, 24, 52, 61, 69, 108, 123, 125, 126, 128, 136, 140, 145, 157, 178–180, 207, 208, 210–212, 224, 225, 226, 228, 230, 237, 239, 240, 241, 242, 243, 285, 288, 294, 297, 298, 321, 323–325
Shiryaev, A. N., 210
Shively, T., 157
Shumway, R. H., 4, 49, 172, 180, 181, 185, 285, 294
Silverman, B. W., 154, 168
Simonato, J. G., 327, 329, 330, 332
Sköld, M., 126, 129, 131, 151
Skovgaard, I. M., 206
Slater, G., 350
Smith, A. F. M., 22, 41, 55, 59, 64, 66, 237
Smith, J., 350
Smith, M., 154
Smith, T. F., 172
Smith, T. M. F., 271
Sorensen, D., 282
Spooner, W., 350
Stabenau, A., 350
Stalker, J., 350
Steel, M. F. J., 157
Stephens, M. A., 94–96, 98–100
Stevens, C. F., 172
Stewart, G. W., 282
Stine, R. A., 187
Stoffer, D. S., 4, 49, 124, 172, 174, 180, 181, 185, 188–190, 197–202, 240, 285, 294
Storvik, G., 236–239, 241, 243, 247
Streibel, M., 104, 114, 115, 317
Stroud, J., 236–239
Stultz, C., 172

Stupka, E., 351
Sögner, L., 126

Tanaka, K., 90
Tauchen, G., 230
Taylor, S. J., 207, 210
Tewfik, A. H., 89
Thombs, L., 187
Thrun, S., 68
Tiao, G. C., 115, 272, 280
Tiller, R. B., 271
Todd, P. H. J., 172
Trimbur, T., 117
Tunnicliffe-Wilson, G., 259

Ureta-Vidal, A., 350

van Dijk, H., 157
van Dyk, D., 123, 126, 132, 133, 143
van der Merwe, R., 62
Vasicek, O., 328, 330
Vastrik, I., 350
Velasco, C., 78

Wahba, G., 154
Walden, A. T., 80, 81, 83, 86
Wall, K. D., 174, 188–190, 197–202
Wan, E., 62
Wang, J., 285
Watanabe, N., 201
Watson, G. S., 93
Wegbreit, B., 68
Wermuth, N., 217
West, M., 4, 41, 44, 59, 69, 154, 157, 237, 285
White, A., 319
White, J. V., 172
Whittle, P., 29, 32, 35, 38
Wilcox, D. W., 270, 271
Wolter, K. M., 271

Xu, X., 210

Yatchew, A., 155, 163
Yeh, R., 347
Young, P., 24

Zhang, M. Q., 349
Zivot, E., 285, 309, 317
Zucchini, W., 340

Subject index

Affine term structure model, 327
Airline model, 271
Anderson–Darling statistic, 93
Antithetic variable, 19, 20
ARFIMA model, 76
ARIMA process, 251
    in state space form, 254
ARMA process, 7, 191, 227, 228, 307
ARMAX model, 172
Ask quote, 208
Autocorrelation function, 231
Autoregression, 27, 28, 198, 228
Auxiliary particle filter, 64, 237

Basic structural model, 251
Baum–Welch algorithm, 342
Bayesian inference, viii, 10, 13, 15, 21, 124, 338
Bearings-only target tracking, 66
Beveridge–Nelson decomposition, 307
BFGS optimisation, 183
Bid quote, 208
Biometrika, xi
BLASTX algorithm, 347
Bootstrap
    state space, viii, 174
Box–Jenkins approach, 7
Box–Pierce statistic, 230
Brownian bridge, 112
Brownian motion, 112, 210

Càdlàg paths, 207
Capital asset pricing model, 299
    time-varying, 290, 305
Certainty equivalence principle, 31, 37
Cholesky decomposition, 281
Clark model, 310
Classical inference, 15
Complex unit roots, 115
Concentrated likelihood, 261
CONDENSATION algorithm, 69
Conditional heteroskedasticity, 110
Conjugate prior, 155
Continuous local martingale, 207

Continuous time process, 169, 208
Continuous wavelet transform (CWT), 79
Control, 29
Control theory, 33
Cost function, 36
Covariance stationarity, 216
Cramér–von Mises distribution, 103, 104, 111, 112, 119
Cramér–von Mises statistic, 93–96, 98, 99
Cross-validation, 156
Cubic spline, 168
CUSUM test, 302
CWT
    see continuous wavelet transform, 79
Cycle, 102
    deterministic cycle, 110, 112, 115
    several cycles, 108
    stochastic cycle, 105, 107
    test for, 112

Data
    global ice volume, 118
    Hudson Bay Company fur data, 117
    nominal interest rate, 175
    Olsen and Associates financial data, 208
    US GNP, 184
    US housing starts, 280
    US retail sales, 271
Data augmentation, 124
Decision theory, 27, 29
Decomposition of time series, 307
Decorrelation property, 83
Deterministic cycle, 103, 107–110, 112, 114, 115
Dickey–Fuller test, 115
Differencing parameter, 76
Diffuse likelihood, 162
Diffuse prior, 287, 288
Diffusion process, 328
    mean reverting, 330
Dilation parameter, 80
Discrete wavelet transform (DWT), 79
Diurnal effect, 207

Doublescan, 347
Drift process, 207
Durbin, James, vii, xi, xiv, xv, 206, 337
Durbin–Watson statistic, 114
DWT
    see discrete wavelet transform, 79
Dynamic linear model, 7
Dynamic program, 29, 32

Easter effect, 277
Econometrics, 206
Ellipsometry, 68
EM algorithm, 13
Empirical Bayes, 152
Empirical distribution function, 93, 97
Ergodicity, 60
Exon, 337
Extended Kalman filter, 45

Filtered cycle, 316, 318
Filtered state estimate, 300
Filtered trend, 314, 318
Filtering, 206, 208, 226
    see also particle filter, 237
Filtering distribution, 238
Financial volatility, 228
Fixed lag particle filter, 238
Fixed point smoother, 256
Forecast error, 264
    confidence interval, 192, 196
Forecasting, 9, 11, 206, 208, 222, 226, 297
Forecasting volatility, 207
Frequency domain method, 77

Gamma distribution, 13
Gaussian likelihood, 259
Gaussian linear state space model, 124
Genscan, 345
Gibbs sampler, 13, 16, 242
Goodness-of-fit test, 93
GPH estimator, 77

Haar wavelet filter, 81
Heavy tail, 212
Hidden Markov model, 337, 338
    generalised, 345
    semi, 345
Hierarchical prior, 161
High frequency finance, 208, 232
High-pass filter, 81

Identifiability
    Regcomponent model, 263
Importance sampling, 19, 50, 297, 323
Initialisation of Kalman filter, 255, 287
Innovation, 287, 296
Innovation filter, 188
Instrumental variable, 206
Integrated variance, 207, 213
Intra-day effect, 212
Intron, 337

Iterated extended Kalman filter, 48

Jitter, 63
Jump Markov linear system, 70
Jump process, 207

Kalman filter, 8, 9, 32, 38, 107, 124, 157, 172, 174, 180, 197, 228, 256, 295, 298
    initialisation, 9, 12, 255
    prediction equation, 297
Kernel density approximation, 64
Kolmogorov–Smirnov statistic, 93, 96
Kolmogorov–Smirnov test, 197
Kurtosis, 184

Lagrange multiplier test, 104, 114, 115, 119
Likelihood
    RegComponent model, 258
Local level, 6, 7, 13, 288
Local linear trend, 6, 7
Locally best invariant test, 114
Locally bounded variation, 207
London School of Economics, xi, 206
Long memory, viii, 76, 207, 212, 231
Low-pass filter, 82
LQG structure, 28, 30

Market microstructure effect, 208
Markov chain Monte Carlo, viii, 14, 124, 143, 179, 237, 242, 244, 297
Markov process, 28
Maximum likelihood, 13, 27, 296
Measurement equation, 172, 287, 296
Measurement error, 76
Meddahi regression, 227
Missing value, 9, 11, 124, 297
Mixtures of distribution, 154
MLE, 174
    bootstrap distribution, 198
Molecular biology, 172
Moving average, 228
Multinomial resampling, 55
Multivariate linear regression, 7
Multivariate model, 287

Navigation, 68
Newton–Raphson optimisation, 326
Non-Gaussian, 40
Non-Gaussian state space, 17, 18
Noninformative prior, 155
Nonlinear, 40
Nonlinear filtering, 38
Nonlinear state space, 17, 18
Nonparametric probit, 153
Nonparametric regression, 152, 153, 156, 169
Nonparametric tobit, 153

Observability, 28
Option pricing, 237
Ornstein–Uhlenbeck process, 210
Out-of-sequence measurements, 70

Outlier, 296
Ox, 24

p^*-formula, 206
Parameter estimation, 9
Parameter learning, 237
Partial linear model, 155
Particle filter, viii, 49, 237
    auxiliary particle filter, 237
    stochastic volatility, 241
    sufficient statistic, 239
    see also the practical filter, 237
Parzen spectral window, 109
Penalised likelihood, 154
Periodogram, 27, 96, 114, 117
Periodogram test, 104
Phillips, Peter C B, xii
Plant equation, 31
Poisson distribution, 17
Portfolio selection, 237
Practical filter, 237
Prediction error
    decomposition, 297
    one-step ahead, 295
    variance, 295
    variance matrix, 298
    vector, 298
Prior distribution, 141
Prior elicitation, 157
Prior hyperparameter, 153
Probit, nonparametric, 153
Proposal distribution, 60
Pyramid algorithm, 81

Quadratic cost, 28
Quadratic variation, 206, 210
Quadruture mirror filter, 82
Quarticity, 211
Quasi-likelihood, 173, 179, 230, 321

Random walk, 6, 288
    with drift, 316
Random-effects model, 127, 132
Rao–Blackwellised particle filter, 65
Realised variance, 206, 210, 212–216, 218–222, 225, 227, 229–231
    asymptotic theory, 215
    logarithmic transformation, 224
Realised volatility, 205
Recursive least squares, 299–301
RegARIMA, 249
REGCMPNT, 249, 256, 262, 272, 281
RegComponent model, 249, 250
    concentrated likelihood, 261
    example, 277
    forecasting, 264
    identifiable, 263
    likelihood, 258
    likelihood maximisation, 260
    seasonal adjustment, 270
    signal extraction, 267

Regression lemma, 10
Regression model
    time-varying, 8
    with ARMA errors, 8
Regularised particle filter, 64
Reparametrisation, 124
Repeat sample survey, 270
Resampling, 54
Residual resampling, 55
Retail sales in U.S., 271
Reuters, 208
Reverse time, 199
Riccati equation, 32, 33
Risk-sensitive criteria, 37

S+FinMetrics, 285
S-Plus, viii, 285, 318, 326
    arima.mle, 310
    Optimization, 298
Sampling error, 270
Scaling coefficient, 81
Seasonal adjustment, 270, 276
    model based, 271
    X-12-ARIMA program, 276
Seasonal component, 252
Seasonal heteroskedasticity, 280
Seasonal unit root test, 115
Seasonality, 110
Semimartingale, 207, 208
Semiparametric regression, 152
Sequential importance sampling, 52, 59
Shift parameter, 80
Signal, 76
Signal extraction, 34, 107, 256, 274, 279
    RegComponent model, 267
Signal to noise ratio, 88, 127
Simulated maximum likelihood, 323
Simulation smoother, 8, 10, 240, 297
Simultaneous localisation and mapping, 68
Skewness, 186
SLAM algorithm, 347
Smoothed disturbance estimate, 296
Smoothing, 206, 208, 226
Smoothing algorithm, 107, 256
SNR
    see signal to noise ratio, 88
Spectral analysis, 103
Spectral density, 27, 97, 106
Spline, viii, 8, 154
Spot variance, 207
Spot volatility, 207
SsfPack, viii, 24, 285, 289, 292, 294
STAMP, 24, 108
State equation, 31, 172
State smoothing, 8, 9, 295, 296
    variance matrix, 296
State smoothing residual, 295
State space form, vii, xiv, xv, 157, 254, 286
    affine term structure model, 328
    ARIMA model in, 254
    bootstrap, viii

bootstrapping, 174
CAPM model, 300
in general, 5, 7, 27, 286, 291
in SsfPack, 288
linear, 172
reverse-time, 199
stochastic regression, 175
time varying regression, 290
Vasicek model, 331
Stationarity property, 83
Stochastic cycle, 104, 107
Stochastic regression, 175
Stochastic seasonal, 116
Stochastic volatility, viii, 18, 69, 178, 206, 207, 211, 318
filtering, 237
Structural break, 296
Structural time series model, 206
Subordinator, 210
Sufficient statistic, 237
Superposition, 228, 229, 230, 231
System matrix, 287, 291
Systematic resampling, 55

Term structure model, 327
Vasicek model, 329
Time invariant parameter, 288
Time series decomposition, 307
Beveridge–Nelson, 307
Clark model, 310
Morley–Nelson–Zivot model, 317
Unobserved component, 310
Time varying parameter model, 124
Time varying system element, 291
Time-varying autoregressive model, 69
Time-varying trading day effect, 276
Tobit nonparametric regression, 153, 169
Tracking, 40
Trading day effect, 112, 272, 276
Transition equation, 287, 296
Trigonometric seasonal, 252

Unobserved components decomposition, 310, 317
Unscented Kalman filter, 48

Vasicek model, 329
Vector autoregressive, 7
Viterbi algorithm, 341, 342

Wald test, 109, 113, 115
Wavelet, viii, 77, 79
White noise, 229
Wiener process, 328
Wold decomposition, 103, 107

X-12-ARIMA program, 276